Aerial Photo-Ecology

Aerial Photo-Ecology

JOHN A. HOWARD

B.Sc. (Wales), M.F. (Minnesota), Ph.D. (Melbourne), Dip.For. (Bangor)
F.L.S. (London), A.M.A.I.C.
Senior Lecturer, School of Forestry
University of Melbourne, Australia

1970

FABER AND FABER

London

First published in 1970
by Faber and Faber Limited
24 Russell Square London WC1
Printed in Great Britain by
Alden & Mowbray Ltd
at the Alden Press, Oxford

I SBN o 571 08592 X

Collotype plates
printed by Cotswold Collotype Co Ltd

TO
M. B. H.
M. M. H.

AUTHOR'S PREFACE

An attempt is made in this book to bring together those topics which are pertinent to the aerial photographic evaluation of the biological environment. In fact, some readers may consider the field of selected topics too wide. Unless, however, the photo-interpreter has acquainted himself with the more important physical factors influencing the aerial photograph he may fail to interpret as fully as possible what he observes on the aerial photographs; and unless he has some training in aerial photogrammetry he may be reluctant or unable to prepare simple maps and to quantify his observations. The book has been divided into five parts to facilitate reference and to enable a particular aspect to be separately studied. For simplicity, the dynamic aspects of integrated photo-interpretation have been kept to a minimum in Part V. When, however, the interpreter has gained proficiency in his field of study, he may find it necessary to give equal attention to both the static and dynamic aspects. Some stereo-pairs (i.e. stereograms) and stereo-triplets include non-stereoscopic areas, as it is often essential for the interpreter to obtain an impression of a larger area. Terminology in the text conforms, as nearly as possible, to the definitions given in the *Manual of Photogrammetry*; and Blakely (1955) has been followed for the names of eucalypts.

The writer is indebted to Mr J. H. Chinner (Head of the School of Forestry, University of Melbourne) for his encouragement during the writing of the book, to other members of the School for their co-operation, to many friends and acquaintances for providing information and to government and commercial companies. In particular, appreciation is extended to the following for giving time to discussion and/or for reading parts of the typescript: Dr D. Ashton (School of Botany, University of Melbourne), Professor T. E. Avery (School of Forestry, University of Illinois, U.S.A.), Mr G. Bervoets (Department of Surveying, University of Melbourne), Professor A. Buchanan (Chemistry Department, University of Melbourne), Mr J. E. Coaldrake (C.S.I.R.O., Brisbane), Mr B. Cole (School of Optometry, University of Melbourne), Mr A. Eddy (Forestry School, Creswick, Victoria), Mr D. Hocking (Department of National Development), Mr F. Horwood (C.S.I.R.O.), Dr I. Langdale-Brown (Department of Forestry and Natural Resources, Edinburgh

AUTHOR'S PREFACE

University), Dr T. Neales (School of Botany, University of Melbourne), Mr C. Ollier (Earth Science Department, University of Papua–New Guinea) and Professor R. Specht (School of Botany, University of Queensland).

*   *   *   *   *

Acknowledgements of permission to publish
aerial photographs and diagrams

Some of the figures in this book have been included through the courtesy of individual authors or government and private organizations. If material has been inadvertently adopted or adapted, where permission should have been obtained or acknowledged, it is hoped the oversight will be excused. The following acknowledgements are made:

Fig. 1.1a (The Director, Science Museum, London); Fig. 1.1b (Cultural Attaché, Burmese Embassy, London on behalf of the Burmese Forester, 1925; Mrs Scott on behalf of C. Scott deceased, see bibliography R. G. Kemp *et al.*; Figs. 3.3 and 4.1 (Professor M. P. Meyer, Forestry School, University of Minnesota); Fig. 5.2 (G. C. Brock, Itek., New York); Figs. 6.2, 9.2, 14.1 (R. Goguey, Lycée H. Fontaine Dijon). Figs. 6.3a, 7.4, 17.5, 17.6, 17.9, 18.6, 19.2, 22.2a (Surveyor-General, Department of Lands and Survey, Victoria); Fig. 6.3b (Reproduction by courtesy of the Department of Lands, New South Wales); Fig. 6.5 (Professor C. Chambers, Botany School, University of Melbourne); Fig. 6.6 (Professor R. N. Colwell, University of California, Berkeley); Fig. 7.3 (Forestry and Timber Bureau, Canberra); Fig. 7.5 (U.S. Department of the Interior, Geological Survey—Part of map, Idaho Garden Valley); Figs. 10.1, 15.2 (Dean S. Spurr, Faculty of Natural Resources, Ann Arbor, Michigan; Ronald Press Company, New York); Fig. 10.7 (Professor E. Thompson, University College, London; Photogrammetric Record); Figs. 12.2c, 13.5b (Hilger & Watts Ltd., London); Fig. 12.2d (S.F.O.M., Paris); Fig. 13.5 (4) (Bausch and Lomb, Rochester, U.S.A.); Fig. 13.6a (K. E. Moessner; U.S. Forest Service, Utah); Fig. 13.6b (Carl Zeiss, Oberkochen, West Germany); Table 14A (I. Langdale-Brown, Department of Forestry and Natural Resources, Edinburgh University); Fig. 17.1 (National Aeronautics and Space Administration, Washington, D.C.; see also P. D. Lowman in bibliography); Fig. 17.2 (C. Ollier, Earth Science Department, University of Papua and New Guinea); Figs. 17.8, 17.9, 19.1, 20.2 (Director, Division of National Mapping, Department of National Development, Canberra); Fig. 18.1 (P. Becket, R. Webster, Oxford University; Lands and Surveys Department, Uganda; Director, Military Engineering Experimental Establishment, Christchurch, U.K.). Fig. 18.4 (R. Ray and W. Fischer, c/o Photogrammetric Engineering); Fig. 18.5 (D. Crossley, North West Pulp and Power, Hinton, Alberta); Fig. 18.7 (R. Muhlfeld, Bundesanstalt für Bodenforschung, West Germany); Fig. 19.3 (J. L. Readshaw and Division

AUTHOR'S PREFACE

of Entomology, C.S.I.R.O., Albury, N.S.W.); Fig. 19.4a (H. Lamprey, Serengeti Research Institute, Arusha, Tanzania; Fig. 19.4c (B. Grimes c/o Nature Conservancy; The Nature Conservancy, London; J. K. St Joseph, Director in Aerial Photography, University of Cambridge); Fig. 19.5a (Conservator of Forests, Forests Department, Western Australia; F. Podger, c/o Forest Experimental Station, Dwellingup, W.A.); Fig. 19.5b (Controller, H.M. Stationery Office, London; see also Brenchley and Dodd in bibliography); Fig. 19.6 (G. Wolff, Institut für Forstwissenschaften, Eberswalde, East Germany); Fig. 20.1 (Department of Lands, Wellington, New Zealand); Fig. 20.3 (Department of Lands and Surveys, Hobart; R. Hurley, Forestry Department, Hobart, Tasmania); Fig. 20.4 (E. J. Swellengrebel, c/o Commonwealth Forestry Review); Fig. 21.1 (Institut de Geographie National, Paris, France); Fig. 21A (U.S. Forest Service; T. E. Avery, Head, School of Forestry, University of Illinois, Urbana); Fig. 21.2 (D. R. Cable c/o Forest Science, Washington, D.C.); Fig. 22.1 (Editor, Commonwealth Forestry Review, London); Fig. 22.2b, 22.2c (R. Parsons, Geography Department, University of Melbourne); Fig. 22.3, 22.4 (U.S. Forest Service; see R. C. Heller *et al.* in bibliography).

Lastly, I am grateful to my wife for encouragement and assistance at all times.

JOHN A. HOWARD

University of Melbourne,
Australia, 1968

CONTENTS

CONTENTS

CONTENTS

CONTENTS

FOREWORD

AERIAL photography's special value is that it gives the viewer a scale of observation not open to him with the naked eye, and not readily deliverable to him in map form. Used effectively, and with a shrewd appraisal of its limitations, it gives the scientist winged feet, and enables him to synthesize regionally where not so long ago he could scarcely see beyond the range of ground vision. A high proportion of the world's vegetation map was, until aerial photography transformed our methods, effectively based on what one could see from river, road or even moving train, or piece together from laborious field surveys on the micro-ecological scale.

A shrewd appraisal of its limitations, coupled with an equally shrewd summary of the effective uses of aerial photography, is what John Howard gives in these pages. He speaks of the photo-community, as if the world's vegetation had actually picked itself up off the landscape and started to grow on the airphotos. I admire such singlemindedness in the user of a scientific technique. Plant geography, he says, has come to treat large-scale distributions, whereas most ecologists work on the local, micro-ecological scale. His own techniques work most valuably between these extremes—and help to bridge the gap that separates them.

Though this is a general survey of the technique of airphoto interpretation as applied to vegetation, it gives major emphasis to forest areas. My sole contribution to Howard's exhaustive bibliography is an account of a reconnaissance survey of the forest and tundra of Labrador-Ungava, conducted when the art of airphoto interpretation was less well advanced. My colleagues and I learned by trial and error the lessons Howard now elegantly presents within a single volume. Had we had access to his knowledge, we could have saved ourselves many hours of false starts, false interpretations and false hopes. Armed with this book, the student who now sets out on such a project would have a five years' start on us.

The practical forester, aiming to cut and then to regenerate forest on his concessions, will find this book useful. He, too, makes use of aerial photography as a standard working tool, and will already have learned some of the drill set out in the following pages. But his students will not have done so, and for them also the text will be good news.

FOREWORD

Another theme can be discerned in what the author has written. He is alive to the value of aerial photography in geographical and geological survey. Except in the most featureless terrain the detail on an airphoto includes much information about landforms, soils and aspect. In structurally vigorous country the strike of sediments and the foliation of gneisses is as clear as the alignment of faults and joint systems. The physical geographer and geologist alike have learned to use the airphoto as a valuable aid to survey. Not merely, moreover, as a base-map, or as a ready means of extrapolation and interpolation; from personal experience I can certify that one occasionally sees on the airphotos patterns that have eluded the eye in quite detailed ground survey. One can be so close to the forest that one cannot see the trees.

The cultural landscape, of course, is the most naked and obvious of the things visible on the airphoto. Not only do the contemporary and recent past uses of the land reveal themselves at once, but the archaeological traces of ancient cultures can sometimes be seen—and airphoto detection is one of the standard working tools in the reconstruction of the land-use systems of the past. So it is not simply the physical and biological scientists who can read this book with profit: human geographers, archaeologists, anthropologists and town planners will find it helpful, too.

As Howard indicates, at the beginning of Chapter 22, the real strength of airphoto interpretation is that it is at root a holistic tool. One can use it for one's special purposes, which may be very limited. But if the horizon is broadened, if relationships between patterns are sought—for example between vegetation and landforms—the evidence remains permanently there for later re-examination. It does not get up and go away.

One final note, not perhaps of despair, but at least of exasperation. Suppose that one has mastered the techniques described in this volume. Suppose, further, that one decides to use them (as I once did) to conduct large-scale regional reconnaissance by team methods. The obstacles are formidable: field keys must be produced by leg work and a lot of subsequent laboratory analysis; and the eye of the interpreter must be trained. The most difficult task remains. It is to assemble, or to get access to, the photographs themselves. It is all very well if one works for some major company, or government, that has just commissioned the photography. But if, instead, one works on photographs previously accumulated by a government, red tape and sheer poor filing make them hard to get hold of. I know of only one country whose national airphoto library is comprehensive, efficient and completely accessible to the research worker—Canada. Perhaps Howard's book will shame some other jurisdiction into copying.

F. KENNETH HARE
President, The University of British Columbia

Vancouver, Dec. 31, 1968

ILLUSTRATIONS

TEXT FIGURES

The text figures are referred to by arabic figures (e.g. 2.2) and appear throughout the text in their relevant chapters.

STEREOGRAMS, ETC.

The stereo-pairs, stereo-triplets, etc., are referred to in the text by **bold** figures (e.g. **6.3**). They have been printed by the collotype process and appear in one section at the end of the book.

MONOCHROME PLATES

Monochrome plates other than the stereograms are referred to in the text by *italic* figures (e.g. *6.5*) and appear in one section between pages 128 and 129.

Acknowledgments are given on page viii.

1

HISTORICAL OUTLINE
AND TERMINOLOGY

No apology is offered for the encumbrance of an historical outline. Equipped with a thumbnail sketch, one is better prepared to argue and advance the cause of aerial photo-interpretation on an ecological basis. F.M. Cornfield once said: 'There is only one argument for doing something, the rest are arguments for doing nothing' (*Microcosmographia Academica*, 1908).

As long ago as 30 B.C., Vitruvius recorded that pigments were bleached by light. In Ancient Greece, Aristotle commented on the quality of light and the principle of the camera obscura; but a camera obscura was not produced in Europe until the sixteenth century. The fact that a pin-hole of light will provide an inverted picture of a bright object was known to the Arabs in the tenth century. The legend goes that Alhazen of Basra was the first to rediscover this phenomenon when he noticed that an inverted picture of a bright landscape was formed on the inside of his darkened tent by a pin-hole of light.

In 1826 Nicéphore Niépce of France, combining the principle of the camera obscura and the bleaching properties of light, provided the first photograph (Anon, 1966). He obtained an inverted image on a glass, copper, or pewter surface coated with silver salts and bitumen of judae. The exposure was between 2 and 8 hours! Previously, Scheele (1742–86) had observed that silver salts were quickly blackened by bluish light. Niépce later worked with Daguerre, who is often credited with the invention of the photograph. By 1839, Daguerre had developed a practical means of photography and the Daguerreotype Camera was being produced. In 1851 Scott Archer in the United Kingdom discovered the wet colloidal process, which reduced the exposure time to 2 seconds; and by 1880 a dry-plate process was being marketed. In 1881, Hannibal Goodwin patented the first transparent flexible photographic film; and in 1899 Eastman introduced the celluloid film for amateur photographers.

Although the development of aerial photo-interpretation has ensued from the development of the camera, it is also closely associated with the development of terrestrial photogrammetry in the nineteenth century and later aerial photogrammetry. Terrestrial photogrammetry resulted from the development of the camera and the application of the principles of perspective drawing.

1

HISTORICAL OUTLINE AND TERMINOLOGY

In 1759, J. H. Lambert had published his findings on the theory of perspective drawing to determine dimensions. The French hydrographer, Beautemps-Beaupré, used his theory to provide maps of the coast of Tasmania. Between 1791 and 1793, whilst sailing in Australian waters, he made free-hand sketches of the Tasmanian coastline from which maps were drafted on his return to France. Between 1810 and 1820, Beautemps-Beaupré published his own findings on the principles of perspective geometry.

The marrying of these principles to the photograph to provide the science of photogrammetry, nevertheless, did not occur until 1861. In that year, Laussedat applied Beautemps-Beaupré's principles to terrestrial perspective views recorded in photographic form, and provided the first orthographic map from photographs of a village near Versailles. Three years previously Laussedat had taken what, as far as is known, were the first aerial photographs. These were taken using a 'glass-plate' camera supported by a string of kites and later by a captive balloon. However, he gave up his aerial experiments two years later, as he experienced difficulty in establishing the scale throughout the photograph, since the technique of ground control had not been developed. Nevertheless he continued experimenting in terrestrial photogrammetry with a photo-theodolite. This has earned him the name of 'father of photogrammetry'.

At the time, as Laussedat's work was of military importance, it was not publicized abroad; and, in consequence, Meydenbauer, an architect, carried out similar work in Prussia by producing orthographic drawings from perspective photographs of monuments. It is doubtful whether there was any military purpose in Meydenbauer's work; but terrestrial photogrammetry was used by the Prussians in the siege of Paris in 1871. The first recorded use of the word 'photogrammetry' in a published article was made by Meydenbauer in 1893 (Whitmore, 1952).

A method of ground control in the form of a 200 m square was developed by Stolze in 1881 in order to establish the height above ground of the camera and the tilt of the aerial photograph. At about the same time the nadir point was successfully located in the photograph by hanging wires; and ten years later Stolze expounded the theory of the floating mark. In 1893, Adams (U.S.A.) outlined the principles of radial line plotting; but these principles, as used today, were not widely applied to photogrammetry until a few years after World War I.

From the end of the nineteenth century onwards, attention was being given to the design of special instruments for photogrammetry, although as early as 1838 Wheatstone had exhibited a reflecting stereoscope in the United Kingdom (fig. *1.1a*) and Brewster had made a lens stereoscope in 1849. For example, in 1902, Fourcade in South Africa suggested a stereoscopic instrument for measuring differences of parallax; but independently von Hubl and Orel (Austria) and Pulfrich (Germany) had developed simple photogrammetric instruments (e.g. Stereocomparator).

In 1910 in Fiji the Stereocomparator was used experimentally for measuring parallax; and by the mid-1930s contour mapping from aerial photographs had

2

attained a satisfactory standard. It is worth while mentioning that, in 1933, Zeiss at Jena introduced the Multiplex aeroprojector. This instrument, with some modifications, has remained popular up to the present time; but in recent years has had to compete with an ever-increasing range of specially designed, highly accurate instruments.

Modern aerial photography for peaceful purposes is deemed to have been initiated in 1913, when an Italian aircraft took photographs of Benghazi town and nearby landscape in Libya (Tardivo, 1913). These photographs were used to prepare a mosaic for geological mapping. With the advent of World War I in 1914, aerial photography was rapidly expanded and military photo-interpretation was introduced and developed. Immediately after the beginning of hostilities, a German airship on war-time patrol was found on capture to contain an aerial camera. At the battle of the Aisne, September 1914, aerial photographs were taken by the Royal Flying Corps, later the R.A.F.; and, by March 1915, maps were being prepared from aerial photographs taken by aircraft and were used for the assault on Neuve Chapelle (Donaldson, 1962). However, up to the time of the Palestine campaign, photographs were taken by the method termed *pin-pointing*. This entailed photographing selected points, which were usually close together; but in the Palestine campaign *strip photography* was used. By 1918, French photographic units, attached to the army, were providing up to 50,000 photographs a week and, prior to the final offensive against the German army, photographs were interpreted for military installations well behind the enemy lines. Also in use by that time was the popular present-day 'film-filter' combination of panchromatic plate and yellow haze filter.

Between the two world wars, aerial photographs came to be widely used throughout the Commonwealth, particularly in Canada, in the Middle East and in the United States. As early as 1920, Thomas writing in *Nature* drew attention to the potentialities of aerial photography for the purposes of archaeology, botany, geography, geology and meteorology; and Blandford and Watson were negotiating for an aerial survey of delta forest in Burma (Kemp *et al.*, 1925). If a date is to be given for the beginning of photo-interpretation in peace-time, then, on available information, it should be 1919 when Ellwood Wilson (1920) used aerial photographs for forest stock-mapping. Half a million acres were photographed at a cost of 2·6 cents an acre. The first peace-time use of strip aerial photography was in Egypt (Thomas, 1920b).

It requires noting, however, that the first known ecological use of aerial photographs occurred during the American Civil War. At Richmond, Virginia, in 1862, the Unionist army used photographs to delineate pinewoods, swamps and rivers. In 1888, Deville, in the course of terrestrial photogrammetrical mapping in the Rocky Mountains, also recorded vegetation. A year earlier, the newspaper *Berliner Tageblatt* had mentioned the taking of aerial photographs of beech, spruce, oak and pine by balloon (see Spurr, 1960).

In 1924, aerial photographs were being taken in the Commonwealth as far apart as Canada, India and Australia. For example, 1,000 square miles of the Irrawaddy delta, Burma, had been photographed vertically (fig. *1.1b*); and in Australia aerial photographs, having a 4 in. by 5½ in. format were taken. In 1928, aerial photographs were used by Bourne for ecological studies in central Africa (see Oxford Forestry Memoir No. 9 and Chrosthwait, 1930). In 1929, Sealey in Canada carried out a forest survey in which tree heights were recorded. By 1930, 19,000 square miles of eastern Australia had been covered by the R.A.A.F., using 7 in., 8¼ in., 10 in. focal lengths and a 7 in. × 8 in. format. In the same year, aerial photographs were taken of New River, south-west Tasmania, by the Royal Navy and north-west Tasmania was flown by the R.A.A.F. A forest officer was aboard the naval aircraft. In 1931, Bourne wrote of the role of aerial photographs in soil survey (Communication No. 19, Imperial Bureau of Soil Science); and in the same year G. W. Leeper used aerial photographs for a soil survey at Mount Gellibrand, Australia. Under Bourne ecological work using photographs was carried out in the United Kingdom. In 1936, a 1/63,360 contour map of Sale, Australia, was prepared entirely from vertical aerial photographs using a principal point traverse and detailed plotting by the Arundel method.

In the United States, aerial photo-interpretation was also being developed, although military photo-interpretation was neglected. For example, in 1928 aerial photographs were applied to geological studies in relation to oil exploration in Oklahoma; in the 1930s specialist geographers were employed by the Tennessee Valley Authority on photo-interpretation and in 1936 aerial photographs were used for a sea-lion census in California.

In World War II, the tactical use of pressurized aircraft in high-altitude aerial photography led to the use and interpretation of smaller-scale photographs (e.g. 1/50,000). Infrared aerial photography was also developed along with faster and finer grained panchromatic films. A concise and objective report on the use of photo-interpretation during the war is given in the *Manual of Photographic Interpretation* (pp. 7–10).

Since the end of World War II, there has been an enormous expansion in the peace-time use of aerial photographs. For example, forest photo-interpretation has been used in Finland since 1945 for taxation purposes. Gradually, the advantage of truly vertical aerial photographs due to their relative simplicity for both mapping and interpretation has been recognized. Australia was completely covered by vertical aerial photographs of varying scales by 1968, being the first continent to achieve this. Canada, the United Kingdom and the United States have been covered by a combination of vertical and oblique photographs. India, on the other hand, although early in the field, has but a small area covered by aerial photographs. Considerable areas of western, eastern and central Africa are covered by photographs at scales between 1/30,000 and 1/40,000, and extensive areas of parts of South America are also covered.

The value of, and application of, small-scale aerial photographs have also been recognized in post-war years for photogrammetry. For example, for planimetric and topographic mapping fewer ground control points are required. In geological interpretation, too, smaller-scale photographs (e.g. 1/25,000) are often preferable to larger-scale photographs. Even in forestry, there is accumulating evidence that under certain circumstances photographs of 1/20,000 may be as useful as at 1/15,840. In the northern hemisphere temperate regions, particularly North America, the best available forest scale is acknowledged as 1/10,000 to 1/15,840. This takes into consideration the economic factor. In many parts of the world, the need to distinguish between aerial photography for photogrammetry and for photo-interpretation is being recognized. For the identification of many tree species from photographs and for photo-ecological studies, photographs at scales of 1/10,000 and larger (e.g. 1/5,000) may be needed in the future. Recent developments include the taking of 1/100,000 photographs for mapping, orthophotography, satellite photography, automatic visual scanning, high-quality colour photographs, the use of computer programming, and remote sensing beyond the photographic spectrum.

AN EXPLANATION OF TERMINOLOGY

The terms aerial photographic interpretation, aerial photogrammetry, aerial photography and aerial surveying are frequently used with considerable variation in meaning. It is therefore desirable to consider the terminology before proceeding. Aerial photographic interpretation is synonymous with aerial photo-interpretation; and obviously the prefix 'aerial' is used to indicate that the photographs were taken from an aircraft flying at a convenient height above the ground. In the ensuing study the word 'aerial' is often omitted, as only aerial photographs are being considered.

A distinction between photogrammetry and photo-interpretation is desirable due to the rapid growth of interest in recent years in the peaceful application of photo-interpretation. Such separation is no more controversial these days than in other fields of the arts and the sciences. For example, statistics is now normally recognized as a subject distinct from mathematics.

Aerial photo-interpretation may be defined as *the art and science of studying and identifying objects formed as images on the emulsive surface of the film and evaluating their significance.* Normally in this process the media is in the form of black-and-white photographs; but occasionally black-and-white positive transparencies, and recently colour photographs, are also used. Both inductive and deductive reasoning are applied in formulating conclusions. Frequently, data is evaluated in a quantitative form; but this cannot be satisfactorily obtained from the photographs by the interpreter unless he has had a basic training in photogrammetry. In photo-interpretation, the measurement of areas and of heights are two important facets of photogrammetry which are frequently used. For example, Spurr (1960) stated that 'photo-interpretation of forest tree-and-stand detail is largely confined to the identifications of forest types and site and the measurement of tree height, crown diameter, crown closure, crown counts and stand area'.

HISTORICAL OUTLINE AND TERMINOLOGY

The term *photogrammetry* is derived from three Greek words meaning light, a drawing and to measure; and this possibly suggests measuring graphically, using light. Thus, both photo-interpretation and photogrammetry overlap in the procurement of quantitative data from the photographs; but, for the former, measurements of the earth's surface are a means to an end, whilst in photogrammetry the measurements are normally an end in themselves, i.e. in the form of a map. This use of the term may be compared with the definition of the American Society of Photogrammetry which has defined photogrammetry as 'the science or art of obtaining reliable measurements by means of the photograph'. In France, Martin (1948) in his book *Cours de photo-topographie* defined photogrammetry similarly as 'une science appliquée qui consiste à utiliser la photographie à des mesures plus ou moins précises dans tous les domaines'. In the United States the term photogrammetry has been extended to cover photo-interpretation (see p. 1, *Manual of Photogrammetry*); but it is hoped, however, due to the increasing importance of interpretation, that the two will be separated in the next few years. The International Society of Photogrammetry has established a separate commission for photo-interpretation (i.e. Commission VII).

The terms *aerial surveying* and *aerial photography* have been used to cover the entire field of the study of aerial photographs, especially in the United Kingdom. 'Aerial surveying' has also been used with this meaning in English language translations of articles in Russian; and is sometimes used in both Canadian and British journals. 'Aerial surveying' seems a better term to use for the combined fields of photogrammetry and photo-interpretation if such a term is required; but it has not been found necessary to use it in the ensuing text. The term aerial photography is applied sometimes to cover the taking of photographs from an aircraft; and it is this second restricted meaning, which will be adopted. Frequently the term is extended to include the subject matter which is given in Chapters 1 to 5 (Part I) (e.g. Commission I of the International Society of Photogrammetry).

At this juncture, it may require emphasizing that an understanding of the elements of pure photogrammetry is an essential prerequisite of successful photo-interpretation; and frequently the worker in the field, who is using aerial photographs, needs to be sufficiently acquainted with photogrammetry to prepare field maps to satisfy his requirements. Part III, Elements of Photogrammetry, is presented with this object in mind.

The approach to pure photogrammetry has been made graphically and not mathematically, as the student will usually have had a biological training and a very limited training in mathematics. An analogy may be made between the way in which the ecologist usually studies statistics and the way in which the mathematician or engineer approaches the subject. Photogrammetry, as outlined by continental Europeans, is frequently difficult for the biologist to comprehend due to their mathematical consideration of the subject. The graphical approach is therefore given, without apologies to the mathematically minded photogrammetrist.

HISTORICAL OUTLINE AND TERMINOLOGY

The term *ecology* is well known to biologists, having a Greek root, *oikos*: a house or dwelling place. It is the study of living organisms in relation to their environment. For over sixty years, two broad divisions have been made, namely plant ecology and animal ecology. With the former is associated phytosociology. Economists and sociologists also use the term 'ecology' when referring to human ecology; but as the greatest interest and study of aerial photographs on an ecological basis is related to vegetation, it is with this aspect that the book is primarily concerned; and the subject will be referred to as *aerial photographic ecology* or *aerial photo-ecology*. The term 'aerial photo-ecology' may be compared with similar photographic terms prefixed by 'photo'. These include photo-geology and photo-geography as used in the *Manual of Photo-Interpretation* (1960).

Within the frame-work of photo-ecology, forest photo-interpretation assumes great importance. There are two reasons for this. First, forest photo-interpretation in relation to forest inventories has been well developed. Secondly, the forest or woodland areas of the world are extremely extensive, comprising about one-third of the land surface or approximately 16·7 million square miles (Roussel, 1962). This results in the recording of trees on many photographs, even though the primary object of the photography is not for forestry. In future, in order to maximize the advantages to be gained by use of aerial photographs in inventories of angiospermous forests, particularly in the tropics, it seems desirable that the forester should be trained in both ecology and photo-interpretation, and have a working knowledge of landform interpretation. This book sets out to bring together aerial photographic topics of interest to the land-resource planner, the forester and the ecologist; and develops a procedure for carrying out integrated surveys of the natural environment.

To provide the photo-ecologist with a better understanding of what he sees recorded on a photograph and to enable him to suggest what steps should be taken to improve the interpretative value of the photographs, it is considered well worth while to examine some of the factors influencing the procurement of the aerial photographs before proceeding to a study of the elements of photogrammetry (Part III) and interpretation (Part IV). For this purpose, selected topics have been presented in Chapters 2–6 (Part I). These cover characteristics of the aerial camera with special reference to the lens and filter, the aerial film, film processing, tropospheric conditions and the reflective properties of the objects being photographed. Surprisingly, little has been published in photographic journals on the phenomena of the reflection of light by vegetation (Chapter 6). Part II covers aerial photography.

The flow-chart (fig. 1.2) will help to summarize what has been said. An examination of this diagram suggests that the approach to aerial photographs, after some initial considerations, can be divided into three fields. These are aerial photography, aerial photogrammetry and aerial photo-interpretation. Each of these is closely related, especially photogrammetry and photo-interpretation. In fact, planimetric mapping techniques are frequently employed in conjunction with photo-interpretation. As

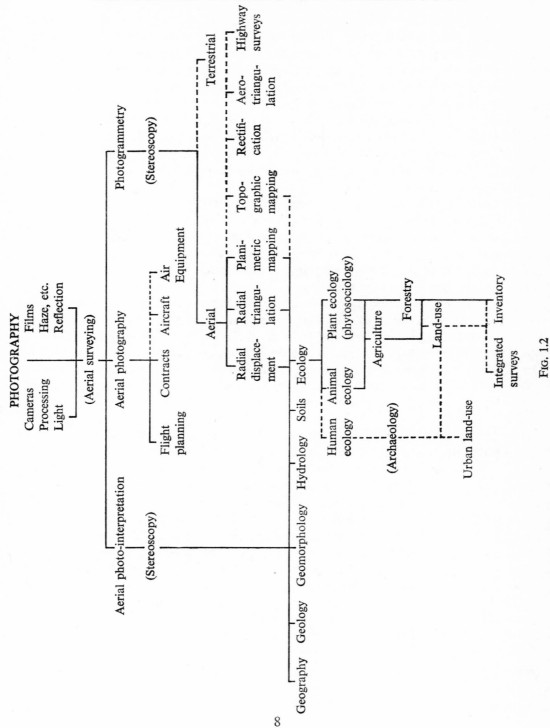

Fig. 1.2

photo-interpretation on an ecological basis is also linked with other field subjects, it is desirable that the interpreter should have some knowledge of the methods used by the geographers, geomorphologists, pedologists and hydrologists. Armed with this additional information, he is more capable of approaching his own field; and will find that forest photo-interpretation has been very widely used and particularly well developed by way of forest inventory. An approach to the study via other disciplines can conveniently be termed *vertically integrated photo-interpretation.*

PART ONE

Factors Influencing the
Aerial Photograph

2

THE AERIAL CAMERA

INTRODUCTION

THE perfect aerial camera has so far not been constructed. Many attempts have been made and, no doubt, many more will be made to provide the perfect camera. There are and have been so many types of camera that to describe and appraise them adequately would need a full textbook. Most cameras are constructed to satisfy a particular requirement, being initially designed for military purposes. As cameras are expensive, it is unusual to purchase a camera for a specific project, and preferable to employ a reputable company with suitable tried aerial equipment to carry out the project. Where a government sets up its own monopoly to provide aerial photographs for a particular purpose, it is usually found within a few years to be unable to provide the exacting photographic requirements for the increasing range of special projects.

To meet the increasing demand for colour aerial photography, it is desirable to have available a camera with the highest quality lenses. For experimental colour photography in Japan in 1962, Maruyasu & Nishio chose the Wild RC-5a and Wild RC-8 fully automatic cameras and the 21 cm (8¼ in.) Aviotar lens. This lens is widely used in continental Europe in conjunction with a 7 in. format, and has the longest focal length of cameras used in topographic mapping at the present time. Brief specifications of the six fully interchangeable lenses are given in table 2.A. The RC-8 camera has a 23 cm by 23 cm photographic format. *Format* refers to the

TABLE 2.A

Name	Focal length (cm)	Angular field (angle of coverage)	Photographic format (cm)
Aviotar	21·0	Normal (60°)	18 × 18
Aviogon	11·5	Wide (90°)	18 × 18
Aviogon	15·0	Wide	23 × 23
Infrater	21·0	Normal	18 × 18
Infragon	11·5	Wide	18 × 18
Infragon	15·0	Wide	23 × 23

picture size. A format of 9 in. by 9 in. is preferred in many countries, excluding continental Europe.

An examination of an aerial camera will show that the principal components are the lens and lens cone, shutter and diaphragm, filter, drive-mechanism, film magazine, camera body and focal plane. Most of these components are readily located in a miniature 35 mm camera. In the aerial camera, the lens, the filter and the shutter

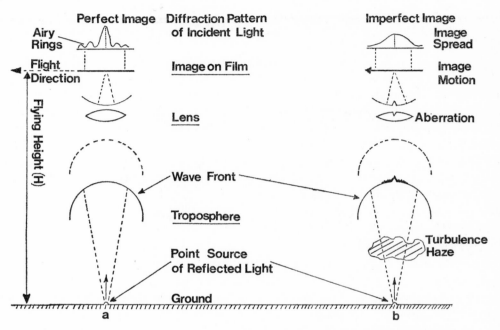

FIG. 2.1. In figure (a), a point source of reflected light on the ground is illustrated as emitting a wavefront which is not deformed on reaching the film surface. In consequence a perfect image is formed. In figure (b), a similar circular wavefront is subject to the principal factors causing wave spread/image distortion; and in consequence an imperfect image is formed on the film in the aerial camera. The wavefront passes through the troposphere, is reshaped by the convergent lens and may be subject to further distortion due to camera movement/vibration. The relevant chapters are reflected light (Chapter 6), haze (Chapter 5), lens aberration (Chapter 2), image motion (Chapter 2) and image quality (Chapters 2 and 4).

are of particular interest due to their influence on the 'interpretation' value of the photographs, and therefore will be separately discussed.

The film providing a single photograph lies at the time of exposure in the focal plane of the camera, and previously and afterwards is stored on two spools contained in the magazine. The magazine also contains a metering device for the correct amount of film to be fed from spool to spool via the focal plane. The film in the magazine of the aerial camera is advanced by an electrically powered drive-mechanism, which also has the function of winding the shutter. The drive-mechanism can normally also be operated by hand.

(a) The lens

The function of the camera-lens is to form a sharp image on the film surface in the focal plane of the camera. By using a high-quality lens, correctly focused, and a suitable exposure for the film the photo-interpreter should be provided with photographs of satisfactory quality. The principal factors influencing the quality of the image, during film exposure, are illustrated in fig. 2.1.

By knowing a little about lenses, the photo-interpreter is better equipped technically in understanding his work and for prescribing specifications for the taking of aerial photographs. Of interest are the characteristics of the lens, and particularly the focal length of the lens. In recent years, the overall quality of lenses has been improved and the maximum angular field of lenses increased.

Modulation transfer functions can be determined for the lenses; and these provide a valuable aid in judging the performance of the optical systems. A *modulation transfer function* is a curve showing the modulation (i.e. measured luminance) of test objects against the spatial frequency of the sine-wave pattern of the test objects. For a full explanation of modulation transfer functions reference may be made to *Fundamentals of Photographic Theory* (Higgins, 1960). By determining the modulation transfer functions separately of each element of the complete photographic system, it would be possible to form different combinations of the separate functions in order to obtain the most suitable for a project. Alternatively, when resolution is used to determine the most suitable combination, it is necessary to test physically each combination.

CHARACTERISTICS OF THE LENS

The lenses of aerial cameras are nominally achromatic; that is, the focus of the lens is the same for all wave-lengths. This, however, is not strictly correct, as in practice it is possible to bring only two wave-lengths to the same focus; and a few degrees change in temperature can produce an aberration of one wave-length. The position of best focus behind the lens varies according to the wave-length of the light entering the lens, as is illustrated in fig. 2.2. In this diagram, the blue, green and red wave-lengths of the white light reaching the lens are seen to focus at different distances along the optical axis. The distance between the position of focus of blue light at about 0·42 microns and of red light at 0·60 microns was about 20 millimicrons for a high-quality lens manufactured in 1944 (Brock, 1952).

From this one may conclude that, in order to provide the best focus for the picture, the lens should be adjusted according to the properties of the film and filter being used; and possibly at least in theory according to the colour of the ground cover. For

C

the former this in fact is done when using infrared film with certain lenses (e.g. Ross 8¼ in. focal length); but for certain recently marketed lenses, corrected for chromatic aberrations and used for modified infrared photography, it is considered unnecessary (e.g. Wild RC-8). These lenses are termed 'universal'. A meniscus filter, as a simple front lens, or a space ring by the magazine have been used to provide the correct focus for mapping in the infrared. Up to the present time, both in black-and-white photography and in colour photography the same lens is normally employed.

A second important characteristic of the lens is that its resolving power in the focal plane is not uniform. The intensity of the illumination reaching the film declines from the centre of the lens outwards. Thus with a lens of 20 in. focal length and a 9 in. by 9 in. photographic frame, the lens may give about 80% of the axial illumination at the edge of the field of view and about 40% of the axial illumination at the corners of the frame (Brock, 1952). It must always be remembered that although the photograph is square the image provided by the lens is circular; and it is only at the extreme corners of the photograph that the image coincides with the image

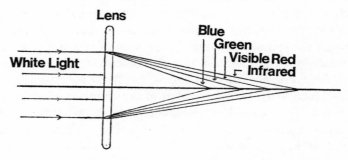

FIG. 2.2. The position of best focus behind the lens varies according to the wavelength. In the visible spectrum blue light comes into focus first and red is the last. Even longer is the focus of infrared waves.

provided by the periphery of the lens. For a lens of 3¼ in. focal length and the same format as for the 20 in. lens, the illumination at the limit of the field of view may be considerably less. Excluding the vignetting effect of the lens mount, the illumination on the negative falls off approximately as $\cos^4 \beta$, where β is half of the angular field.

The falling-off in illumination outwards from the centre of the focal plane will influence the density of the images recorded in the negative and can result in poor-quality photographs, especially towards the edges of colour photographs. Sometimes, where there is considerable shade towards the edges of the photographs, as in the case of forest photography, details are lost to interpretation in the negative and cannot be recovered by careful printing. The falling-off of illumination towards the edges of the field of view is not normally too serious for black-and-white 9 in. format photographs taken with a 12 in. lens or an 8¼ in. lens; but it can be troublesome in photographs taken with a 4½ in. lens or a 6 in. lens as these lenses are wide-angle. If, however, a smaller photographic format (e.g. 7 in. by 7 in.) is used the problem of illumination differences is greatly reduced. It requires noting at the same time that, for photographs of the same scale, a smaller format results in more expensive photography.

As a lens aperture is reduced in size, the resolving power of the lens increases. By reducing the aperture of the diaphragm, the light is focused through the central part of the lens, which is virtually free from aberrations. An *aberration* is a defect in the optical image due to the fact that no lens system can form a perfect image.

Further reductions in the aperture of the lens to pin-point size would result in deterioration of image quality due to *diffraction*. This may be explained with reference to physical optics. It is known that light from a point source passing through an aberration-free lens will produce an image with a diffraction pattern as shown in fig. 2.3. From this figure, representing readings outwards from the centre of the light

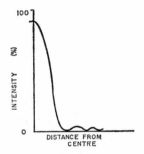

FIG. 2.3. Incident light from a point source forms 'Airy rings', the centre disc of which may contain 85% of the illumination and the outer rings 7% and 3%.

disc, it is seen that the pin-point of light after passing through the lens comprises a central disc of light surrounded by dark rings of light of much lower intensity. For example, the central disc may contain 85% of the light from the point source and the second and third rings only 7% and 3% (Brock, 1952). This spreading of the pin-point of light is further exaggerated when forming an image on the shiny surface of the film, due to light-scattering in the film (e.g. *halation*). Two point sources of light close together will not resolve separately as images unless the centres of the diffraction patterns of the images are separated by a distance equal to the radius of the first ring. The radius of the central dark ring (Airy disc) varies with light quality and the F-number of the lens for objects at infinity.

Resulting from these phenomena, the size and nature of the images recorded on the film surface are limited. Thus, on medium-scale photographs of forest areas the small light areas, produced by the reflection of light from several leaves, spread outwards to form one 'high-light' in the photograph as the individual leaves are not resolved. Halation appears to be the cause of the 'high-lights' produced by the tips of black spruce (Losee, 1951). A second effect of halation (Losee, 1951) is to brighten

17

the grey tones, when they are not sufficiently numerous to overcome entirely the dark tones surrounding them. Halation is partly responsible for the lighter tone on some photographs of the images of the conical crowns of conifers on their sunward side. This is unfortunate for interpretation, as it may reduce the tonal differences between conifers and hardwoods.

FOCAL LENGTH

The photo-interpreter is concerned with the focal length of the lens both for providing photographic details in a form most suitable to his work and for determining the nominal scale of the photographs in relation to calculations of area, distance

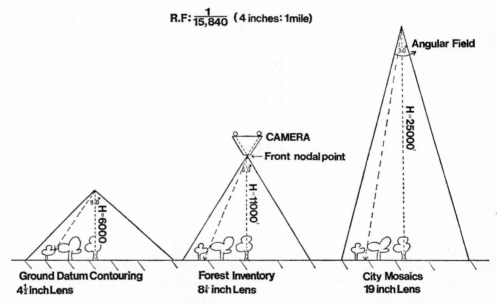

FIG. 2.4. For the same scale photography and for identical trees, the ground/ground vegetation is more frequently recorded using longer focal lengths; but focal lengths in excess of 12 in. will introduce other adverse factors. Focal lengths of 6 in. and 8¼ in. are most commonly used in forest surveys. Flying heights of 6,000 ft, 11,000 ft and 25,000 ft will provide photographs of about the same scale with focal lengths of 4½, 8¼ and 19 in.

and height of objects. Smith *et al.* (1960), working in western Canada, concluded when measuring the crown closure of tree crowns and the height of the trees that photographs taken with a 12 in. lens at a scale of 1/14,000 were preferable to photographs taken with a 6 in. lens at 1/15,600 and 1/28,000. Similarly in the Snowy Mountain region, eastern Australia, W. A. Mueller (personal communication, 1963) favoured a fairly long focal length for 'heighting' forested country as the ground can

18

be more frequently seen between the crowns of the tall eucalypts. Probably, at a scale of 1/8,000 to 1/10,000, best overall results are obtained using a lens of 8¼ in. focal length. If a 6 in. focal length is used, less ground is seen and the crown closure appears greater.

The effect of using lenses of three different focal lengths over tall trees is illustrated in fig. 2.4. It is seen from this diagram that, for the same scale photography and for trees at the same distance apart, the base of the trees is more frequently recorded using a longer focal length. In general photogrammetry, wide-angle lenses are favoured today, not only because the cost of photography is thereby reduced, but also because the wider angle of coverage gives a more exact orientation of images. In the United Kingdom, 3½ in. and 6 in. focal lengths are most commonly used.

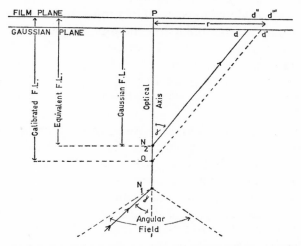

FIG. 2.5. A single ray of light (solid line with arrows) is shown as entering the lens of the aerial camera at the front nodal point (N_1) and emerging at the rear nodal point (N_2) before being incident on the film plane.

As shown in figs. 2.4 and 2.5 the *angle of coverage* or *angular field* is the maximum angle subtended at the front nodal point of the lens by rays passing through. These rays will provide the images recorded in the corners of the photograph. With an 8¼ in. focal length, the normal angular field is 60° to 75°. Cameras with a lens of short focal length, e.g. 6 in., have a wider angle of coverage, e.g. 90°. Some modern aerial cameras have a field of coverage of 110° to 115°. In Russia, Mikhailov (1961) reported use of wide-angle lenses of 100° to 120° for forest projects; and according to Steiner (1963) the focal length for interpretation varies between 36 mm and 500 mm (1·4 to 19·7 in.).

Recently L. Bertele has designed a lens with an angular field of 120° and a focal length of 29·3 mm. This has been used for taking photographs at representative fractions of 1/20,000 or smaller. The radial and tangential distortion is ±10 microns and the axial resolution 110 lines (Gruner, 1964). It is yet to be seen whether this high-quality lens results in a reorientation of techniques and instrumentation such as followed the advent of the Zeiss Topogon lens in 1934, which increased the angular field from about 65° to 90°. Lenses used in war-time at considerable altitudes have focal lengths of 20 in. or more; and are often termed *telephoto-lens*. In New Zealand, a camera with a focal length of 19 in. and a narrow angular field (30°) is used to provide photographs for preparation of mosaics, and when the effective areas of the photographs are to be enlarged and used as maps.

In view of progress made in modern high-precision photogrammetry, it is desirable to appreciate that approximations have been made in formulae derived from simple geometric optics. As an example it will possibly suffice to consider only a single ray and to represent it diagrammatically, bearing in mind that a large number of rays are directed by the object considered to be at infinity towards the camera and are directed through the lens to form image points on the film. In fig. 2.5 a ray from a bright object and directed towards the camera lens passes through the frontal nodal point of the lens (N_1) and forms angle α with the optical axis. It then continues through the lens to the rear nodal point (N_2) before passing out of the lens at angle α to form an image in the Gaussian plane at point d. The nodal points are the points in the lens through which all rays pass.

In commonly used formulae, N_1 and N_2 are assumed to coincide at N_2; whilst the point to which the image points are joined to obtain the corresponding object space directions is more satisfactorily placed on the optical axis between the two nodal points but close to N_2. This point (O) is termed the *interior perspective centre*. The image is not formed on the Gaussian plane but on a curved surface due to curvature of the field of the lens. Hence the emulsion surface of the film is not placed in the Gaussian plane but at a short distance behind it to provide the best average definition and consequently the focal length is increased, being no longer equal to the *Gaussian focal length*. The distance from the rear nodal point to the film plane of best average definition is termed the *equivalent focal length*. The focal length may require further adjustment to compensate for radial distortion of the lens, no lens being aberration-free. The focal length from O is termed the *calibrated focal length*. Radial distortion will result in point d being recorded at d' in the Gaussian image plane or at d''' in the film plane (focal plane). The distance $d''d'''$ is termed the *radial distortion* or simply the distortion of the lens; and can be determined from the following formula:

radial distortion $= r - f \tan \alpha$, where α is the angle of the ray to the optical axis, f is the equivalent focal length and r is the radial distance from the principal point to point d'''.

Focal length of hand stereoscopes. For calculations involving a simple hand stereoscope, the basic lens formula may be applied directly in calculating the focal length:

$$-1/f = 1/v - 1/u \quad \text{(but object distance is negative)}$$

where the image distance is u, the object distance is v and f is the focal length. If the image and object distances are known, then the magnification (M) is given by the formula:

$$M = u/v.$$

When the object distance is so far away that it may for practical purposes be considered to be at infinity, the image distance is equal to the focal length; and this provides a simple way of determining the focal length of a hand stereoscope by measuring the distance between the lens and the image of a bright object formed on a sheet of white paper.

Example. A bright object is 16 cm. from a lens and forms a clear image on the opposite side of the lens at a distance of 25 cm. What is the focal length and what is the magnification?

$$-1/f = 1/v - 1/u = 1/16 - 1/25$$

$$f = 44 \text{ cm.}$$

$$M = u/v = \frac{25}{16} \cong 1 \cdot 6.$$

(b) Shutters

A shutter is a mechanical device in the camera which when set in motion allows light passing through the lens to reach the emulsion surface of the film for a predetermined time. The shutter functions as a light valve controlling both the time in which the light can pass to the film and the quantity of the light that passes. If the shutter regulates only the time, then a separate *diaphragm* is required to control the quantity of light passing in unit time. The time the shutter is open depends primarily on the external light intensity, the speed of the film, the properties of the filter, the ground speed of the aircraft in relation to flying height above the ground and the aircraft's angular velocity due to yaw (swing), roll and pitch. The term *exposure* should preferably be limited to cover the product of the intensity of the illumination multiplied by the time it lasts. Some photographers use the term also to include the act of exposing a section of film.

The ratio of the focal length of the camera lens at infinity divided by the lens diameter is termed the *F-stop* or exposure number (e.g. $F/4$, $F/5\cdot6$, $F/8$); and is used to achieve the best exposure setting of the camera in conjunction with the film speed, filters and light conditions. For example, Maruyasu & Nishio (1962), using Fuji negative colour film and apertures of $F/4$, $F/5\cdot6$ and $F/8$ and shutter speeds of 1/125 and 1/250 seconds at altitudes between 750 m and 4,000 m (focal length 135 mm) concluded that the most acceptable prints were obtained with a setting of $F/8$ and 1/250 seconds. Most black-and-white photographs are taken with shutter speeds of 1/150 to 1/300 seconds and $F/4$ to $F/11$. In the United Kingdom, an aperture of

$F/6\cdot3$ and a speed of 1/300 is commonly used. A capital F has been used with stop numbers, as f refers to focal length.

There are basically three principal types of shutter. If the shutter is between the lens and film, it is termed a *focal plane shutter*. This may be considered as a curtain with a slot cut out of the centre. To provide the correct exposure the slot moves across the film at a predetermined speed. A second type of shutter is fitted between the component lenses, being termed a *between-the-lens shutter*, and consists of a 'rosette' of thin metal leaves which open and close to provide the exposure. This type of shutter is probably the most popular today. It is sometimes termed a rotary shutter. Finally there is a *louvre shutter*, which consists of parallel thin metal strips opening and closing like a venetian blind. This type of shutter is normally placed in front of the lens. It is not often used these days. To complete the description of camera-shutters, the *shutterless camera* must be mentioned. As the name implies there is no shutter. The film is rapidly fed from spool to spool across the focal plane in the direction of flight and at a speed related to the ground speed of the aircraft (Mignery, 1957).

Usually for photo-interpretation the between-the-lens shutter provides the best negatives. The characteristic of this type of shutter is that light passes to all parts of the unexposed film, instantly at the end of the exposure. Cluff (1952) pointed out that the focal plane shutter, which is very popular in the United States, introduces positional errors of varying magnitudes of points being photographed and that such errors are eliminated with a between-the-lens shutter.

Focal plane. This is best described as the plane within the camera on to which all light rays, passing through the lens of the camera, come into focus. Two types of focal planes are used. Particularly in older cameras, the focal plane is a sheet of high-quality glass having optical qualities as near as possible to the lens. The fact that the glass focal plane cannot be provided aberration-free has resulted in preference for an 'open' focal plane, as described below. In addition, glass focal planes have the disadvantage of causing scratches on the exposed surface of the film if they are not perfectly clean; and under dry conditions a build-up of static electricity on the glass surface results in tree-like lines developing on the film due to electrical discharge. These defects will show up in the processed film and contact prints used in interpretation.

The 'open' focal plane is simply an air space enclosed by a metal frame. The film occupying this space is held in position and against a metal plate at the back of the film by air-pressure or a partial vacuum. Sometimes on photographs of water surfaces off the Australian coast the pattern of the pressure plate has been observed in the print.

Attached to the metal frame, enclosing the focal plane, are four metal markers. These are located at the mid-point along the sides of the frame or in the corners of the frame. At time of exposure of the film, the outline of the markers is recorded on the film; and the images of the markers are known as the *fiducial marks*. These fiducial marks are used by the photo-interpreter to locate the geometric centre of the photograph (see Chapter 9).

(c) Filters

Filters are strictly not part of the camera, being external accessories which are varied according to the type of film used and to a lesser degree according to environmental conditions and purpose of the photography. Primarily filters are fitted to the front of the lens to eliminate the effects of haze, which severely scatters the light at the blue end of its spectrum. The scattering is in accordance with Rayleigh's law, there being very little diffused light in the red and infrared part of the spectrum. The choice of filter also enables the contrast between certain features of the terrain to be increased, by allowing those parts of the spectrum most suited to the film and requirements of the photography to reach the emulsion surface of the unexposed film. As explained later, in colour photography, filters are used to control the quality of the light towards sunrise and sunset by bringing it to a standard colour-temperature for the film. In the U.S.S.R., special shadow-filters are reported to be used with wide-angle lenses (e.g. 100°) and colour film (Mikhailov, 1961). These cast a shadow on the image plane, the density distribution of which is the reciprocal of the light distribution in the field of view.

Most filters are made of high-quality coloured glass, although sometimes dyed gelatine between two plain sheets of glass or simply dyed gelatine are used. It is important that the optical qualities of glass filters should equal the optical qualities of the lens so that resolution is not reduced. Aberrations are minimized by careful grinding of the filter, in manufacture, to a flat surface. Nowadays gelatine filters are used only in experimental work and with colour films, as they are fragile. For colour photography, the gelatine filters are sometimes used in pairs. One acts as a haze filter to correct for excessive blue and the other provides a balance of colour rendering of the ground features. If too much of the blue light is eliminated, blue objects may appear black on the processed film. The large-scale manufacturers of colour aerial film normally enclose filters with the film, as the colour-balance filter needs adjustment to each batch of film manufactured. Mikhailov (1961) reported that in the U.S.S.R. colour CN-film is used without filters or with only a faint yellow filter (JS-3) for flying altitudes up to 10,000 ft.

The effect of the most commonly used 'film–filter' combination is shown diagrammatically in fig. 2.6. It is seen that a minus-blue filter (Wratten 12) placed in front of the film absorbs all the violet and blue light. The exclusion of the blue light increases the contrast in black-and-white photographs and reduces bluish tinges in colour photographs and transparencies. Ultra-violet light, invisible to the eye, is also excluded, as the emulsion surface may be sensitive to it. It will also be observed from the diagram that although the extreme red light and infrared light passes through the filter, the emulsion surface is not sensitive to these wave-bands. With aerial

23

photography below a flying height of about 2,000 ft, it has been found unnecessary to use a haze filter with panchromatic film and some lenses.

If, now, the panchromatic film is replaced by infrared film, the far-red and infrared light would be absorbed by the emulsion but the light below about $0.50\,\mu$ would be reflected. This combination of film and filter has been termed *modified infrared*, and is frequently used in North America for separating hardwood species from conifers. If the minus-blue filter is replaced by a green filter, the blue-green light will also be reflected by the filter and only green light and longer wave-lengths will reach the film. If this is further replaced by a red-coloured filter, all the spectrum excluding the red and infrared will be reflected by the filter. Since this would require an extremely long exposure for panchromatic film, being about four times greater than the normal

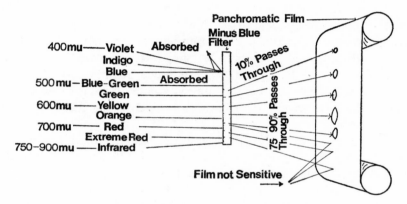

FIG. 2.6. *Film–filter combination.* Solar energy on reaching the filter is absorbed at the blue end of the spectrum, but most of the remainder passes through the filter. Panchromatic film, however, is not sensitive to the infrared.

exposure time, red filters are only used with infrared films. With infrared film, the use of a red filter doubles the exposure time in dry conditions; and in damp or misty conditions an even longer exposure may be needed as mist readily absorbs infrared energy. As the red filter excludes much of the visible spectrum, nearly all diffused light will be excluded. This can be a serious disadvantage in photo-interpretation, as the recording of details in shadow is improved by the light of the shaded areas which mainly belong to the blue end of the spectrum. Infrared photography is useful in urban areas subject to long periods of haze.

In colour photography, the three-layer emulsion of the film is so balanced that the colour of the objects can best be recorded at a colour temperature of about 6,000° K. Under cloudy conditions, the tropospheric colour temperature may reach 10,000° K. and just after sunrise and before sunset it will be about 1,750° K. Although colour aerial photographs are not taken under these extreme conditions, it may be necessary to raise or lower the colour temperature of the light coming into the lens to that of the standard colour temperature when intermediate conditions exist which are considered suitable for the taking of aerial photographs. A filter of the blue colour system will raise the temperature, and a filter of the amber system will lower the temperature. In the United States, a filter to cut out all light below about $0.42\,\mu$ is used for about 90% of aerial colour photography; and a 'peach-shaded' filter to cut out light below $0.38\,\mu$ is used for early-morning and late-afternoon photography (Swanson, 1964). Maruyasu & Nishio (1962) recorded that there was an increase in the dark blue or purplish tint of processed colour film with increased altitude at a constant exposure. This suggests that the change in colour depends not only on haze but on other atmospheric considerations as well. In Europe, satisfactory results have been obtained with a Wild lens and no 'UV' filter for altitudes below 15,000 ft, as the lens excluded light below $0.42\,\mu$.

A further type of filter to be considered is the *polarizing filter*. This can simply be described as a polarizing screen, used in front of the lens, to eliminate undesirable reflections of polarized light. It is not a colour filter. Light is polarized by water surfaces and sometimes by waxy leaves. According to wave theory natural or non-polarized light is said to vibrate in all planes perpendicular to the direction of propagation; whilst some reflected light and light passed through a polarizing screen vibrates in a single plane, i.e. is plane polarized.

3

THE FILM

A PHOTOGRAPH is basically a perspective record of length, as depicting size or area and actinic brightness. It is erroneous when considering the aerial film to think in terms of portrait and of landscape photography, as the wrong conclusions may be formulated. For example, a low-contrast aerial photograph may be preferred if it contains fine detail within micro-areas. Such a tone of print would be highly undesirable for landscape photographs, as gross areas are often preferred in sharply contrasting black-and-white. A greyish appearance is sometimes essential for interpretation if resolution and detail are to be at a maximum. In landscape photography it is often preferable to use a long exposure time and a high aperture number (e.g. $F/22$) to improve the depth of field, but for aerial photography this is impossible due to the ground speed of the aircraft, and depth of field is not applicable.

In the last two decades there has been continuous improvement of aerial films, particularly in increasing film speeds without loss of photographic quality. There is little doubt that in the next decade, the black-and-white film will be increasingly replaced by the colour film, as colour processes are improved and costs are reduced. For example aerial colour film was about six times slower than the fastest aerial black-and-white film used under similar conditions. Recently, the U.S. government has introduced colour aerial photography for routine work in coast and geodetic surveys. The reason for this step is that colour photographs for photogrammetry are now of a quality equal to black-and-white and the photographs have the added advantage of increased and easier interpretation.

(a) Black-and-white

The black-and-white film is sometimes termed monochromatic, but, etymologically, achromatic is the more correct. Black-and-white films consist essentially of two parts. These are the emulsive surface layer termed the emulsion, which is sensitive to light, and the supporting back for the emulsion, termed the film base. An aerial film is about 1/200 of an inch thick.

The two important constituents of the unexposed emulsion are the gelatine, which

26

holds the light-sensitive materials, and the light-sensitive materials themselves, which are silver halides in granulated form. The granules usually have diameters between 1/10,000 and 1/25,000 of an inch; but on films requiring a very short exposure (i.e. fast films) the granules may be much larger. In general the fastest films have larger granules; and the final prints will be more grainy and will normally record less fine detail due to light-scattering in the emulsion.

The silver halides in the 'emulsion' are basically sensitive only to the blue, violet and ultra-violet light. The precise sensitivity depends on the particular halide and the exposure. Fast emulsions contain silver bromide, and sulphur compounds are also

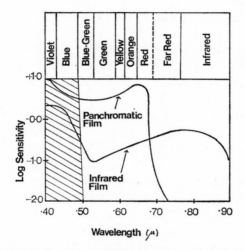

FIG. 3.1. A comparison of the two sensitivity curves shows that, in association with a minus-blue filter (hatched area), the panchromatic (black-and-white) film has a much greater sensitivity in the visible spectrum (blue-green to far red) than the infrared (black-and-white) film; but only the infrared film is sensitive to infrared solar radiation. The filter prevents the shorter wave-length radiation (below $0.5\,\mu$) from reaching the film.

added to increase the sensitivity to light. To extend the range of spectral sensitivity of the film certain dyes are added. These dyes do not act as light filters, but are absorbed by the silver halide crystals.

Black-and-white films can be conveniently divided into two broad classes according to whether this sensitivity is in the visible spectrum only or in the visible and near infrared. The former are termed *panchromatic*, recording from the violet to the deep red at about $0.75\,\mu$, and the latter are termed *infrared*, recording to about $0.90\,\mu$ in the near infrared. A third class, *orthochromatic* films, were popular for aerial photographs before World War II. At exposure these films recorded in the violet, blue, green and yellow-green parts of the spectrum up to about $0.60\,\mu$.

Panchromatic films. Panchromatic film was widely used in World War II and, with

27

continuing improvement, has remained the most popular type of aerial film. The object in sensitizing panchromatic film is to provide an emulsion with a spectral sensitivity similar to the spectrum recorded by the human eye (i.e. $0.40\,\mu$ to $0.75\,\mu$). In fig. 3.1, a comparison of the curves shows that the panchromatic film has the greatest sensitivity in the violet-blue and orange-red bands of the spectrum, whilst the infrared film has greatest sensitivity below $0.45\,\mu$ and in the near infrared. The hatched area indicates the wave-lengths excluded by a minus-blue filter. The high sensitivity of the films at the blue end of the spectrum is to be expected due to the inherent spectral sensitivity of the silver halides. The graph also shows that the sensitivity of the panchromatic film in the visible spectrum is considerably higher than infrared film. It is unfortunate that neither film has a peak to its sensitivity in the visible spectrum between $0.50\,\mu$ and $0.60\,\mu$, as this is the part of the spectrum with greatest variation in reflection from mature leaves. If the spectral reflectance graphs (fig. 6.4) are compared with fig. 3.1, it will be seen that the maximum sensitivity peak of the panchromatic film coincides with the trough of the reflectance curves of the tree species between $0.65\,\mu$ and $0.725\,\mu$.

A.S.A. exposure indices should be used with caution for indicating the speed of aerial films, as the index was designed only for landscape and pictorial monochromatic photography in the visible spectrum. With the ASA system, the exposure index is designed to provide a high safety margin for under-exposure. In practice this had resulted in undesirably high densities; and in order to overcome this problem the ASA index of speeds has now been doubled. Recently Kodak have commenced using an 'aerial exposure index' for aerial films. The A.E.I. is defined as 'the reciprocal of twice the exposure at the point on the toe of the characteristic curve where the slope $= 0.6\,\gamma$' (Kodak Techbit No. 2, 1965). The ASA standard speed is expressed in terms of the exposure to obtain a density of 0.1 above fog density.

Infrared films. The term infrared strictly covers wave-lengths beyond the visible spectrum up to about $100\,\mu$. Consensus of opinion differs on the exact boundary of deep red and the near infrared. It is variously quoted between $0.725\,\mu$ (e.g. Brock, 1952), and $0.75\,\mu$ (e.g. Spurr, 1960). In the present discussion it is accepted as $0.75\,\mu$. Recently 'thermal maps' have been prepared using wave-lengths emitted by sensory devices beyond $2.0\,\mu$; but a limitation for normal photographs occurs at about $1.2\,\mu$. At the present time infrared films at $1.2\,\mu$ are known to be unstable and are used only for special indoor spectroscopic work. Currently aerial infrared films have a maximum range of sensitivity at $0.90\,\mu$ to $0.975\,\mu$. Clark (1949) stated that there are fundamental reasons why sensitivity of the emulsion of aerial films cannot be increased much beyond $0.90\,\mu$.

Infrared film has three disadvantages when compared with panchromatic film. Normally infrared film requires storing at about 55° F. until a day or so before use (Brock, 1952). Many lenses, especially of older manufacture, must be recalibrated before being used with infrared film. Infrared film requires a longer exposure than panchromatic film. Brock (1952) quoted that infrared film with a tricolour red filter required about four times the exposure of panchromatic film with a minus-blue filter. However, the exposure can often be halved for green vegetation, due to its high reflectivity. The exposure should not, of course, be reduced for leafless trees. Growing grass may provide up to 40% reflectance of infrared light compared with less than 10% reflectance in the visible spectrum.

The purpose of using infrared film in certain work is not so much to penetrate haze as to provide greater contrast. Northern hemisphere conifers normally record dark compared with hardwoods. With infrared film, leafless trees against growing grass are recorded in contrast similar to that of leafless trees against snow when panchromatic film is used.

Several writers (e.g. Nyyssonen, 1955; Spurr, 1948; Welander, 1952) concluded that an estimate of tree height and stand height is as accurate with infrared film, using a minus-blue filter, as with panchromatic film and a minus-blue filter. Haack (1962), working in Alaska, found under the conditions prevailing there that for forest types, stand sizes and tree heights were more readily identified on photographs from infrared film. No doubt this is partly due to the need to recognize separately conifers and hardwoods. However, for measurement of shadow-length infrared film is considered to be unsuitable (Nyyssonen). This is to be expected, as in accordance with Rayleigh's law there will be very little diffused light in the infrared part of the spectrum for recording shaded areas. In northern Sweden, Welander (1952) also favoured infrared film for species identification.

Colwell (1960a), in comparing photographs from infrared film and panchromatic film in North Carolina, considered that for an area of agricultural and non-agricultural land, panchromatic film was superior for showing shallow water, individual trees, moist depressions, cultivation marks, the boundary between marsh grass and upland grass, forest roads, differences in crop maturity, bare ground surrounded by forest trees, and ripple marks. Infrared film was superior for showing the boundary between two marsh grasses, bodies of fresh water, individual orchard trees, an admixture of dark-toned conifers to light-toned hardwoods, standing grain and stubble and drainage swales and fallow ground.

Meyer & Tranlow (1961) working in the conifer–deciduous forests of northern Minnesota with Kodak Tri-X film and Infrared film found that, although there was some improvement over normal film–filter combinations for summer photography, polaroid filter photography was inferior to panchromatic film with minus-blue filter taken in the autumn and infrared film with a minus-blue filter taken in the summer. They used polaroid variable colour filters favouring tonal relationships in the yellow and green parts of the reflected spectrum. On the American west coast, conifers being dominant, panchromatic film and the minus-blue filter are commonly used. Ryker (1933) preferred panchromatic film with a green filter (Wratten 55, 59 or 64) and was able to distinguish four species at a scale of 1/9,600. Jensen and Colwell (1949) in California concluded that panchromatic film with a minus-blue filter was best for classifying forest, grasslands and soils. Hardwoods in autumn, with their range of colours between yellow and red, record nearly white with a minus-blue filter and panchromatic film.

(b) Colour films

A further choice of film for photo-interpretation is provided by the increasing number of colour films now being marketed. Aerial colour film was used as long ago as World War II for camouflage detection; but for many years the use of colour film in photo-interpretation was curtailed due to the slow speed of this type of film, e.g. 8 ASA to 10 ASA. Anscochrome (25 ASA) was the first multi-layer reversal type of colour transparency film with a speed approaching monochromatic films. Recently Ansco has marketed a much faster film: Super Anscochrome Tungsten film (100 ASA). Agfa markets Aerial Agfacolor (40 ASA) for exposures of 1/250 to 1/400 seconds at $F/4$ to $F/5 \cdot 6$; and Gevaert has aerial film at 25 ASA (Gevacolor). Kodak has recently marketed an improved Aerial Ektachrome from which colour prints can be provided and also manufactures Ektachrome Infrared (i.e. camouflage detection/false colour film). Fujicolor, marketed in Japan, has a speed of 20 ASA. In North America most colour film is supplied by Eastman Kodak. In Western Europe and certain Commonwealth countries colour aerial film is supplied by Agfa, Ansco (Ilford), Gevaert and Kodak. As mentioned in section (a), ASA indices are not strictly applicable to black-and-white aerial film; and similarly the ASA numbers given above for colour film should also be considered as only a rough guide.

Normally reversal film is used, so that after processing, interpretation is carried out on transparencies. For colour prints, negative film is preferred, as the production of prints from transparencies is a more costly process, requiring the making of a negative from the transparency, and the quality of the print is often inferior. As mentioned elsewhere, special filters may be required with each batch of reversal film at the time of exposure. Reversal and negative films have an emulsion of three light-sensitive layers, each layer being sensitive to a particular part of the visible spectrum. Whereas panchromatic colour film is designed to reproduce the colours of objects faithfully, three-layer infrared film is designed to embrace the contrast of objects having different reflective properties in the infrared, which are not conspicuous to the human eye. In the U.S.S.R., two-layer film, termed spectrozonal (i.e. SN-film) is used in addition to three-layer colour film (i.e. CN-film). The three layers of Ektachrome Infrared (sometimes termed camouflage detection) are sensitized to green, visible red and infrared respectively. The combined effect of the light-sensitive layers is to record, in the processed film, green vegetation as red, other green objects and dead trees as blue and 'visible' red objects as yellow or brown. The spectral sensitivity of a reversal film is shown in fig. 3.2a. The use of a minus-blue filter results in blue objects recording as black images on the transparencies.

Mikhailov (1960) recorded that spectrozonal or false colour film in the U.S.S.R. greatly facilitated the interpretation of detail (see also Belov, 1959). In this type of film, the lower layer, which is panchromatic, has a peak at 0.65μ; whilst the upper layer has its peak sensitivity at 0.60μ and 0.735μ (fig. 3.2b). After processing, the upper layer is rendered in light blue and the lower layer in magenta. The speed of this film and the CN-film were reported to be about 100 ASA and 65 ASA respectively. It was also claimed that as low-density filters are used with this type of film, photographs can be taken at higher altitudes or under less favourable conditions. Mikhailov was of the opinion that three-layer Russian film is to be preferred for geological and many soil studies as images are more readily identified in their natural colours; and SN-2 film is preferable for taking photographs of wooded areas and possibly soil types in desert areas, as it facilitates the recognition of detail.

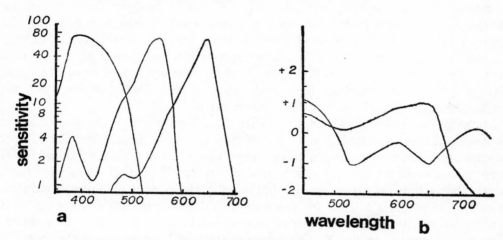

FIG. 3.2. (a) Panchromatic (colour) film is a three-layer film and has three well-distributed peaks of sensitivity. The film is not sensitive to infrared solar energy. (b) Spectrozonal (false-colour) film is a Russian two-layer film which is sensitive to solar radiation in both the visible spectrum and part of the infrared spectrum (see text).

Worthy of comment are the precise controls required in manufacture. For example, consider a panchromatic colour film similar to that illustrated in fig. 3.2a. There are three colour layers to be controlled in manufacture, which is carried on in complete darkness. The colour sensitivity of each layer is different, and the tolerance in thickness for each layer requires to be kept within 0.15μ! The layers may have an average thickness of 5μ and additional layers are required to provide halation protection, optical filtering and physical-chemical separation of the three emulsion layers.

Aerial reversal colour films have certain limitations when compared with black-and-white films. The latter have a wider tolerance to over- and under-exposure than reversal films. For all colour photography environmental conditions are more exacting. The hours of daylight suitable for colour aerial photography are fewer, the times of the year are shorter and the number of suitable flying days are less, due to

D   31

the effect of haze. The colour temperature of sunlight may change throughout the day from about 1,800° K. at sunrise to 5,000° K. or 6,000° K. at noon. For comparison the colour temperature of a tungsten bulb is 2,500° K. to 3,400° K. When using black-and-white panchromatic film it is only necessary to adjust the exposure to the quantity of available light after excluding the ultra-violet end of the spectrum; but in colour photography the adjustment has also to be made according to the quality of the available light.

In favour of colour, despite the higher cost, is the enormous increase in combinations provided by hue, value and chroma, as compared with tonal differences only in black-and-white photographs. Using the Munsell colour notation, *hue* refers to colour (red, yellow, blue, green, purple), *value* to its brightness or dullness, and *chroma* to its strength (sometimes termed intensity saturation), (see fig. 3.4). For some purposes in Scandinavia, small-scale infrared colour photography (e.g. 1/40,000), has been found as satisfactory as much larger scale black-and-white photography in the visible spectrum. In the Munsell classification, there are five basic colours, subdivided into two intermediate hues, and these are each further divided into ten smaller divisions (i.e. giving a total of 100 hues).

Maruyasu & Nishio (1962) reported that the interpretation of shaded areas on photographs was impossible when infrared film was used, limited and difficult for panchromatic but easy for colour. Haack (1962) recorded that colour transparencies brought out minute tonal differences, and differences in ground detail were more clearly shown than on panchromatic photographs. Heller *et al.* (1963) reported that, at scales 1/3,960 and 1/1,180 for nineteen Boreal tree species, more species were identified correctly using Super Anscochrome Colour film than when using monochromatic Aerographic Plus X. The accuracy achieved between interpreters was the same for the colour film but not so for the black-and-white film.

On colour prints and transparencies, visibility through the top canopy layer of the trees is often improved sufficiently to recognize the next tree layer or shrub layer or herb stratum. Rocks, rivulets and dead trees also stand out clearer, but a dry green meadow and swampy grassland may record the same hue although in black-and-white their tones are different (for films recording in the visible spectrum). In Malawi, colour prints proved better for distinguishing land-use classes than black-and-white. In Alaska (Haack, 1962), however, it was concluded that for forest inventory there was insufficient gain in accuracy to warrant the use of colour film. Low oblique transparencies taken with a miniature camera have been used with varying success in entomological and pathological studies in North America, Malaya and Australia.

Experimental work in North America with camouflage detection film (i.e. infrared colour film) has indicated that this type of film can be useful in determining the vigour of tree stands, as the tone varies conspicuously according to the amount of new foliage and chlorophyll activity. In East Germany, spectrozonal film was preferred for detecting the vigour of forest stands and detecting disease (Wolff, 1966). The use of black-and-white infrared film and colour infrared film has enabled early detection to be made of *Puccinia* infection of oats (Colwell, 1956). In Jordan (1966), colour infrared film was found to be useful under desert conditions for locating ancient settlement, due to the conspicuous pattern provided by the shrubs growing on the sand-covered ruins.

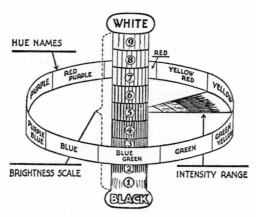

FIG. 3.4. Munsell colour wheel showing the three dimensions of colour: hue, value (brightness) and chroma (intensity).

The first colour diapositives for photogrammetry were provided by Eastman Kodak in 1963 by laminating Kodak colour duplicating film on ¼ in. glass. The colour balance, contrast, granularity and resolution were all reported to be highly satisfactory and the metric stability to equal black-and-white emulsions for aero-triangulation (Swanson, 1964b). The standard error on a single grid intersection co-ordinate on colour plates was under 2 μ after a linear least square fit for correspondence.

(c) Film–filter combinations

Due to the large number of film–filter combinations existing, it is economically impracticable to determine the best combination for a specific project by trial photographs from an aircraft with various films and filters; but the selection of a number of practical combinations can be helped by means of spectral studies (see Chapter 6), or by examining the photographs taken with a cluster of miniature cameras in the aircraft. In many parts of the world a restriction of choice results from what films and equipment are locally readily available.

The film–filter combination has a pronounced effect on the tone and contrast of black-and-white photographs (fig. *3.3*) (see Meyer & Tranlow, 1963); and on hue and colour contrast characteristics (see Colwell, 1960a). Within reasonable limits, the film–filter combination seems to have no effect on the stereoscopic parallax characteristics and very little effect on image sharpness characteristics (e.g. Colwell, 1961).

Film–filter combinations, utilizing the shorter wave-lengths, however, usually provide the best details in shadow, as mentioned previously, and, where by other combinations the ground would be obscured by shadow, heights of trees are easier to assess. Combinations favouring the intermediate wave-lengths are frequently the best for soil mapping, and combinations favouring long wave-lengths give the best images of water courses and are best for detection of swamps and shore-lines.

A selection of film–filter combinations is given in table 3.A. The data provided in the table is based on information extracted from published articles relating to the interpretation of natural resources, especially forests. Column two shows the approximate wave-length below which the solar spectrum is excluded from passing through the filter. Whenever possible an appropriate Wratten number has been shown.

TABLE 3.A
Film–filter combinations

Filter colour	Excludes wave-lengths below	Equivalent Wratten number	Film	Remarks
Peach	0·39		Colour film	Late-afternoon colour photography
	0·39/0·41	—	Panchromatic	Ultra-violet filter (UV)
Light yellow	0·42	3	Colour film	e.g. land-use surveys
Medium yellow	0·45/0·475	8	Panchromatic	Light haze—popular photogrammetric combination
Deep yellow	0·52	12	Orthochromatic	Sea-water penetration
Deep yellow	0·475/0·50	12	Panchromatic	Minus-blue filter—most popular combination
			Infrared	Modified infrared—some forest surveys
			Ektachrome IR	Colour photography (medium haze)
Orange	0·57		Panchromatic	River-water studies
Orange	0·52	15	Colour infrared	Plant vigour/disease
Light red (tricolour red)	0·56/0·60	25	Infrared	Heavy haze
Deep red ('black')	0·70	88A, 89B Ilford 207	Infrared	Infrared photography; some hardwood/conifer forest surveys
—	Polarizing 'filter'	—	Panchromatic	Water sub-surfaces
Red	0·62	?	Spectrozonal	e.g. plant vigour/pathology

4

FILM PROCESSING AND PRINTING

(a) Negative processing

To obtain the black-and-white aerial photographs for interpretation from the exposed film two distinct operations are involved, namely negative processing and printing. It will be recalled that taking a photograph consists basically of exposing the emulsion of the film for a period long enough to record a faithful image of the object. As the image on the exposed film is not visible to the human eye until development is complete, it is termed a *latent image* and is formed by the action of reflected sunlight on the silver halide particles of the emulsion. In this process, the electrons of the silver halide have their energy raised by the absorption of photons in accordance with the concepts of modern quantum mechanics. During the ensuing processing the unchanged silver halide grains are washed out in solution whilst the remaining grains forming the image are converted by the developer into metallic silver and remain in the emulsion. For details, reference should be made to Mees (1953).

As some objects are better reflectors of light than others, the density of the developed film will vary from that of a clear film to black according to the proportion of the grains reduced to metallic silver. Since bright objects reflect the most light in the visible spectrum, they will be represented by the darkest layers of silver on the developed film; and conversely darker objects are recorded as the lightest layers of silver on the film. On developed infrared film, green foliage appears dark, as green leaves are normally the highest reflectors of light in the near infrared. At printing, however, the tones of the images are reversed; and the foliage that appeared dark in the infrared negative will be recorded in light tones on the photographic print. In developed colour negative film, the colours of the objects will be represented by images in complementary colours.

Normally in large laboratories the developing is by the *continuous processing method*. The exposed film is passed through developer, washed in water and then fixed before being washed again in water to remove excess chemicals. Finally, it is dried, by exposure to warm air. Sometimes, instead of a continuous process, the exposed film is treated as a unit in one solution before being transferred as a unit to the next solution, or alternatively the film remains *in situ* and the solutions are changed. This method is termed *batch processing*. In black-and-white photography. the developer is basically a reducing agent, which converts silver halide to metallic

35

silver. The developer also contains an alkali to adjust the pH, a preservative (e.g. sodium sulphite) and potassium bromide, which may act as an anti-foggant.

Reversal processing. In addition to negative processing there is reversal processing, as used in the production of colour transparencies. In processing colour films to provide colour transparencies the film is first developed to a negative silver image as mentioned above. Then the silver is bleached and the remaining halide is exposed to light so that it is completely fogged. Finally, the film is immersed in a colour developer, which converts the halide in each layer to its appropriate colour.

As mentioned in Chapter 3, the colour is provided by three layers. Usually the layer sensitive to blue light is on top, the layer sensitive to green light is in the middle and the bottom layer is sensitive to red light. In the reversal processing the three layers take on complementary colours. The red-sensitive layer becomes the blue-green portion of the *positive transparency*, the middle green layer will provide the magenta layer and the top blue layer will provide the yellow layer.

(b) Printing

The black-and-white photographic print is obtained by exposing a light-sensitive printing paper in contact with the negative to artificial light for a specified time. The printing paper then contains the latent image which is developed, fixed and washed to provide a black-and-white photograph. The fixer usually contains hardening agents. Finally, the print is sponged to remove excess moisture from the surface and dried in warm air.

The three principal methods of printing black-and-white photographs include *projection printing* used only for enlargements. Normally best-quality prints are made from negatives by *contact printing*; but in recent years *automatic dodging* has become increasingly popular for economic reasons and as quality and techniques have been improved. In contact printing a battery of lamps are manually operated so that variation in the density of the negative can be compensated for in the printing process by providing varying light intensities. Occasionally it is desirable to make two contact prints, one light and one dark, from the same negative to record its entire contrast range. Alternatively, a single diapositive may be preferred, as it will contain almost the same amount of detail as the two prints. A *diapositive* is a positive print on a transparent medium, which is usually glass.

Fleming (1963) pointed out that the automatic dodging printer has the advantage of accommodating on one print the range of densities occurring on the negative. Normally, when contact printing, the range of densities of the negative are too great to be reproduced fully on a single photographic print. In automatic electronic dodging, the negative is scanned by a spot of light emitted from a cathode-ray tube, the light intensity varying in proportion to the density of the negative. A disadvantage is that when the scanning spot moves suddenly from a high density to a low density, and depending on the contrast of the printing paper and the response of the electronic circuit, the sharp transition may cause over-exposure on the print in this zone. Fleming (1963) termed this *'edge halo'*. Some photo-interpreters consider that on prints produced by electronic dodging there may be a loss of important subtle changes in the tone of vegetation and there may also be a loss of shadow provided by the forest canopy. Possibly within a small area on the photograph, the tonal range is contracted, but there should be no loss of micro-detail.

Often, stereoscopic examination will show that the greyish toned photograph contains more details of the negative than the 'contrasty' photograph in which details will have been lost in the distinctive white and black zones, due to the tones of interest being displaced to the 'toe' or 'shoulder' of the characteristic curve. *Glossy prints* with a high contrast have the advantage of a greater density range, but *matt prints* can easily be written on and the reflected glare from the surface is not so trying to the eyes. Nowadays, a *semi-matt* or semi-glossy or glossy matt surface is generally preferred as a compromise between the two types. Glossy implies a shiny surface.

In black-and-white printing methods, the paper on to which the image in the negative is transferred comprises a white base, a silver bromide or silver chloride emulsion and (most commonly) a glossy surface. A polyester base is preferable to a cellulose base as it is very stable. In Western Europe, including the United Kingdom, bromide papers are used, whilst in North America chloride papers are customary. Neither can be said to be superior. In both bromide and chloride papers there is a wide range of contrasts from which to choose as best suiting the negatives and the desired finish of the print, For example, Stellingwerf (1966) preferred for maximum tonal differences a soft bromide printing paper, an exposure of 0·9 seconds and a development time of 2 minutes.

In making colour prints, as in black-and-white printing, the exposure time of the print paper is important. The print paper is three-layered. In the actual printing, filters are used to correct for offending colour casts. For example, if the unwanted overall cast is blue, a blue filter is used with the white light source. Mikhailov (1961) claimed that it is preferable to print on normal three-layer colour print paper when using Russian two-layer film as the yellow layer of the former improves the tones. On photographs of Malawi, greater tonal differences were observed when printed in black-and-white from colour negatives than from black-and-white negatives.

Certain defects encountered in aerial photographs are concisely covered by Tupper *et al.* (1952) and Fleming (1963). These are given below:

Blurred areas—These are caused by lack of contact between the paper and negative in contact printing, or in projection printing by the use of a poor lens, improper focusing or vibration during exposure.

Fingerprints—Photographic paper should always be handled by the extreme edges and corners, and the hands should be clean to avoid contamination.

Abrasion, streaks and scratches—Fine scratches may be due to abrasion in removing the paper from the box. Larger scratches, creases and cracks are due to careless handling.

Air bells—These are small bubbles which may form on the emulsion surface when the paper is first immersed in the developer. They prevent the chemical solutions from coming in contact with the emulsion. They are less likely to occur if the print is immersed, emulsion side up, with a quick sliding motion and agitated well during processing.

Fog—This may be caused by excessive development, inadequate safelights, light leaks in the printer, and use of over-age paper or of paper stored under adverse conditions, such as excessive temperature and humidity.

Streaks—These may be caused by immersing the paper too slowly so that portions of the print are developed appreciably longer than others.

Irregular white spots—These may be caused by dust on the surface of the negative during exposure.

Irregular dark spots—These usually result from pin-holes or scratches on the negative.

Flat prints—These are caused by overexposure and underdevelopment, or by using a paper of too low contrast.

Excessive contrast—This is caused by using a paper of too high contrast to suit the negative.

Density too high or too low—Excessive density is caused by overexposure or over-development; inadequate density is due to underexposure or underdevelopment.

Blisters—These may be caused by temperature differences between successive processing solutions, or by permitting a strong spray of water to strike the print continuously at one spot.

Brown spots—These may be due to rust particles in the water, to insufficient rinsing after development, or to the use of exhausted developer.

Yellow stain—This usually does not appear until some time after the print is dried. It may come from using an exhausted developer or fixing solution, insufficient washing, using over-age paper, or drying the prints on cloth which has been previously contaminated by inadequately washed prints.

Fading—This may be due to insufficient fixing, exhausted fixing bath, or inadequate washing.

Double image—Over even-toned areas, e.g. water surfaces, a double image is sometimes observed. This double image is said to be the result of the anti-vignetting coating on the front of the camera-filter reflecting light into the lens.

Drying marks—These are caused by uneven drying, usually because droplets of water have been left on one or both sides of the film. They may be avoided to a marked degree by rinsing the film in a wetting agent after washing.

Pressure-bar marks—Long lightish coloured streaks may occur on the print due to uneven pressure on the film as it is wound through the developers.

Fork-like or finger patterns—These are caused by static electric discharge when the film is unwound rapidly under dry conditions at the time of taking the photographs. These days, this defect has almost been eliminated, as most manufacturers seal their films at time of packing at a relative humidity of about 60%.

Dark streaks—Across the surface of the negative may be recorded shadows of vapour trails.

Small overexposed circular zone—Possibly the no-shadow zone (see Chapter 6).

(c) Important physical characteristics of the image

Finally, before considering tropospheric conditions and factors influencing the reflection of light from objects being photographed, it is desirable to draw attention to certain important physical properties or characteristics of the film and print. Some of these have already been mentioned and the importance of others will be appreciated as Parts III and IV are read. The quantitative study of the response of photographic materials to light has been given the name '*sensitometry*', and was adequately covered in relation to aerial photographs by Brock (1952). Unfortunately Brock's book (*Physical Aspects of Aerial Photography*) is now out of print, but an excellent reference work from the point of view of general photography covering the topics now to be mentioned in *Fundamentals of Photographic Theory* (1960).

Density. The developed film has a surface of varying degrees of darkness, the tones of which can be measured scientifically and expressed as a number called the density. The higher the density number, the darker is the developed film. The density is evaluated on the basis of the proportion of light which can be transmitted through the developed film or negative. It is strictly a measurement of the opacity of the negative in common logarithms; and is defined as the common logarithm of the reciprocal of the transmission (Brock 1952). Thus if $1/10$ of the light passes through the negative, the *opacity* is 10 and the density is ($\log_{10} 10$). Similarly if $1/100$ of the light passes through the negative, the opacity is 100 and the density is 2. In formula form this may be expressed as

$$D = \log_{10} \frac{I_o}{I_t},$$

where I_o is the intensity of the incident light in the negative and I_t is the transmitted light through the negative.

Density is a factor important to the aerial photo-interpreter, as a photograph from a negative with a high density will lack a considerable amount of detail. When photographs of a forest area are taken in an east–west direction, it will often be observed that the density of the negative varies considerably from the north side to the south side. In colour infrared photography it may not be possible to accommodate satisfactorily in a single photograph desert surfaces and seascapes, nor tropical forest and urban areas.

Contrast. Closely related to density is contrast, which again affects the value of a photograph for photo-interpretation. Contrast can be defined as the actual difference in density between the high-lights and low-lights or shadows on a negative or photographic print. Density measures the darkness of a negative, whilst contrast is not concerned with the magnitude but the difference in the densities. Long developing time and active developers are associated with the maximum contrast in the negative. The *reflection density* of the photographic print may be expressed as the brightness of the processed emulsion (B_1) in relation to the paper base (B_2), i.e.

$$\text{reflection density} = \log \frac{B_2}{B_1}.$$

Tonal contrast between an object and its background is principally influenced by the light reflectivity of each, the spectral sensitivity of the film, spectral transmission through the filter, scattering effect of the atmospheric haze, developer, developing times and characteristic of the printing paper.

It is known from optics that a grey object against a white background appears darker than against a dark background. Trees of given species having a background of the same species, as for example, regeneration at the edge of a wood, are frequently less conspicuous on photographs than trees of the same species mixed with a species appearing darker, e.g. black walnut (light-coloured) and beech (dark-coloured) in eastern North America.

Possibly contrast is not as significant in forest photo-interpretation as sometimes stated. Meyer & Hugo (1961) used absolute stereoscopic parallax difference and the number of tree crowns capable of being counted on unit areas of photographs, as criteria of quality when examining different degrees of contrast on variable contrast paper. The same areas were examined at five stages of contrast on variable contrast paper. The variable contrast prints as a group were found to be equal in quality and interchangeable with the single contrast prints. Variable contrast prints were found to decrease in quality as contrast approached maximum or a minimum in terms of the number of tree-crown counts as shown in fig. 4.1. There was a significant quadratic relationship between photograph sets A, B and C. Photograph 'D' had the lowest contrast and photograph 'A' had the highest contrast. The point of maximum quality at 'B' corresponds to a slightly flat print.

The significance of subject contrast is illustrated by the fact that a 3 in. white line on a dark road may be conspicuous on a photograph at 1/10,000, using an $8\frac{1}{4}$ in. lens, although the resolving power of the object being recorded is about 2 ft. Similarly, on a 1/10,000 photograph sheep may be seen in a field, being white contrasted against a dark background, whilst the crown of a sapling, although larger, may not be conspicuous due to its having a tone similar to the ground. Nyyssonen (1955) recorded that on photographs of Saarijarvi at 1/10,000 to 1/15,000, the power lines themselves were distinctively visible, although they should normally be indistinguishable. This was due at least partly to *'image spread'*. That is the spreading on the photograph of the image of brightly illuminated ground objects. Lack of contrast often makes the identification of understory species difficult or impossible on black-and-white photographs; whilst identification may be possible by distinctive colour hues and colour values in colour transparencies and colour prints.

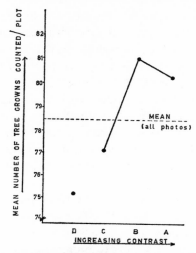

FIG. 4.1. Meyer and Hugo used the number of tree crowns capable of being counted on stereo-pairs of photographs as a criterion of quality when examining the different degrees of contrast. Maximum quality (B) corresponds to a slightly flat print.

Characteristic curve. A useful aid in studying contrast is the '*characteristic curve*'. This curve is obtained by plotting density against the logarithm of the exposure. Typical characteristic curves for matt (soft) and glossy (contrasty) printing papers are shown in fig. 4.2. There are also characteristic curves for the film.

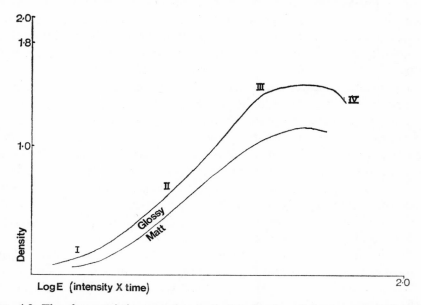

FIG. 4.2. The characteristic curve (see text). I is the toe of the curve and II is the gamma (γ) or slope of the curve.

41

The S-shaped curve is of similar basic shape for all films and for print papers, although the exact shape of the curve will vary with the type of the emulsion and the way in which the film is developed. The slope of the straight part of the S-curve is known as *gamma* (γ); and is used to describe the contrast of the negative or photograph. The steeper the curve, i.e. the higher the gamma, the greater is the contrast. Thus high contrast paper has the S-curve which is the steepest; and the paper which is the least contrasty has the lowest gamma. Brock (1952) cautioned against the indiscriminate use of gamma as a measure of contrast in aerial photography; and suggested that the employment of the entire characteristic curve is the best way of evaluating sensitivity. Film speeds are based on the density at the toe of the curve (see Chapter 3).

The negative density range should not exceed the log exposure-scale of the positive material on which a contact print is made. For contrasty subjects, such as snow and dark clumps of trees or deeply gullied country in winter, a paper with a low gamma (e.g. 0·70) may give the best results for interpretation in black-and-white; but a high gamma (e.g. 1·20) may be necessary for photographs in haze or at flying heights of 20,000 ft or more above the datum plane. A gamma of about 1·4 is frequently used for medium and high altitude colour photography.

Graininess. This is used to describe the visual non-homogeneity of the black silver particles on the negative or photograph. If a small area of the negative or print is examined under high magnification it will be seen to comprise clumps of black silver particles of varying size and tones. This physical aspect is known as *granularity*; and the impression given is termed *graininess*. Graininess depends principally on the type of emulsion, density of the negative, the developer and development time. Fast emulsions are generally more grainy and energetic developers provide greater contrast. Granularity may be reduced by using special developers. These, however, may give insufficient contrast and necessitate increased exposure.

Graininess, resolving power and sharpness limit the magnification at which photographs can be examined and can limit the recognition of small details. Fine-grained, high-resolution films are to be preferred for photo-interpretation, but this conflicts with the need for fast films to eliminate image movement caused by the ground speed of the aircraft. A compromise on the choice of film has therefore to be made.

Sharpness. This is closely associated with graininess, contrast and density. The closer the image edge on the photograph corresponds to the edge of the object photographed, the sharper the photograph is said to be. If the sharp edge of a bright object is photographed against a dark background, the image will probably appear to be sharp; but when highly magnified it will be found to have a density gradient or *edge gradient*. This is primarily caused by the scattering of light in the film emulsion at the time of exposure due to its turbidity. Sharpness as a term is now often restricted to the impression received by the observer and *acutance* applied to its physical characteristics in the photograph. Details on a photograph are more easily recognized if the images are sharp. Sharpness is reduced by vibrations of the aircraft at time of film exposure, by flying at low altitudes and by flying at very high speeds. The focus of the lens system and aberrations in the lens system also influence sharpness.

Sharpness can be important to the interpreter in providing valuable information, especially on small-scale photographs. The size, shape and pattern of the edge of the image formed on the photograph are significant. For example, unbroken, straight or continuous curved lines immediately suggest man-made objects; whilst continuous but irregular curved lines indicate natural features, e.g. drainage lines. Close erratic patterns are usually provided by vegetation.

Fog. When an unexposed film is developed under ideal conditions, the negative should be transparent. Normally, however, the film after development has a more or less uniform low density. This physical characteristic is termed fog. Fog is associated with the developer, length of developing, incorrect storage of the unexposed film, long storage of unexposed film, accidental exposure of the film and occasionally by faulty chemical properties of the emulsion. Fog is greater on fast films and may reduce the value of the photographs for interpretation. On aerial negatives a fog density of 0·1 to 0·2 is generally tolerated. These fog values are higher than values considered tolerable for ordinary photography.

Resolution. This refers to the resolving power as applied to the lens, the negative and the photograph; and may be expressed as either the maximum number of lines per millimetre that can be seen as separate lines in the image plane or in terms of combined modulation transfer functions. There is no problem in combining the transfer functions once they have been determined (e.g. Hempenius, 1964), but accurate techniques need developing for their determination. At the present time lenses are being tested commercially under laboratory conditions, using optical transfer functions. The direct measurement of the resolving power of a system, using test cards (or test targets) and expressed in number of lines per millimetre, is relatively simple and is comprehensive, as many factors are included in a single measurement (e.g. gamma, grain, granularity). Unfortunately, a standard resolution target cannot be used to provide a point on an M.T.F.-curve. Sine-wave targets are needed, since the standard targets represent square wave functions of lines/bars versus distance.

In the United Kingdom, resolution is usually measured with Cobb cards, having a size difference of 10% between cards. On each test card there are lines, the length of which are three times their width. The separation distance between two lines equals the width of a line. A similar type of test is used by the U.S. air force (3-bar, edge gradient targets). Macdonald (1958) suggested that test cards should possibly be based on area and not linearity. He observed that the area of objects just detectable on the negative is inversely proportional to the resolution. Possibly targets providing a maximum of 30 lines per millimetre is sufficient when testing the film-camera combination.

As mentioned earlier, aberrations in a lens vary across the surface of the field, being least towards the centre. The resolving power will therefore be at a maximum towards the centre and least at the maximum lens-angle; and the resolving power of the camera-film system may be higher for a faster film (e.g. Kodak Super XX) than a slower film (e.g. Kodak Plus X). That the resolving power of the system with the faster film should be greater is contrary to what one might expect; but this is due to using a smaller aperture (i.e. a larger F-number).

Brock (1952) commented that the resolving power using an infrared film is about 40% of what it is with panchromatic film. This is partly due to using longer wavelengths; and partly due to the inferior correction for aberrations of the lens in the infrared wave-band, which is not overcome by a change in focal length.

It is essential to appreciate that due to lack of resolving power it is impossible to identify on photographs of normal scales leaf arrangements on the twig, the ends of branches and the tips of the crowns. A 5 in. leaf on a photograph at 1/10,000 would be only 1/5,000 in. long if it could be resolved!

On 1/15,840 photographs of a *Pinus radiata* plantation at Isandula in Tasmania, the images of trees about 13 ft tall and having a crown width of under 3 ft above the bracken did not resolve. In theory the crowns would record as 0·058 mm on the negative and would require a theoretical resolving power of a film with 18 lines per millimetre. Under laboratory conditions, Schwidefsky (1959) obtained 37 lines per millimetre with panchromatic film but, in practice, resolution was reduced due to a number of major factors. For example, Schulte (1951) reported Kodak Super XX (Aerographic) as having a *rated resolving power* of 30 lines per millimetre but in actual photography the *field resolution* was reduced to 10 to 14 lines. Assuming a resolution of 0·004 in. (10 lines per millimetre) in a positive, the best resolution obtained for tree images on a semi-matt print were:

Item	Scale
Hardwood leaves (4 in.)	1/1,000
Individual leafy branches, e.g. oak (5 ft long)	1/15,000
Individual conifers with narrow crowns, e.g. spruce, balsam fir (15 ft diameter)	1/45,000
Individual hardwoods (30 ft diameter)	1/90,000

Losee (1951)

5

TROPOSPHERIC CONDITIONS

AN examination of a few of the more important physical aspects of the troposphere in which the photographs are taken, is needed before considering aerial photography, photogrammetry and photo-interpretation. The photo-interpreter, who has spent time examining tropospheric conditions and the reflection of light from the surface being photographed is better equipped to obtain the maximum information from the photographs. For those wishing to pursue their studies of the physical aspects of the troposphere further than the introduction now to be given, reference should be made to Brock (1952), Gates (1962) and Middleton (1952). The troposphere is that part of the atmosphere below about 30,000 ft.

Pertinent information relating to the present study may be conveniently grouped under 'the quality and quantity of solar radiation', 'the scattering of light' and 'the influence of haze'. Weather conditions, such as cloud and mist, need little or no consideration in relation to peace-time aerial photography, as these conditions are normally avoided at the time of taking the photographs. The effect of other weather conditions are considered better dealt with under 'Planning Aerial Photography'. The phenomenon of reflected solar radiation is so very important that it will be discussed separately in Chapter 6.

The Quality and Quantity of Solar Radiation

As mentioned elsewhere, consensus of opinion does not agree on the exact limits of the wave-length of the visible energy spectrum; and therefore in order to avoid a metaphysical discussion, Crittenden's definition (1923) of light has been accepted. He considered light (i.e. the visible spectrum) as radiant energy evaluated in proportion to its ability to stimulate our sense of sight. The justification and the necessity for this definition lie in the fact that the stimulation of the sense of sight varies from zero at a wave-length of about $0.4\ \mu$ to a maximum at $0.555\ \mu$ and back to zero at about $0.72\ \mu$ (Middleton, 1962). It is customary to place the average wave-length of white light at about $0.57\ \mu$.

The visible spectrum comprises only a small part of the electro-magnetic spectrum.

45

This is shown diagrammatically in fig. 5.1. The electro-magnetic spectrum includes all energy which moves with the constant velocity of light in a harmonic wave pattern. It is classified by wave-length and represented by the general formula:

$$f = \frac{v}{\lambda} . a,$$

where f = wave frequency, v = wave velocity, λ = wave-length and a = a constant varying with the measurement units used.

The near infrared wave-band, by convention, is considered to extend to $1.5\,\mu$; and therefore, as mentioned in Chapter 3, includes the spectral range of infrared photographs. The middle infrared wave-band extends from $1.5\,\mu$ to $5.5\,\mu$; and as new

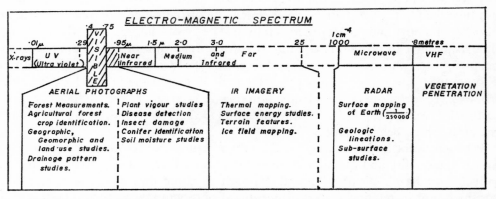

FIG. 5.1. The visible spectrum ($0.4\,\mu$ to $0.75\,\mu$) comprises only a small part of the electro-magnetic spectrum. At the earth's surface, approximately 97.5% of the incident solar energy can be in the spectral range $0.40\,\mu$ to $2.0\,\mu$. The remainder is nearly equally divided between the ultra-violet and mid-infrared. Passive sensing by aerial photography uses the solar energy between $0.4\,\mu$ and $0.95\,\mu$ for the recording of images. Active sensing by infrared imagery in the mid-infrared and by radar is still at the experimental stage of development for peaceful purposes.

infrared detecting and recording devices are developed, this wave-band may be of interest to the photo-interpreter. It appears that plants virtually cease to reflect solar energy at about $2.4\,\mu$ (see Chapter 6). Van Wijk & Ubing (1963) commented that about 98% of the radiant energy entering the atmosphere is contained in the wave-length interval between $0.25\,\mu$ and $4.0\,\mu$. In fact, when the sun is at an elevation of 60° to the horizon, possibly 98.7% of the solar radiation at sea-level can occur below $1.90\,\mu$. A spectral breakdown of this figure in calories per square centimetre per minute is given by van Wijk and Ubing (after Moon, 1940) under conditions of a turbidity factor of 4, a water vapour constant of 2.0 cm and 300 dust particles per cubic centimetre. These are 0.008 ($<0.40\,\mu$), 0.302 (0.40 to $0.70\,\mu$), 0.301 (0.70 to $1.50\,\mu$), 0.056 (1.50 to $1.90\,\mu$) and 0.01 ($<1.90\,\mu$).

Approximately half of the solar radiation incident on the earth's surface in the troposphere consists of light, and most of the remaining energy is in the infrared (see fig. 6.4a). Only a small percentage is in the ultra-violet below $0\cdot40\,\mu$. As pointed out by Gates (1962), the median of the solar energy spectrum occurs at about $0\cdot7\,\mu$. It is important to consider this factor when choosing a film–filter combination and when evaluating spectral graphs made in the laboratory. In the infrared part of the spectrum, the reduction of solar energy at the earth's surface is due to water vapour and, to a lesser degree, due to carbon dioxide and ozone. The absorption bands of these constituents are at about $0\cdot85$, $0\cdot95$, $1\cdot1$, $1\cdot4$, $1\cdot9$, $2\cdot4/2\cdot8$ and $6\cdot3\,\mu$ for water vapour, $4\cdot3$ and $15\cdot0\,\mu$ for carbon dioxide and $9\cdot6\,\mu$ for ozone. As the photographic range does not exceed $0\cdot95\,\mu$ only the absorption band due to water is of interest; but infrared sensing requires attention to be given to the absorption bands of carbon dioxide and ozone. Radiation provided by terrestrial objects is termed 'earth glow'; and extends from about $1\cdot0\,\mu$ to $80\,\mu$, with a maximum at about $10\,\mu$. At about $3\,\mu$, the reflected infrared energy of objects on the earth's surface falls below the emitted radiation of the objects.

THE SCATTERING OF LIGHT

In 1899, Rayleigh correctly referred the colour of the atmosphere to its actual cause, and demonstrated that the atmosphere in its purest form and containing nothing else but molecules of permanent gases scatters light as the inverse fourth power of the wave-length. This is referred to as *Rayleigh's law*. Rayleigh also pointed out that the molecules of the sky alone are sufficient to provide scattering and to produce a blue appearance in the absence of any small particles. This blue light is more or less plane polarized; and the direction of the electric vector is perpendicular to the direction of the incident beam.

In the laboratory, the principle of Rayleigh's law may be demonstrated as a simple experiment in physics by passing dust-free air through a black tube having the sunlight focused through a lens and observed at a peephole. If, now, a fine cloud of say iodide vapour is introduced, the light loses its blueness according to the increase in size of the particles. If the beam of sunlight is focused into a vacuum, the beam becomes invisible to the human eye. If very small particles such as those of sodium vapour in hydrogen gas are used at low pressure, the colour provided will be violet of great intensity.

The deficiency in the atmosphere of the scattering of long wave-length radiation is shown by making an exposure of vegetation with infrared film so that the lower-quality light is excluded. The vegetation will be resolved whitish, but the sky remains black (R. Wood, 1959). If a distant whitish view is observed through a nicol prism, the light will appear blue in increased splendour, being termed residual blue (R. Wood, 1959). This is due to the effect of polarized light with its vibrations parallel to the z-axis. The intensity of the light scattered along the z-axis varies as the inverse

E

eighth power of the wave-length, so that the residual blue is purer (Rayleigh, *Phil. Mag.*, **12,** 81).

Several types of instruments have been used to measure the scattering of light. These include polar nephelometers for measuring the scattering function at more than one angle (see Waldram, 1945; Brock, 1952; Middleton, 1952); polar nephelometers for measuring the scattering function at a fixed angle and instruments used in integrating the scattering function. By using the first type of nephelometer to record the intensity of the white light scattered at varying angles, graphs with characteristic curves as shown in fig. 5.2. can be obtained (after Brock, 1952). An examination of this curve will show that the scattering of light in relation to the

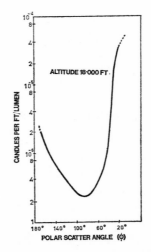

FIG. 5.2. The curve shows that the scattering of light in relation to the direction of incident light is at a minimum when $\phi = 90°$ and should be at a maximum in the direction of the light and towards its source (i.e. 0°, 180°).

direction of the incident light should be at a maximum when $\phi = 0°$ and $\phi = 180°$ and is at a minimum when $\phi = 90°$. When $\phi = 90°$, the polarization of light is nearly complete and diminishes to zero at 0° or at 180°.

R. Wood (1959) commented that with very small particles the intensity of the scattered light is the same in both directions, but as the particle sizes in the atmosphere increase, the intensity of the scattered light increases rapidly in the general direction in which the light is travelling (i.e. 0° in British publications) as compared with a similar direction backwards (180°); and is proportional to the square of the volume or the sixth power of the diameter. Work carried out at Woomera (South Australia), using a nephelometer over a 12 month period has indicated that at ground level scattering is greatest in August and lowest in February (Crosby & Koerber, 1962). As the area has a Mediterranean type of climate, particle size is presumably larger in the dry summer weather of February; and the reduced scattering may be explained by reference to Rayleigh's law and Wood's comment.

The photo-interpreter will be interested to learn that the maximum polar scatter ($\phi = 180°$) in the direction of the incident light coincides with an area of over-exposure often observed on aerial photographs and which has been referred to as the 'hot spot', 'shadow point' or 'no shadow point' (see Chapter 6). It occurs on the side of the photograph farthest away from the sun; and it will be found that the sun's disc, the lens of the camera at the time of exposure and the point on the ground where the maximum brightness appears all lie in the same straight line. The angle of maximum scatter remains constant irrespective of the altitude of the aircraft.

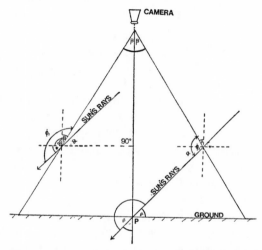

FIG. 5.3. For a known elevation of the sun and angular field of the aerial camera, it is possible to determine the maximum polar scatter angle and elevation at which a reflex reflection zone will be recorded on the photograph (see text). The direction of two rays from the sun is given by the arrows. The sun's elevation is θ, and angular field of the aerial camera is 2β. P is the ground principal point.

For a known elevation of the sun and a given angular field for a camera it is quite simple to calculate the maximum and minimum angles of polar scatter for light entering the camera-lens. This is shown diagrammatically in fig. 5.3 and is a useful method for determining the sun's elevation which results in an area of no-shadow (reflex reflection) on photographs for a known angular field of the camera. In the diagram, for simplicity, only two rays of sunlight have been shown. If the angular field is 2β and the sun's elevation is θ, then the maximum (ϕ_1) and minimum (ϕ_2) scatter angles are:

$$\phi_1 = 90 + \beta + \theta; \phi_2 = 90 - \beta + \theta.$$

Example. For a camera with an angular field of 60°, what is the minimum elevation (θ) of the sun before it will produce an area of over-exposure on the photograph?

Answer:
$$\phi_1 = 180° = 90° + 30° + \theta$$
$$\theta = 180 - 90 - 30 = 60°$$

THE EFFECT OF HAZE

From what has been said in the last section, it will be appreciated that the intervening atmosphere between the object reflecting the light and the aerial camera may be considered as a turbid medium in which some reflected light from objects on the ground and some solar radiation are scattered in all directions. The light entering the aerial camera will comprise not only the reflected light from the objects being photographed, but also scattered solar radiation. The latter has been referred to in German literature as '*Luftlicht*' (i.e. air light). To the human eye the combination of the two sources of light results in a uniform brightness rather than a differential brightness, being illustrated by the fact that distant objects appear less contrasty than near ones. Hence, when examining vegetation on aerial photographs, the brightness differential of the vegetation is less than when viewed close at hand.

Middleton (1950) commented that if the atmosphere were pure the visual range along the ground would be more than 220 miles; but in practice the visibility is much less and varies according to the direction of the prevailing wind and the density and size of particle the wind contains. In eastern Canada, for example, the visibility at ground level has been reported as about 45 miles under ideal conditions with winds from the polar regions; but the visibility is reduced to as little as $2\frac{1}{2}$ miles with dust-laden winds from the Gulf of Mexico.

Particles in the atmosphere include ash from industrial cities, forest fires and volcanoes, particles from atomic explosions, fungal spores, bacteria, dust from outer space, soots, ice crystals, snowflakes and salt nuclei from the sea in the form of water droplets. Volcanic dust and similar inert material are commonly of a diameter of $1\ \mu$. Liquid droplets frequently have radii of 10^{-6} to 10^{-1} cm. Wright (1936) deduced from the ratios of charged and uncharged particles that the radii fall between $3\cdot0$ and $4\cdot5 \times 10^{-6}$ cm and number from 21,000 to 72,000 per cubic centimetre in the vicinity of London. Earlier, Owens (1926), estimated that in dense fog there were 21,750 particles per cubic centimetre compared with only 200 in fine clear weather.

Of importance to aerial photography is the gross effect of the suspended particles or haze on both the visual range and the colour of the distant objects. Visual range is measured by referring to the extinction of light, scattering and visual range itself. A valuable review was given by Middleton (1952). He also provided a number of nomograms, based on Tiffany data, showing the visual range in yards of circular objects on the ground as seen from an aircraft; and has pointed out that when the contrast between an object and its surroundings become less than 2%, the object is no longer visible. Waldram (1945) considered that the reduction in contrast is a subjective phenomenon similar to that of a reduction in sharpness as may be demonstrated by introducing haze between the object and the observer. Hugon (1930) drew attention to shadow as influencing the loss of detail in relation to visual range. A number of authors (e.g. Hall, 1954) have drawn attention to haze as influencing the density of negatives so much that they are unsuitable for photogrammetrical mapping.

Hendley & Hecht (1949) found that distant objects approach the hue of 0·475 μ irrespective of their value, hue or chroma. Dark distant objects appear blue whilst distant objects of high positive contrast look pale orange (see Middleton, 1952). No doubt distant hills appear blue in summer due to the scattering of the blue light by the haze. This is particularly noticeable when the prevailing wind contains salt nuclei and few dust particles. The greatest purity in the blue horizontally is achieved within 10 to 30 miles; but on the way to this, the hue passes close to the achromatic point and is a very unsaturated blue green (i.e. greenish grey) at about 3 miles.

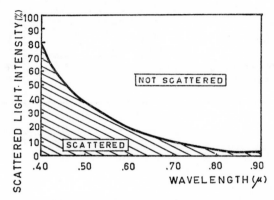

FIG. 5.4. The scattering of light in the troposphere according to wave-length is maximal at the blue end of the visible spectrum (i.e. 0·40 μ) and minimal at the red end. For this reason a 'minus blue' filter is often used in aerial photography.

The C.I.E. system of colour differentiation is based on an aperture of 2° angular diameter; but as pointed out by Middleton (1952) the mixture data changes markedly by the time the angular field is reduced to 15′ of arc. Probably the ubiquity of red and green as separate colours on aerial photographs is due to their invariance with changing size. Small areas of red and orange tend to be achromatic only at about the visible range. This possibly explains why orange is such an excellent colour for 'ground-control' markers and why on colour photographs the partial defoliation of orange-red twigs, the flush of evergreen new growth and autumnal tints are so conspicuous and readily recorded. Maruyasu & Nishio (1962) commented that the effect of haze on blue hues in colour photographs is increased at a fixed ratio within the altitudinal range of 3,000 ft to 13,000 ft. The scattering effect of haze according to wave-length is shown in fig. 5.4. Colman (1948), Middleton (1950) and others have observed that the effect is more conspicuous than indicated by Rayleigh's law towards the ultra-violet end of the spectrum.

Haze also affects the quality of the incident solar radiation. As the angle of the sun from the zenith is increased by latitude, season or time of the day, the light reaching the ground has to pass through increasing thicknesses of atmosphere. This varying thickness acts as a filter and affects the quality and quantity of the solar radiation as shown in table 5.A (Lundegardh, 1931). When the sun is at its zenith, about 50% of the visible spectrum is in the blue-green wave-band; and, as the angle of the sun decreases, the maximum quality moves through yellow to red.

TABLE 5.A

The effect of haze on the quality of solar radiation

Spectrum	Time of day			
	Noon	2 p.m.	4 p.m.	6 p.m.
Red	33%	34%	34%	41%
Yellow	34%	34%	34%	35%
Blue	33%	32%	32%	24%
Achromatic light (white light)	100%	88·7%	63·6%	20·4%

6

THE REFLECTION OF LIGHT BY VEGETATION

INTRODUCTION

THE energy exchange between the earth's surface, its atmosphere and outer space is influenced by the reflective properties and the distribution of the earth's vegetation. A photograph of a forest taken from an aircraft is a permanent record of the radiative properties in the visible spectrum or the near infrared of the trees, their background vegetation and the exposed surface of the earth. The photograph provides a record of a small but very important part of the electro-magnetic energy spectrum. To obtain a complete record of the energy exchange or the reflected and emitted energy between 0.29μ and 2.0μ would require supplementary recordings by remote sensing devices other than the camera.

Within the photographic spectrum, the amount of energy reflected by leaves will depend on their group texture, vigour, age, angle of incidence of the solar energy, wave-length, geographical location and season. In addition, the tone and colour of the image on the photograph will be influenced by other factors including atmospheric conditions, film–filter combination, lens quality and processing techniques. Geographic location, expressed in terms of aspect and slope, will have considerable effect on the tone or hues of the images. For example, the annual radiation index at a latitude of 40° N. is about 48 on level ground but declines to about 32·8 on a north-facing 40° slope and increases to 57·8 on a south-facing 40° slope (Lee, 1963).

As pointed out by Colwell (1956), spectral analysis, using a suitable spectrophotometer in the laboratory, is useful in predicting the correct film–filter combination for obtaining recognizable tone or colour differences on photographs of soil types, mineral types, plant species and forest and agricultural crops attacked by insects or disease.

In this chapter, the characteristics of reflection and the optical and spectral properties of leaves and vegetation will be examined in relation to the aerial photograph, as an understanding of the reflective properties of the plant is highly desirable if maximum use is to be made of the photograph in field studies. In the course of discussion, reference will be made to earlier publications, which, although not associated with photo-interpretation, have contributed to the study of the optical and spectral properties of leaves.

Extensive studies relating directly to reflection as recorded on the aerial photograph may be considered as beginning with Krinov's work in Russia between 1934 and 1938. The Russian-language publication of his work was translated into English in 1953, but it has since been out-dated by the findings of other Russian workers (e.g. Pronin, 1949; Vinogradova, 1955, Belov & Arcybasev, 1957), who used improved equipment (see Steiner, 1963).

In Sweden, Backstrom & Welander (1953a) studied on panchromatic photographs the reflective characteristics of nineteen tree species in relation to age and site and concluded that neither was important. They also observed that hardwoods reflect more light than conifers and that spectral differences between species is at a maximum in the spring. In New England, Branch (1948), also pointed out that hardwoods with their broad leaves tend to be good reflectors, and that this is especially the case on the side from which the light is falling. Conifers, on the other hand, usually register darker because of the greater absorption of light between their thin 'leaves' and their more diffuse reflectance. In California, Colwell has carried out a number of detailed studies, including the detection of disease using both colour and infrared photography (Colwell, 1956, 1960a, 1961a; Colwell et al., 1963). Other workers contributing towards the study of reflection in relation to the aerial photograph include Fischer (1960, 1962), Walker (1961), Langley (1962), Olson (1962, 1963), Fischer & Gray (1962), Wolff (1966) and Rossetti & Kowaliski (1966).

(a) Optical properties of reflective surfaces

It is convenient to commence by defining several terms in accordance with the C.I.E. vocabulary of international lighting (1957). *Reflection* refers to the backward reflection of radiation by a surface without change of frequency of the monochromatic components of which the radiation is composed. Transmission refers to the passage of this radiation through the medium of the reflecting surface; and *absorption* refers to the transformation of the radiant energy by the intervening medium. *Mixed reflection* is the sum of the two components: the *diffuse reflection factor* and the *specular* or direct reflection factor. *Reflex reflection* refers to the light reflected close to the direction in which the light is incident.

The ratio of the luminous flux (radiation) reflected by a body to the incident flux is termed *reflectance* (U.S.) or *reflection factor* (U.K.). Meteorologists use the term *albedo* to express the ratio of radiation reflected from a body to the total radiation on it. *Spectral reflection* refers to the reflectance of incident light for given wavelengths. If the thickness of the reflecting medium is such that any increase in thickness will provide no increase in reflectance, then the term *reflectivity* is used instead of reflection factor. Similarly *transmittance* and *transmission factor*, *absorptance* and *absorption factor* are the comparable terms relating to *transmission* and absorption,

OPTICAL PROPERTIES OF REFLECTIVE SURFACES

but *transmissivity* and *absorptivity* imply the transmittance and absorptance for a body of unit thickness. A medium which absorbs completely all incident radiation irrespective of wave-length or direction of incidence is termed a *black body*. A cavity with these properties is referred to as *black cavity*.

A specular or mirror surface in nature is provided by a sheet of still water and sometimes erroneously illustrated as occurring on leaf surfaces (see fig. 6.1a) (Howard, 1966b). The light incident on water will be reflected in accordance with the

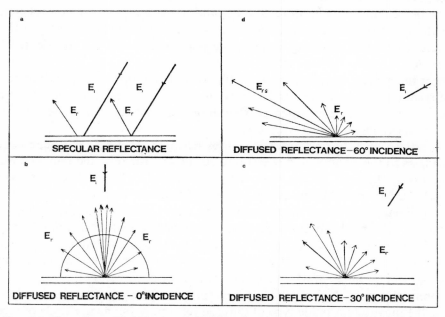

FIG. 6.1. The reflected light by leaves is diffused and not reflected specularly upwards. (a) Specular reflectance which has not been observed for the leaves examined. (b) The characteristic diffused patterns of reflectance for normally incident light on a leaf of *Eucalyptus regnans*. (c) The same leaf as (b) but the angle of incidence is 30°. (d) As in (c) but the incident energy is at 60°. Note the development of spread reflectance and a specular component (E_{rs}) which will not be directed towards the camera.

sine law, such that the angle of incidence equals the angle of reflection. Thus if the sun (angle of subtense 0° 30′) is at an angle of 45° to the horizon, it will be recorded on the photograph as an over-exposed spherical area at a reflection angle of 45°. The image of the solar disc reflected from calm water has a subtense angle of about $2\frac{1}{2}$°. Geometrically, the specular reflection zone is located on the photograph at the intersection of the line subtended at 45° to the plumb-line and passing through the lens of the camera. As soon as a sheet of water becomes disturbed by wind action, the incident sunlight will be scattered at differing angles and provide a much larger area of over-exposure. When the area of reflection of the sun is not within the angular

55

field of the camera-lens, the image of the water surface will record very dark on the photograph, if the depth of the water is considerable. In the infrared, even a thin sheet of water always records black on the photograph, irrespective of the angle of reflection.

As most surfaces of interest to the photo-interpreter have a luminous reflectance of a diffusing nature rather than that of a mirror surface, it is well worth while considering momentarily a perfectly diffusing surface. A perfectly diffusing white surface receiving an illuminance of π lumens per unit area will have a luminance of one candle per unit area in all directions; and hence any unit area on a hemisphere about the point of reflection will have the same luminance. Thus, in fig. 6.1b, if the leaf were a perfectly diffusing surface, the luminance could be represented by the semi-circle.

One of the nearest approaches to a perfectly diffusing surface is provided naturally by smoked magnesium oxide; and somewhat surprisingly, the reflection of some leaf surfaces also provides a diffusing pattern approaching that of magnesium oxide for perpendicularly incident light (fig. 6.1b).

This pattern of diffusion, however, varies between species of the same genus. At small angles to the normal, and even up to 30° or 45°, the reflected light is diffused in all directions, although, as the angle increases, *spread reflection* becomes characteristic (6.1c, 6.1d). Eventually, as the angle of incidence becomes large, e.g. 60° or 75°, a specular component may develop (fig. 6.1d). When a surface produces not only diffuse or spread reflection but also a specular component, it is referred to as a *mixed reflecting surface* (6.1d). Wallis (1943) concluded that the light reflected from leaves varies only slightly with the angle of incidence, but as his studies did not go beyond an angle of 45° incidence, he is only partly correct.

As pointed out by Keitz (1955), many surfaces for perpendicular incident light provide the magnesium-oxide type of diffused reflection, but as the angle of incidence departs more and more from the normal, spread reflection is observed. Even surfaces which appear smooth may have a rough microstructure. In a leaf, a rough microstructure is probably provided by the internal cell walls, although the leaf appears smooth due to the window-like cuticle. Leaves also often have a rough external micro-surface provided by waxes on the cuticle surface. The wax-pattern is discernible under an electron microscope (fig. *6.5*).

When the light is incident obliquely on a shallow trough-like micro-surface, the reflection spread is greatest in the direction away from the incident light. When the troughs are deep or the surface is visibly rough or broken, then the luminance is often greater as seen from the direction of the incident light. It seems likely that the leaves, whilst providing the first type of surface individually, provide also as an assemblage of leaves and branches the second type of reflective surface.

It will be appreciated that irrespective of the angle of incidence on a leaf, the light is diffused upwards towards the aircraft fairly uniformly, and hence as far as the interpretation of vertical photographs is concerned, we may consider all vegetation as being diffusing irrespective of the angle of incidence. Secondly, it will be appreciated that the luminance may appear very different when viewed on the ground and when seen from the air.

On the low oblique photograph, the recorded reflected light may be quite different towards and away from the incident solar radiation. Thus when the sun's elevation is low, micro-relief may be accentuated by the long shadows and spread reflection may be recorded on the film if the camera is pointing in the direction of the sun. Goguey (1966) has used these conditions as an aid to archaeological photo-interpretation. In fig. 6.2, the two photographs were taken on the same day within a few minutes of each other (July 1964); but photograph (a) was taken with the camera pointing towards the sun and photograph (b) with the sun behind the camera (note the reflex reflection zone with the aircraft's shadow in the centre: A). The geometric designs recorded in the photographs are part of a Bronze Age settlement near Dijon. None of these ancient earthworks was recognizable on the ground. (1) is an ancient tumulus of sand and gravel, which has been totally eroded. The earthwork (4) is not visible on photograph (a) and earthworks (5) and (6) are not recorded on photograph (b). On both photographs, the outline of an ancient wide ditch (forming a right-angle) is conspicuous. The arrow-like images in photograph (b) are telegraph poles and their shadows.

Contained luminance. In addition to the light being freely diffused outwards into the atmosphere by the first layer of leaves on the tree, the light will be diffused also in all directions by the other layers of leaves; but little of the light from the second and third layer of leaves will be recorded by the camera in the aircraft as it is contained within the foliage. The term *contained luminance* (or contained solar energy) is suggested to describe this phenomenon as recorded on the aerial photograph and as produced by the crowns of trees and other broken or rough surfaces. Losee (1951) used the term *contained shadow*.

Leaves and needles produce innumerable small shadows which, at the scale of aerial photographs, do not record separately as a mosaic of black and lighter tones of reflected light from the leaves, but produce an overall darker tone on the photograph than would be produced by a continuous reflecting surface. With hardwoods of the northern hemisphere, the leaves usually overlap more than the conifers and therefore provide fewer shadows and form a relatively smooth surface. In consequence, hardwoods record lighter than conifers on aerial photographs. This is particularly the case in the infrared. However, at midday in midsummer, shadows being least, hardwoods and conifers are almost indistinguishable at the latitude of south-eastern Canada (Losee, 1951).

Low recorded reflectivity resulting from contained luminance is also helpful in classifying undergrowth. The taller and thicker the undergrowth, the darker, in general, is its tone on the panchromatic photograph. Losee (1951) suggested that *Sphagnum* bogs appear light toned due to a very low moss cover, providing a relatively smooth surface. Bogs with shrubs record darker. Often tall undergrowth under eucalypts appears dark although the leaves have a normal reflectance as given by spectrophotometer studies. As a rule, the reflectance of a surface decreases with an increased wetness in the visible spectrum. However, in individual cases, Krinov (1947) observed that moist surfaces were brighter than dry surfaces, but dry *Sphagnum* had a reflectance 2·7 times greater than wet moss.

Reflex reflection. This may be defined as reflection in which the path of the returning light lies close to the direction in which the light is incident, whatever may be the angle of incidence at the reflecting surface (C.I.E., 1957). Work carried out by the writer has shown that a fine powder surface, such as smoked magnesium oxide, has a much higher luminance factor within 2° of the normal than at, for instance, 10° or 45°. The writer has also observed the same phenomenon for leaves of several species of eucalypts and for pine needles. For a mosaic of leaves or needles, the phenomenon is enhanced as the area of reflex reflection coincides with an area of little or no shadow.

The increased luminance factor for the eucalypts is sufficient to record as a tonal difference on the photograph if the sun's elevation is high enough. As pointed out in the last chapter this occurs when a hypothetical straight line subtended from the sun's disc through the centre of the lens intersects the ground within the area covered by the angular field of the camera.

Various names have been given to this zone of over-exposure as observed on the photograph, including no-shadow point, shadow point, hot spot, flare point, back-scatter point and diffuse reflectance zone. The latter is used to distinguish it from the specular reflection point or zone as recorded by a water surface. The term *reflex reflection zone* is preferred by the writer and will be used for the remainder of this book. Of historic interest is the use of the term 'no-shadow zone' in 1925 (Kemp *et al.*). If the hot spot or flare point is produced by factors other than those producing reflex reflection, then the zones may not coincide.

The specular reflection zone, the principal point and reflex reflection zone will all lie in the same straight line across the plane of the photograph; and the specular and reflex reflection zones will be at equal distances from the principal point. This will be obvious by reference to fig. 5.3.

A study of large numbers of photographs representing many types of surfaces in Australia and New Zealand, recorded on panchromatic, infrared and colour film, has shown that choice of the film-filter combination does not eliminate the presence of the reflex reflection zone; and neither is its presence affected by the flying height or type of surface, except water. At Mount Disappointment, Victoria, the reflex reflection zone was observed on photographs at several scales, including 1/3,500, taken at an altitude of 2,400 ft. Over deep water, the reflex reflection zone is not recorded (fig. **6.3b**), although the shadows of the plane may be seen as a small darkish spot on the sea bed under shalllow water (fig. **6.3a**). The fact that a reflex reflection zone is not recorded over water indicates that it is not primarily a phenomenon of the troposphere. On photographs taken in cloudy conditions, as in military operations, a zone of back-scatter or a 'halo' may sometimes be observed on the clouds below the aircraft.

(b) The spectral and optical properties of the leaf

In the early days of camouflage for military purposes, it was mistakenly thought that green paint could conceal a building. It was not appreciated that the spectral reflectance of green paint in the photographic spectrum is often different from the spectral reflectance associated with the green chlorophyll content of leaves.

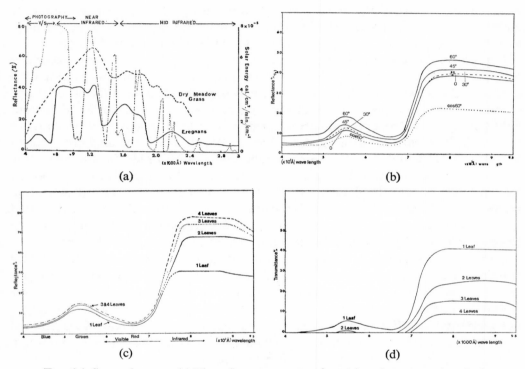

FIG. 6.4. Spectral curves. (a) The reflectance curves of samples of green eucalypt leaf (solid line) and dry grass (broken line) collected in late summer 1965. On the aerial photograph (panchromatic) the grass is light toned and the eucalypt is dark toned. The dotted line (and right y-axis) represents the incident solar energy. (b) The reflectance of the leaf of *Eucalyptus regnans* when the angle of the incident energy is varied. The dotted line represents the cosine corrected solid line for 60°. (c) The reflectance curves of mats 2–4 leaves compared with a single leaf (*E. regnans*). Note in the visible spectrum there is little change in the reflectance with increased layers of leaves and the reflectivity is the same for 3 and 4 leaves. In the infrared, up to 4 leaves contribute considerably to reflectivity. (d) Transmittance curves (*E. regnans*) showing the change in transmittance with mats of 2–4 leaves and a single leaf. The spectral range in (a) extends over most of the solar spectrum; and in (b), (c) and (d) the spectral range coincides with the photographic spectrum (visible and infrared).

The overall reflectance spectrum with constant incident energy from 0.4μ to 3.0μ (left y-axis) is given in fig. 6.4a (Howard, 1966a) for a green leaf of *Eucalyptus regnans* and dry grass. Water absorption bands occur at about 0.85μ, 0.95μ, 1.10μ, 1.40μ, 1.90μ, and 2.40μ to 2.70μ. Incident solar energy is represented by the dotted line (right y-axis). If reference is made to Chapter 5 and studied in conjunction with fig. 6.4, it will be observed that the reflection of solar energy is of minor importance beyond 2.0μ, and the green plant acts virtually as a black body beyond about 2.4μ.

The marked change or break in the level of reflectance at the commencement of the infrared (i.e. $0.72 \mu/0.75 \mu$) is not surprising and may be explained as follows: in the visible spectrum, the incident energy stimulates electron excitation within the atom whilst in the infrared, the vibration is in the form of symmetrical stretching, asymmetrical stretching and deformation of the hydroxyl bonds of the molecules.

Using a Dunkle-Gier spectrophotometer it is possible to determine the effect of leaf orientation on the reflection of incident energy; and either this instrument or a Beckman DK 2 (with integrating sphere) are suitable for examining the effect of one or more layers of leaves on the transmittance and reflectance of normally incident light (Howard, 1966b). Spectral curves in the photographic range 0.4μ (4,000 Å) to 0.95μ (9,500 Å) are given in fig. 6.4b, 6.4c and 6.4d. When the angle of incident energy exceeds about 45°, the spectral reflectance as recorded in an integrating sphere increases considerably (e.g. 60° solid line, fig. 6.4b). This result is supported by the work of Shulgin, Khasanov & Kleshnin (1960), who obtained increasing reflectance with increasing angles of incidence. It will be appreciated, however, that in the case of leaves under field conditions, a cosine correction must be made for the angle of solar incidence (e.g. 60° dotted line, fig. 6.4b); and as the reflectance is now less than for leaves more or less normal to the incident light, trees with vertically hanging leaves would be expected to record darker toned crowns on the photographs (e.g. many eucalypts). In the case of the reflectance of light from more than one layer of leaves, it will be observed that the number of leaf layers in the visible spectrum is not very important (fig. 6.4c) in increasing the reflectivity. In the case of transmitted light, the transmittance in the visible spectrum (fig. 6.4d) is low after passing through a single leaf layer and is negligible after the second layer. This suggests that the interpreter in advance of aerial photography might obtain some idea of the likelihood of obtaining photographic tonal differences by examining the orientation and distribution of the top leaf layer (first-order surface) of the leaves of the principal species.

Tageeva & Brandt (1960) have provided one of the few references on absorptance according to the angle of incidence. They found, for thin leaves (e.g. *Lactuca sativa*) that absorption remained more or less constant for angles between 15° and 90°, although reflectance increased with decreasing transmittance. Tageeva and Brandt also found that, for thick leaves (e.g. *Ficus elastica*), the reflectance remained relatively low, irrespective of the angle of incidence. In table 6.A, percentages are provided showing the absorptance of light in the photographic spectrum for four eucalypts (Howard, 1966b).

SPECTRAL AND OPTICAL PROPERTIES OF THE LEAF

LEAF STRUCTURE

The leaf is a complex structure, as amateur botanists will know (fig. *6.5a*). The incident light is first in contact with the waxes, hairs or cuticle. If the waxes are conspicuous on the cuticle of the leaf, then the reflectance will be increased by as much as 15% in the visible spectrum. Shulgin *et al.* (1960) commented on the difference in reflection with different conditions of the leaf surface (e.g. a high reflectance for waxy *Begonia peltata*). It seems likely that increased waxiness accounts for the increased reflectance in the visible spectrum observed with greater elevations. Barber (1955) has reported clinal variations in Tasmanian eucalypts, and has observed that surface waxes may increase with altitude and influence reflection by about 10% for *E. vernicosa* (personal communication, 1965). Obaton (1944) obtained twice as much reflectance for mountain species as for flora of the plains. Billings &

TABLE 6.A

Spectral absorptance of normally incident light by single leaves
Values are given as percentages of incident light

Species	Wave-length (Å)					
	4,000	4,500	5,350	6,550	7,500	9,000
E. regnans	97	95	89	95	24	11
E. radiata	94	91	82	94	20	10
E. obliqua	94	93	85	96	23	12
E. gonicalyx	95	93	84	95	19	9

Morris (1951) recorded that in the infrared, the difference in reflectance between alpine woodland and desert groups of plants was not so marked as in the visible spectrum. However, abundant wax on the mature leaves of mesophytic tree species is probably a rarity, and, in consequence, should be considered as a minor factor in influencing reflectance as recorded on the photograph. Leaves tend to lose most of their wax with age, and the accumulation of foreign matter may then be as important as the remnant wax in increasing the reflectance by 2% or a little more (Howard, 1966b). As shown by micrographs using an electron microscope, the pattern provided by the waxes varies greatly even between species (fig. *6.5a, b*) (e.g. see Chambers & Hallam, 1964).

Billings & Morris (1951) concluded that the hairs or scales of desert and sub-alpine herb species were correlated with higher reflectance in the visible region but not necessarily in the infrared region. Desert species reflected the greatest amount of visible radiation, followed by the sub-alpine, and then west-facing and shaded species. Desert species in the infrared reflected about 60%, although *Prunus andersonii* (desert peach) reflected 70%. Schulte (1951) found, when comparing elms from

Minnesota and Indiana, that the leaves of the former reflected slightly less light on their under surfaces; but for a white poplar, the top surfaces had a reflectance of 12%, whilst the white ventral surface had a reflectance of 50% in the visible spectrum. Obaton (1941, 1944), found that in the mountains, the epidermis of the herb species is often covered with hairs, giving the white appearance and higher reflectance in the longer wave-lengths, but the coefficient of reflection was not correlated with hairiness, e.g. *Gallium pyrenaicum*, 0·26 (glabrous): *Myosotis pyrenaica*, 0·26 (woolly). Shull (1929) had observed that a visibly shinier cuticle did not have a higher reflectance than *Morus rubra* (red mulberry) and the hairy surface of *Verbascum thapsus* (common mullein) had little effect. Gates & Tantraporn (1952) re-examined mullein and concluded that the hairs scatter and trap the light, particularly in the infrared. *Mullein* and *Asclepias syriaca* possessed zero and very low reflection from 3μ to 25μ. If a leaf surface without hairs has a high reflectance, then the presence of hairs will most likely diminish the reflectance (Gates & Tantraporn, 1952).

Several workers (e.g. Pokrowski, 1925; Shull, 1929; W. Clark, 1949) quote for the northern hemisphere species that the lower surfaces of leaves reflect considerably more light in the infrared and visible spectrum. Gates & Tantraporn (1952) suggested that this is probably due to the lack of palisade cells on the lower surface of the leaf. As early as 1895, in probably the earliest work relevant to reflection, Bonnier recorded that the coefficient of reflection as compared with magnesium oxide/carbonate doubled for plants of the same species at high elevations (in Obaton, 1941, 1944). He observed that, with increasing elevation, the palisade tissue increased two or three times in thickness and that the spongy tissue decreased. If, however, a juvenile dorsiventral eucalypt leaf is used, the reflectance of the under surface will be greater in the visible spectrum and less in the infrared, and examination of the leaf in the cross-section would show fewer layers of palisade cells on the lower surface of the leaf.

It seems highly likely that all the living cells reflect somewhat similarly irrespective of size, shape and arrangement, and reflectance per cent possibly coincides with the differences in the refractive indices of the cell wall and middle lamella and as modified by chlorophyll (plus carotene.) The reflectance from vascular bundles, however, is higher than that from other cells (e.g. 2%). It is unlikely that the intercellular spaces are responsible *per se* for the reflectance, as suggested by some investigators (e.g. Clark, 1949; Wolff, 1966) since the tissue may be absent or nearly absent without altering to any great extent the reflectance (fig. 6.5a). The following hypothesis may be formulated. The incident light in the visible spectrum is mostly absorbed by the chlorophyll at the blue and red ends of the visible spectrum, and partly absorbed by the carotenes. The unabsorbed light is then reflected by the cell-walls of the palisade tissue and the mesophyll, and some of the reflected light is further absorbed by the chlorophyll before 'escaping'. In the infrared spectrum, the transmittance and reflectance is much higher (about 40% to 50%) and only a small percentage is absorbed (about 10%). Variation would appear to be associated with the infrared absorption by the water molecules and by increased reflection in the cuticle. It is again thought that the cell walls act as reflectors. Colwell (1963) has pointed out that bubble-like surfaces are often good reflectors in the infrared.

Carotenes commence absorbing in the ultra-violet ($0 \cdot 36 \, \mu$), attain a maximum at about $0 \cdot 5 \, \mu$ and absorb again strongly in the medium infrared. Seybold & Weissweiler (1943, in Rabinowitch, 1951) drew attention to the possibility of red pigments in leaves (i.e. phytocyanins) increasing the absorption of incident light.

Shull (1929) suggested that the leaf pigment and epidermis of *Quercus robur* and *Rhus glabra* influence reflectance in the visible spectrum. Obaton (1941, 1944) concluded that neither the epidermis, its attachments (i.e. hairs), palisade tissue and the chlorophyll, nor the air spaces of the mesophyll (parenchyme lacuneux) confer the property of reflecting so much radiation, but it is probably due to the climate modifying the tissue, particularly the cell walls ('parois de ses cellules'). He also suggested that the high reflection in the infrared is not due to the waxy epidermis, nor to the cuticle of the leaves. He found that fresh cuticle has a very low reflectance and a high transmittance.

As early as 1913, Coblentz recorded the reflectivity of a large number of surfaces including plant species both in the visible spectrum and in the infrared below 1,000 Å. Also in 1913, Willstatter and Stoll carried out extensive examination of chlorophyll of various species of different families under a wide range of ecological conditions, and concluded that the green pigment in leaves (i.e. chlorophyll a and b) is the same for all green plants and that the carotenoids, yellow in colour, are masked by the chlorophyll and are usually also present in leaves. They suggested that solar radiation in the visible and infrared passes through the epidermis and palisade cells, diffuses in the parenchyma of the mesophyll, and is then reflected back towards its source.

Surprisingly, subsequent investigators up to Dinger (1941), Clark (1949), and Gates & Tantraporn (1952), did not extend the infrared reflectivity studies. Dinger (1941), having examined the reflectivity and transmission of leaves in the near infrared, found it to be at a maximum between $0 \cdot 80 \, \mu$ and $1 \cdot 30 \, \mu$, as chlorophyll and carotenoids are very transparent to infrared.

Fluorescence and overtones. Although incident light in the chlorophyll absorption bands is associated with fluorescence (i.e. re-radiation at longer wave-lengths), studies by the writer have indicated that the intensity of fluorescent light is too small to be significant to aerial photography. Fluorescence also occurs at shorter wave-lengths beyond the photographic wave-bands. For example, stilbenes occurring in leaves of eucalypts fluoresce at $0 \cdot 32 \, \mu$. Essential oils, found also in the leaves of eucalypts, absorb strongly below $0 \cdot 30 \, \mu$ and beyond $3 \cdot 0 \, \mu$ and in the medium infrared at about $3,000 \, \text{cm}^{-1}$.

Overtones or the re-emission of incident energy at twice the fundamental wave-lengths were recorded for several species of eucalypt between $0 \cdot 60 \, \mu$ and $0 \cdot 80 \, \mu$ and incident radiation at $45°$ between $0 \cdot 30 \, \mu$ and $0 \cdot 40 \, \mu$; but like fluorescence, it does not appear important, at least at present, in photo-interpretation and image analysis.

F

Phenology. The changes in the value, chroma, hue and appearance of vegetation in relation to the time of the year can be particularly helpful in the identification of ripening crops, tree species and sometimes grazing areas for wildlife and domestic stock. The most obvious example is the separation of deciduous and evergreen trees in spring and autumn.

The order in which different tree species flush each spring or at the beginning of the wet season, and their tones or hues, may provide valuable clues to the species; but this is not always reliable as the time of the leaf flush can be locally altered in some regions by climatic conditions, site, genetic variation, age of trees and vigour of the trees. In Wales, observations indicated that there is sufficient lapse in time between the early flush of birch in the spring to distinguish it in a matrix of oak, which flushes much later. Again, for example, in eastern Canada, white birch flushes earlier, and large-toothed aspen late.

The rate at which leaves expand also determines their photographic appearance in the first weeks after the buds have burst. The leaves of the elm, for instance, grow slowly, which prolongs the bare appearance of the tree after flush has started. Twigs, young flowers and fruits can also give tree species a characteristic colour or tone.

Particularly for colour photography, the chroma and hues of the leaves are important. Young greenish leaves, still immature in size, have a high reflectance (about 10% to 20%) in the visible spectrum with a peak at about $0.535\,\mu$. If the leaves are reddish due to the presence of phytocyanins, then the peak usually occurs in the vicinity of $0.60\,\mu$ or as a plateau extending to $0.535\,\mu$. Old green leaves reflect less in the visible spectrum than young green leaves, but more in the infrared. The actual grey tone or colour recorded on the photograph for the foliage of an individual tree will depend not only on the spectral luminance of the surface provided by the leaves but also on their arrangement and on the grey tone or colour of the ground vegetation and soil surface.

According to Swellengrebel (1959), individual species in British Guiana could not be identified, mainly due to the variation of crowns of the same species on the photograph. Some trees had old leaves and appeared dark, while others of the same species were just getting new leaves, which recorded a lighter tone on the photographs. This is a photographic problem in other regions of evergreen forest, particularly if the trees have naked buds. The response of a species with naked buds is often sudden to conditions favouring flush.

(c) Remote sensing

Since World War II, remote sensing has been expanded, which includes not only passive sensing using aerial cameras, but also active sensing using radio-frequency receivers, radar (e.g. Crandall, 1963), magnetometers, gravity meters, seismometers and scintillation counters. *Remote sensing* refers to the acquisition of information about an object or phenomenon which is not in contact with the information-gathering device (Parker & Wolff, 1965). A combination of these devices is termed a *multiband sensor*. A simple example is provided by simultaneous panchromatic and infrared photography. Recently, a multi-lens camera has been developed to provide nine simultaneous photographs using different film–filter combinations. This has been used for evaluating aircraft landing sites through to studying the tonal differences of the vegetation and terrain background (Parker & Wolff, 1965). Colwell (1963) and Harris & Woodbridge (1964) drew attention to the applications of multiband sensing for interpretation and mapping.

In fig. 6.6, a normal photograph and infrared imagery of the same area are shown. *Infrared imagery* is created by recording emitted energy from objects on the earth either passively or by recording the 'echoes' of a sensing machine in the aircraft. Photography is a passive process of image recording which relies entirely on the reflected solar energy; but it is possible by using Laser and Maser sensing devices to identify unknown substances on the ground or close to the surface of the ground by the specific absorption and reflection bands of their molecules. Laser and Maser are abbreviations for light-wave and micro-wave amplification by stimulated emission of radiation. Laser also provides a precise method for the aerial measurement of distance and bearings.

The physical limits of passive sensing of vegetation appear to be between $0 \cdot 29 \, \mu$ in the ultra-violet and $2 \cdot 40 \, \mu$ in the infrared (see fig. 6.4a). Already in research, active sensing with infrared imagery beyond $1 \cdot 0 \, \mu$ has been used for mapping forest fires in the United States (Olson, 1963), for the delineation of thermal areas and mapping ocean currents, and radar has been used for continuous tracing of terrain features at very small scales, for recording soil characteristics and for mapping ice fields. Multiband sensing is also playing an increasing role in satellite probes of the surface characteristics of the moon and planets.

Due to recent advances in the development of sensing devices for military purposes, it is now possible to discriminate in the mid-infrared between objects on the earth in the temperature range 275° K. to 325° K. ($5 \, \mu$ to $25 \, \mu$). Attempts are also being made to develop the ultra-violet band as a long-range sensing tool, using wave-lengths down to $0 \cdot 24 \, \mu$ (Monteith, 1959; Olson & Cantrelli, 1965; Parker & Wolff, 1965); but it is unlikely that active sensing below $0 \cdot 29 \, \mu$ will be widely used due to the poor transmitting qualities of the atmosphere below this wave-length. As pointed out by Colwell (1963), cataloguing and analysing infrared spectra make it possible to identify unknown substances in their specific molecular bands. Within the photographic spectrum it is only possible to determine an object's gross reflectance properties by noting the contrast differences exhibited on the photograph through the use of filters having different transmittance characteristics.

PART TWO
Elements of Aerial Photography

7

PLANNING AERIAL PHOTOGRAPHY

INTRODUCTION

PERTINENT aspects of aerial photography or the taking of photographs from an aircraft need consideration, as it may be necessary to provide flight specifications for a special purpose. Important to the planning are the objectives of the operation and the influence of these objectives on the scale and focal length, cost, time of flying, the preparation of the flight map and flight contract and ground control. It is convenient to examine the taking of aerial photographs before considering photogrammetry and photo-interpretation, as this places the topic in order of the natural sequence of events.

(a) Scale and focal length

When the objective or objectives of the operation are known prior to flying it is possible to prescribe specifications so as to improve the interpretation-value of the photographs. If we assume that the objective is the provision of photographs for ecological studies and forestry, then the objectives will immediately influence the choice of photographic scale and focal length of the lens. It will be recalled that focal length and other characteristics of the lens were discussed in Chapter 2. Emphasis will be placed on the objective for forestry, as most photographs of greatest value to the ecologist have been taken initially for forest inventory or for a combined forest inventory-mapping project. However, the overall approach and most of the discussion will be of interest to the photogrammetrist and interpreters in other branches of natural resources.

SCALE

For forest studies, the photographs will normally be at a representative fraction of 1/10,000 to 1/20,000 and occasionally as small as 1/30,000 or as large as 1/5,000. Stellingwerf (1966) found no significant difference in the identification of tree species in Holland at scales of 1/10,000 and 1/20,000 and Lackner (1964) obtained similar results at scales of 1/6,000, 1/10,000 and 1/15,000 in Austria. In regional forest surveys in Maine and Nova Scotia, photographs at 1/31,000 were reported being as suitable as 1/15,840 (Young *et al.*, 1963). In Washington state photographs at 1/60,000 have been used successfully for locating marketable stands of timber (Kummer, 1964). In satellite photography the scale is likely to be between 1/1,000,000 and 1/2,000,000. For infrared imagery, conventional scales are at present used. In Finland, Nyyssonen (1955) concluded that the normal scale limit in temperate forestry is between 1/5,000 and 1/20,000, but due to cost one should aim towards the use of 1/20,000 photographs. Within the scales used in forest studies, the term 'large scale' is frequently used with reference to photographs between about 1/5,000 and 1/10,000, 'small scale' for photographs at about 1/20,000 or a little less and 'average' or 'normal' scale for the range of about 1/10,000 to 1/20,000. These terms will be used in this context in the ensuing chapters; and the term 'very small scale' will apply to photographs in excess of about 1/40,000. Much of the Commonwealth in the tropics (e.g. Nigeria, Ghana, Malawi, Zambia, Uganda) has been covered by photographs at 1/30,000 or 1/40,000 primarily for mapping; but the photographs in this scale range have proved useful for forest inventory, including the identification and delineation of important plant formations. In Ceylon 1/15,840 photographs have been used in conjunction with 1/40,000 photographs for fairly detailed forest-type mapping and ecological studies including the recognition of plant associations. In the U.S.S.R. the commonest scale for forest photographs has been 1/25,000 (Mikhailov, 1961). In Tasmania, a scale 1/23,000 to 1/24,000 has been used for both mapping and forest photo-interpretation, but 1/15,840 is preferred by forest interpreters. At 1/15,840 it was possible to stratify the forest into several site-quality classes by tree height but it was not usually possible to identify tree species. In the United States, forest-type mapping is normally carried out on photographs at a scale of 1/10,000 to 1/15,840 and occasionally at 1/20,000. In the Landes, France, 1/20,000 photographs have been used for forest inventory. Several workers have used 1/10,000 to 1/20,000 photographs in the tropics (e.g. Hannibal, 1952; Wheeler, 1959). In Japan, colour transparencies and colour photographs at 1/20,000 have been used for forest inventory (Maruyasu & Nishio, 1962).

A maximum for large scale photographs of about 1/5,000 is determined normally by economic considerations, although at times the largest possible scale may be fixed by other factors. Thus, for forest research at Mount Disappointment, Victoria, using an 8¼ in. lens, a Lockheed aircraft and a scale of 1/4,000, the ground speed of the plane (150 m.p.h.) combined with the time required to advance the film did not allow more than 45% endlap. For research using very large scale photographs, a scale of 1/600 has been used in Ontario (Sayn-Wittgenstein, 1963) and 1/1,200 in the United States (Avery, 1958). Avery used twin 35 mm cameras and Lyons (1964) twin 70 mm cameras mounted on a boom projecting from a helicopter. The method should prove useful in ecological work and forestry.

It is also desirable to comment on choice of scale for puroposes other than forestry and ecology. For example, for geographic, geological, soil and hydrologic surveys, scales at 1/20,000 to 1/30,000 or even smaller are often preferred. Revertera (1961) indicated the value of small scale photographs. In South Australia for photo-geologic studies 1/40,000 photographs are used. In Japan, 1/20,000 and 1/5,000 to 1/10,000 have been recommended for special geomorphic studies (Maruyasu & Nishio, 1962). Colour photographs are in increasing use in geological and mineral surveys and are helpful in soil-mapping regions where the ground is clearly seen (e.g. semi-arid zones). Scales between 1/12,000 and 1/35,000 have been used in the U.S.S.R. (Simakova, 1964) for soil studies.

For agricultural, pathologic, entomologic and some land-use studies, large to very large scale photographs seem to be preferred, although some information can be obtained from small scale photographs. For example, near Tabora, in Tanganyika, rice, maize and ground-nuts could be identified on photographs at a scale of 1/30,000 by pattern, tone and texture in conjunction with local knowledge (see Chapter 19). In Japan, 1/12,000 colour transparencies have been used recently in land-use studies (Maruyasu & Nishio, 1962). In Western Australia, black-and-white photographs at 1/5,000 have been used for studying the die-back of *Eucalyptus marginata* (jarrah). In experimental studies of fungal diseases of oats in California, panicle suppression was discernible on photographs at 1/1,000 to 1/2,000 (Colwell, 1960). Disease and degree of infection could be detected by tonal differences at 1/2,000 and 1/4,000. For the identification of tree species, Aldrich (1966) mentions that there was little gain in accuracy with scales larger than 1/1,584.

For topographic surveys, scales have been continually reduced. Early photographs were taken at 1/15,000 to 1/30,000, followed in the 1950s by scales of 1/30,000 to 1/50,000, and recently scales of 1/50,000 to 1/80,000. Photography at 1/90,000 and 1/120,000 has been used in Sabah and Pennsylvania respectively; and in northern Australia scales of 1/80,000 to 1/90,000 have become popular, using a Wild A9 camera and a 3½ in. focal length. At 1/90,000, each photograph covers more than 150 square miles of the earth's surface. Even at this scale individual trees can be distinguished stereoscopically when they are large and widely spaced. As the scale of the photographs is reduced so are the number of control points for a given ground area.

FOCAL LENGTH

Having decided on scale the next step is to decide on the focal length of the lens to be used. This will be influenced by several factors, including the type of aircraft available. The type of aircraft determines the maximum ceiling for flying. Oxygen equipment or a pressurized aircraft is necessary above 20,000 ft. Flying conditions may fix the lowest safe flying height. In many parts of the world, particularly the tropics, flying is often 'bumpy' below 5,000 ft. Avery (1958) has described the taking of large scale photographs by helicopter at 200 ft to 600 ft.

The object of taking the aerial photographs also influences the focal length to be used. At high altitudes in war-time, focal lengths of 20 in. to 36 in. were favoured;

but for most peace-time projects focal lengths between 3½ in. and 12 in. are used. For forest and ecological studies, the most popular focal length is 8¼ in. with a 9 in. by 9 in. format and a 6 in. focal length with a 7 in. by 7 in. format. A 12 in. focal length is used in the U.S.A. occasionally for taking photographs for the mapping of rugged terrain at 1/15,840; and may be preferable for colour transparencies to provide a more uniform exposure from the centre to the edge. For maximum exaggeration of the stereoscopic image, a 6 in. lens with a 9 in. format or a 4½ in. lens with a 7 in. format is to be preferred, being suitable for small trees, possibly shrubs, agricultural crops and the mapping of flat or undulating terrain. Wing points are more accurately located with a short focal length, but mountains are undesirably exaggerated. Often, therefore, a compromise has to be made. If the trees are very tall, e.g. 200 ft, and the scale is large, then as the accommodation of the eyes is exceeded at only a short distance from the centre of the photograph, a 6 in. lens is unsuitable; and an 8¼ in. or a 10 in. focal length may be preferred. In western Canada and Australia, 12 in., 10 in. and 8¼ in. focal lengths have been found the most suitable for assessment of crown closure. The longer focal length may also help in ecological work when it is desired to see the understorey as often as possible. A 6 in. focal length may result in crown closure being over-assessed. The U.S. Forest Service compromises in selection of lenses by using normally an 8¼ in. lens for both mapping and interpretation. In Finland Nyyssonen (1955) recommended a lens having an angular field of 60° for forest studies.

Representative fraction. The ratio of focal length of the lens in feet to flying height above ground datum in feet is termed the *representative fraction* (R.F.), being also expressed as the ratio of the distance on the photograph (*d*) to corresponding distance on the ground (*D*), i.e.

$$\text{R.F.} = \frac{f}{H\text{-}h} = \frac{d}{D}.$$

That this is correct can be easily proved from similar triangles.

The 'R.F.' ratio is commonly termed 'scale' as has been used in this text; but scale may also be expressed as the number of feet or chains on the ground corresponding to 1 inch on the photograph. Thus a representative fraction of 1/15,840 corresponds to a scale of 20 chains to an inch and 1/8,000 corresponds to 10 chains or 660 feet to an inch. It is convenient to remember that there are 63,360 inches in 1 mile and 792 inches in a chain.

Direction of flight. Excluding differences in cost, other factors will also influence the choice of flight direction. For example, let us consider the influence of shadow. On a photograph of open ground, particularly at higher latitudes, the general recorded pattern of shadows on the ground will depend on the direction and elevation of the sun. Overall, one side of the photograph looks light and the other side dark. Fig. 7.1 illustrates the division of light and shadow as it occurs in a vertical aerial photograph of 'even-aged' forest at midday (after Fleming, 1963). Being in the northern hemisphere, the camera records mostly the shadowed side of the trees on the southern half of the photograph and the sunlit side on the northern half of the photograph. The density of the negative will be a record of this and the print can only be partially corrected for it by the technique known as dodging. On mosaics, made from assembled

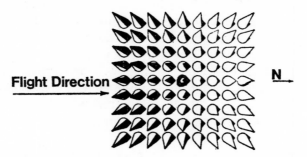

FIG. 7.1. In the northern hemisphere, the camera records mostly the shadowed side of trees on the southern half of the photograph. Overall, one side of the photograph will look dark and the other side light. The diagram represents a number of uniform-sized trees which are equally spaced on the ground, and illustrates how they would record with their shadows (black) on a large scale vertical photograph. The tree crown at the principal point has been scribed with a thick line.

photographs, the shadow-effect results in dark bands corresponding to the southern edge of the photographs in the northern hemisphere; and can result in incorrect photo-interpretation. In fig. 7.2, the shadow-effect is shown in plan-view for flight strips flown north-south. It will be observed that by overlapping the photographs or using only the effective areas of the photographs, it is possible to eliminate much of the shadow-effect. For this reason, as the end lap is considerably greater than the

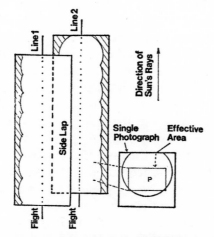

FIG. 7.2. The diagram represents two north/south flight strips being used to provide a mosaic by matching the effective areas of contiguous photographs (see text).

side lap of photographs, it may be preferable to fly the area north/south and not east/west. If the photographs are to be used later for mapping, it is sometimes an advantage to have the flight lines parallel to the national grid in an east/west direction.

For the photography of Connor's Plains, Victoria, the photographs were taken in the form of a 'T', which provided photographs both east–west and north–south. From these there seemed little to be gained for improved interpretation by direction of flight. A question requiring investigation is what is the effect of direction of flight on the stereo-model in association with its shadow. For stereoscopic examination of very large scale photographs, the effect of movement of the individual tree crowns on parallax measurements can possibly be minimized by flying at right-angles to the prevailing wind, as the movement is then in the direction of the y-axis (Aldred, 1964).

(b) The cost

Refer to Chapter 8, 'Economic Considerations'.

(c) Time of flying

TIME OF DAY

Both the time of day and season of flying need to be carefully considered in order to obtain the most satisfactory photographs for the objective; and require consideration in relation to the sun's angle to the zenith or horizon, the lens of the camera and the topography. Important limiting factors to the timing of the aerial photography, weather conditions excepted, are the avoidance of reflex reflection and shadow-length.

For colour film, the differences in colour temperature due to the altitude of the sun above the horizon need to be taken into consideration and, if necessary, a correcting filter used (see Part I). For example, in the U.S.S.R., Mikhailov (1961) says that colour photographs of wooded country are not taken unless the altitude of the sun is at least 20° above the horizon, in order to avoid the solar spectrum from affecting the colour balance of the emulsion. In Japan, Maruyasu & Nishio (1962) recommended that colour aerial photographs should not be taken when the angle of the sun is less than 45°, due again to the marked differences of colour temperature. *Reflex reflection zone.* Several aspects of this phenomenon have been examined in Chapters 5 and 6. On the equator, the task of excluding it from photography is at a minimum in June and December when the sun is on the Tropic of Cancer and on the Tropic of Capricorn; and in temperate regions it is at a minimum when the sun is in the opposite hemisphere. In temperate regions at lower latitudes the sun is at the critical angle for normal angle lenses (e.g. 50° to 70°) only between about 11.30 a.m. and 1.30 p.m. in the four summer months. If wide-angle lenses are used in temperate latitudes, then to avoid serious reflex reflection, especially near the tropics, photography in summer should be avoided or flying should be carried out in early morning or late afternoon. The effect of long shadows may, however, preclude the latter suggestion.

The Forestry and Timber Bureau, Canberra, have provided a useful booklet of charts which cover latitudes from the equator to 60° S. for focal lengths of 6 in., 8¼ in., 10 in. and 12 in. It provides information as to the hours throughout the year which are to be avoided to obtain photographs free of reflex reflection, and includes the chart shown in fig. 7.3. In Canada, solar nomograms have been published (Fleming, 1964).

Sims (1954) commented that the repeated area of no-shadow may cause visual confusion. For forest studies relating to the height of trees, the area of reflection on each photograph cannot be interpreted satisfactorily and tree heights cannot be accurately assessed. At time of viewing a pair of photographs stereoscopically, the reflex reflection area is seen with one eye, and the other eye sees a normal image of the trees and shadows. At first the inexperienced photo-interpreter may have the impression that he is viewing the adjoining photographs in normal stereo-vision; but by closing one eye at a time he will observe that only one satisfactory image can be seen (fig. **6.3b**). Under suitable magnification, the shadows of the trees will be seen to be directed inwards to the centre of this area on the one print and on the other print the shadows conform to a normal pattern. This is due to the radial displacement of the crowns on the photographs exceeding the radial displacement of the shadows. A similar effect may be observed on photographs for clouds and their shadows on the ground.

In the tropics, if photographs are taken when the sun is at or near its zenith, the reflex reflection zone and the photographic nadir will be close together, and in consequence the transferring of the principal point will be difficult. The inconvenience caused by the reflex reflection is not only avoided by choosing the right time of flying in relation to season, but is also minimized by using as long a focal length as possible, by a 7 in. format and by having the maximum side lap and end lap between photographs. Normally, the presence of such areas on the photograph is more of a problem to the interpreter than the photogrammetrist. The photogrammetrist usually finds no difficulty in using black-and-white photographs containing a reflex reflection zone, provided the photographs are at very small scale.

Solar specular reflection zone. **Fig. 7.4** (Eildon Weir, Victoria) shows both the solar specular reflection zone located at the water's edge and opposite to this and at an equal distance on the opposite side of the principal point the reflex reflection zone. The specular reflection zone is caused by the smooth water surface reflecting the sun's rays upwards in accordance with the sine law i.e. $\dfrac{\sin i}{\sin r}$ = a constant. Sin i is the angle of incidence of the rays on the water and sin r is the angle of reflection of the rays from the water, both angles being measured from the normal to the earth's surface. If extensive areas of water are being photographed, there is a risk that on the

75

photographs over-exposed areas in the vicinity of the solar specular reflection zone will conceal details of the shore line, shoals and even entire small islands. On rough water, the solar reflection zone may be of considerable size due to wave action

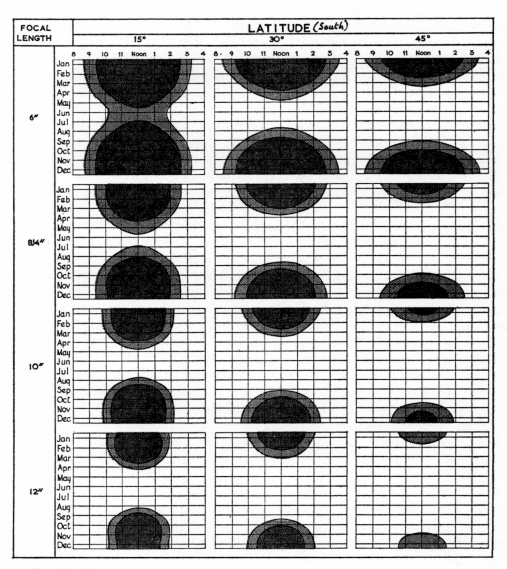

FIG. 7.3. Charts showing the time of the year and time of the day at three latitudes in the southern hemisphere at which photographs can be taken to avoid the 'no-shadow zone' (i.e. reflex reflection). Note also the influence of the focal length of the lens. The lighter (i.e. outer) shaded areas indicate that the zone occurs within the diagonals of the aerial photograph and the darker (i.e. inner) shaded areas that the zone occurs within the width of the photograph (9 in. format).

(e.g. Fleming, 1963). This effect is also shown in fig. **7.4** (left-hand photograph) due to a breeze rippling the water surface. The effect of solar specular reflection can be minimized as for reflex reflection by flying when the sun is at a fairly low angle to the horizon, by increasing the focal length, by increasing the photographic overlap and possibly by careful selection of the filter. Under American conditions, with 80% forward overlap the loss of stereoscopic coverage due to the effect of specular reflection from the water surface should be negligible (Fleming, 1963).

Shadow-length. This is important to the interpreter, being directly related to the angle of the sun to the horizon. When planning a photographic project, shadow-length requires to be considered both in relation to the topography, the shadows of the trees and the combination of both. It has been observed in northern Australia that when several hours have elapsed between the taking of the photographs of adjoining flight runs, it may be difficult to obtain a 'firm' stereoscopic model for mapping due to the different direction of the shadows on the contiguous photographs of the two flight strips (Hocking, personal communication, 1964). Welander (1953), working in the north of Sweden, concluded that the length of shadow on the print should not exceed 1·5 times the height of the tree and that 1·0 gives excellent prints. The shadow-length of trees will be further discussed in Part IV, 'Apparent Height'.

If valleys are steep sided and run east–west, long shadows may obscure much valuable detail at certain times of the day and even for an entire season at high latitudes. For forest photography in New South Wales, the time of the year and day best suited to flying, after considering shadow and reflex reflection, are often determined graphically. For example, for Eden at latitude 36° S., curves for 35°, 40° and 45° were required to provide the minimum sun's elevation for terrain varying from level to rugged in order to avoid excessive shadow. A fourth curve for the sun's elevation of 59° was superimposed so as to avoid a reflex reflection zone on the photograph, when a lens with a 60° angular field is used. Finally, due to the valleys running east–west, additional curves were added for time of the year to take into account the sun's bearing from Meridian North.

TIME OF YEAR

The season of flying to satisfy the objective of the photography is influenced by solar altitude, periodicity of leaf-flush and leaf-fall, and occasionally by time of flowering, and by atmospheric conditions.

Solar altitude. The effect of the solar altitude was discussed in the last section, and will therefore be only briefly commented upon here. In mountainous areas at higher latitudes, one or more months may be precluded due to the long shadows or deep shadows obscuring details in the valley bottoms. Also, the length of shadow provided by the trees on the photograph may exclude certain months. For example at lower latitudes in Sweden and a shadow length of 1·5 times, the season is between May and August. Similarly, one or more months may be unsuitable due to appearance of reflex reflection on each photograph, irrespective of the time of the day. Thus for a latitude north of 45° S., photography for forest interpretation may be precluded, using a 6 in. lens and 9 in. format, from about the middle of November until the middle of January, as shown in fig. 7.3. The period is even longer as one enters the tropics.

Phenology. Phenology plays an important role in determining the season unsuitable for flying. In the cool temperate regions of the northern hemisphere and where there are two distinct seasons in the tropics, phenology is important in relation to time of flying. It may or may not be desirable to fly an area when species are leafless, depending on the objective (e.g. forestry or geology) and composition of the forest (i.e. deciduous and non-deciduous species). Extensive areas of the northern hemisphere, some tropical woodlands (e.g. East Africa and India), are completely deciduous, and some forests are semi-deciduous (e.g. parts of northern Australia). In Britain the deciduous period extends from late October to about April; in central and eastern Africa from about November to March and in eastern Canada October to May. If, however, a deciduous species is growing in intimate mixture with a conifer, then winter or dry-season photography may provide a means for identification of each. In North America, the very definite leaf shades in the autumn, which last only a few weeks, can be critical for recognition of species from photographs, e.g. sugar maple, red oak. Similarly by studying the season of leaf-flush and flowering it may be possible to separate species on photographs. As just mentioned, this critical period may only last a few weeks; and it is essential to ensure that the photographs are taken in this period. Backstrom & Welander (1953) from their spectral reflectance studies in Sweden concluded that the best time for photography of pine, birch, spruce, aspen, oak and beech was in the spring and not in the summer.

Specht & Rayson (1957) and Burbridge (1960) have drawn attention to the fact that many of the plants indigenous to southern Australia have a growth phase out of rhythm with the present Mediterranean-type climate. These plants make their major growth during the drier months of summer. Burbridge describes two distinct periods of growth, one in spring and one in late summer.

If the photographs are required for the study of tree form, then in deciduous forest winter with snow on the ground may be the best time. Also if the photographs are required for the study of the ground vegetation and the ground surface, the photographs will be taken when the trees are leafless. In New South Wales, near Bega, it was possible to identify *Eucalyptus Baxteri* from photographs taken at time of flowering (H. Macdonald, personal communication, 1963). Ground observations at Boola, Gippsland on *Eucalyptus polyanthemos* suggested that this species could also be identified at the time of flowering.

Atmospheric conditions. Excessive haze following long periods of drought or following fires at the end of the dry season will also influence the time of the year for photography. For example, in the tropical woodlands of Tanzania, Zambia and Malawi there is usually so much haze from the annual grassland fires in August and September that the taking of sharp photographs is impossible. Similar conditions also occur in northern Nigeria. Heavy haze not only reduces resolution but also makes it difficult or impossible for the pilot to locate normal-size ground control marks. In years of bad fires, haze is a limiting factor to the taking of photographs in Australia. In Victoria, smoke haze in February frequently excludes aerial photography.

The presence of clouds and cloud shadow on the aerial photographs usually calls for their rejection and often makes certain known cloudy periods of the year unsuitable for aerial photography. This is not a serious problem in mainland Australia, but it is in Tasmania. In western Britain, reasonably long cloudless periods are uncertain and usually occur only in February, November, June and September; but February and November are unsuitable due to the long shadows caused by the sun never rising high above the horizon. In Holland, there are possibly 20 to 25 days a year suitable for black-and-white photography and considerably fewer for colour. In the Canadian south-west, there are about ten flying days in July, August and September and about two days a month from November to March. Tropical rain forest areas such as Malaysia, Guiana, New Guinea and the West African coast, are difficult to fly throughout the year; and it may be necessary to fly below the cloud cover. In the high forest region of Ghana, suitable conditions for photography are limited to a few weeks in January and February at the most due to cloud cover. In Victoria, clouds during winter months preclude flying in June and July and possibly August and again in October. In addition, shadows preclude flying in May, June and July in the mountainous areas.

It is interesting to note that in experimental work at 35,000 ft at very high speeds, e.g. Mach 1.4, certain environmental effects may degrade the quality of the photographs (Nielsen & Goodwin, 1961). These include metric distortion caused by refraction of light rays by the flow field surrounding the aircraft, loss of resolution by scattering of light by the turbulent boundary layers around the aircraft, loss of contrast in the photograph between the object and its background by the presence of luminous air in the flow field and the effect of scattering in the atmosphere (see Chapter 5).

(d) Flight map

Before executing the aerial photography it is often necessary to prepare a flight map for the contractor and it is usually desirable to have a contract covering essential technical specifications. The terms flight plan and flight map are considered to be synonymous. It may also be necessary to establish ground control points before taking the photographs.

A fundamental condition of flight planning is to provide adequate photographic coverage with the minimum number of photographs to satisfy the objective of the flight. The flight map of the area to be flown shows the boundaries of the area, the location of the starting and finishing photographs of each flight line and the location of the flight lines to one another and to important objects on the ground. Part of a flight map is shown in fig. *7.5.* It will be observed that the flight lines are in the direction of the longest sides of the area and that to minimize the number of photographs being taken, the outermost flight lines to the left and right are within the boundaries of the area being photographed and not on the boundaries. Also the flight lines were arranged so that the high peak, Scott Mountain (Idaho), is not towards the edge of a photograph. If the peak were towards the edge of the photograph, radial displacement would be exaggerated and be more difficult to correct during map preparation, as discussed in Part III. It would also, possibly, be necessary to have the flight lines closer together to provide an adequate side-lap. On each flight line, the average *datum plane* (or datum level or vertical control datum) of the land over which the flight line passes has been shown.

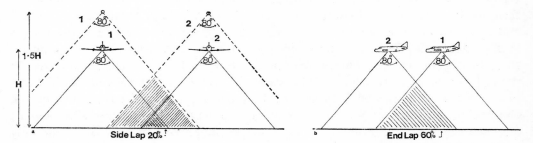

FIG. 7.6. (a) An aircraft is shown flying away from and towards the reader on flight lines (1) and (2). Using a camera with angle of coverage of 80° and a flying height H, there will be a side-lap of 20%. If the flying height is increased to $1.5\,H$ (and the focal length is 12 in.), then the side-lap is increased to approximately 46%. Alternatively, the distance between flight lines can be increased to retain the 20% side-lap. This will reduce the number of flight lines in a fixed area and will also reduce the cost. (b) Two consecutive camera stations on the same flight line are shown for an aircraft. The closer the camera stations are together, the greater will be the end-lap.

The term *datum plane* is strictly not the same as datum level, but in practice for surveys of limited area they can be considered as synonymous. A datum plane is any level surface, perpendicular to the direction of gravity, which is taken as a surface reference from which to calculate elevations. A *flight line* is a line drawn on the map to represent the track of the aircraft. During actual flight the aircraft may easily diverge from this track. Allowance for this divergence needs to be made when preparing the flight plan.

It is useful, in the margin of the map, to annotate the following: name of area, flight altitude above the datum plane, altitude above sea-level, the focal length of the camera-lens, end-lap, side-lap, and shooting time in relation to the angle of the sun above the horizon.

The simple calculations needed in preparing the flight map are provided below as an exercise. This should be carefully worked through. The *overlap* is the amount one photograph overlaps the area covered by another, expressed as a percentage. The overlap between adjoining photographs in the same flight line is termed *end-lap*, being usually 55% to 65%; and the overlap between adjoining photographs in adjacent parallel flight lines is termed *side-lap*, being usually 15% to 45%. The relationship between flying height and side- and end-lap is shown in fig. 7.6. It is suggested that Chapter 9 ('The Aerial Photograph') should be studied before working through this exercise, unless the reader is already familiar with the geometry of the vertical photograph.

Exercise

1. *Object of aerial photography*
   To provide (a) topographic map at R.F. of 1/20,000 and contour interval of 50 ft, and (b) photographs for general forest type interpretation, and for geological studies.

2. *Scale*
   $$12'' = 20,000 \text{ ft}$$
   or $$1'' = 1,667 \text{ ft.}$$

3. *Photographic coverage*
   Size of photographs $= 9'' \times 9''$.
   Therefore $9'' = 15,003$ ft.

4. *Direction of flight lines*
   The flight lines will be north–south to conform to the shape of the area, which has its shortest axis in an east–west direction.

5. *Boundary coverage*
   This should be 25% of photographs on the eastern and western boundaries

   i.e. 25% of 15,000 ft = 3,750 ft; in $1/10'' = \dfrac{3,750}{1,040} \cong 3 \cdot 6$ (see No. 7).

6. *Side-lap* (approx.)
   Average = 30% (see technical specifications).
   Reciprocal of 30% = 70% = Net coverage.
   Net coverage = (70% of 15,000 ft.) $\cong$ 10,500 ft.

7. *Bar scale on map*
   $$5 \cdot 07'' = 10 \text{ miles.}$$

   Therefore $1'' = \dfrac{10}{5 \cdot 07}$; $\dfrac{1''}{10} = \dfrac{1}{5 \cdot 07} = \dfrac{5,280'}{5 \cdot 07} = 1,040$ ft.

8. *Grid measurement*
   10 minutes $= 4 \cdot 20'' = 1,040 \times 4 \cdot 20$ ft.

   1 minute $= \dfrac{1,040}{10} \times 4 \cdot 20$ ft.

9. *Width of flight plan in feet* (A')
   $6 \cdot 3''$ on map $= 10 \times 1,040 \times 6 \cdot 3$ ft $= 65,520$ ft $\cong 65,500$ ft.

10. *Width A'*

Less allowance for boundary coverage (i.e. net width A'').
Net width $(A'') = 65,500$ less $(3,750 \text{ ft} \times 2) = 58,000$ or
$6 \cdot 3'' - 0 \cdot 72'' = 5 \cdot 58''$.

11. *Number of flight spaces*

(From 10) Net width $(A'') = 58,000$ ft.
(From 6) Net photo coverage $= 10,500$ ft.

$$\text{Flight spaces} = \frac{58,000}{10,500} \cong 5 \text{ or } 6.$$

12. *Flight coverage*

Width per flight strip.

(a) $\dfrac{\text{Net width } (A'')}{\text{No. of flight spaces}} = \dfrac{58,000}{5} = 11,600.$

(b) $\dfrac{58,000}{6} = 9,670.$

13. *True side-lap*

$$\text{Side gain} = \frac{\text{Flight coverage} \times 100}{\text{Photo width coverage in feet}} = \frac{11,600 \times 100}{15,003} = 77\% \text{ (a)}.$$

or $\dfrac{9,670 \times 100}{15,003} = 64 \cdot 5\% \text{ (b)}.$

Side-lap $=$ reciprocal of $77\% = 23\%$ (a).
Side-lap $=$ reciprocal of $64 \cdot 5\% = 35 \cdot 5\%$ (b).
As the terrain is rough will choose *35·5%*, i.e. *6 flight spaces*.

14. *Datum plane*

Elevation readings at 1″ distances along flight line

Flight line number	1.	2.	3.	4.	5.	6.	7.
1.	6,000	6,500	5,500	6,500	7,500	7,000	6,500
2.	5,000	5,000	6,000	7,500	7,500	7,500	6,500
3.	4,500	5,000	5,000	7,500	6,000	6,500	6,000
4.	5,000	5,500	6,000	6,500	6,500	5,500	5,000
5.	5,000	5,500	7,000	7,500	6,000	5,500	5,000
6.	5,500	7,000	7,000	7,000	6,500	5,500	6,000
7.	6,000	4,500	7,000	6,500	7,000	6,500	5,500
8.	6,500	6,500	8,000	7,000	6,500	6,500	6,000
Total:	43,000	45,000	51,500	56,000	53,500	50,500	46,500
Mean:	5,400	5,700	6,400	7,000	6,700	6,300	5,800

Grand mean: 6,200 ft.

15. *Height of aircraft above datum*

Refer: Technical specifications.

82

Focal length $= 6'' = \frac{1}{2}$ ft. R.F. $= \dfrac{1}{20,000} = \dfrac{f}{H}$.

Therefore height $(H) = \frac{1}{2} \times \dfrac{20,000}{1} = 10,000$ ft and flying above sea-level $= 16,200$ ft.

16. *Width of flight space in inches*

1 flight space $= 9,670$ ft $= \dfrac{9,670''}{10,400} = \dfrac{93''}{100}$.

17. *End-lap*

Average 60% (55% to 65%) see technical specification.

End-lap $= \dfrac{60}{100} \times 15,000 = 9,000$ ft $= \dfrac{9,000''}{10,400} = \dfrac{87''}{100}$.

(e) The flight contract

A contract is the basis of business deals and there is no reason why aerial photography should be an exception. A contract legally involves an offer, an acceptance and a consideration, e.g. money payment, for flying carried out. Most written contracts are conditional; and in a contract for aerial photography it is advisable to include the items summarized below under 'Technical specifications' as given in the *U.S. Forest Service Handbook*. This set of specifications has been selected on account of its concise yet adequate coverage. Suitable specimen contracts are also provided by the U.S. Geological Survey standard contract and the Australian standard specifications provided by the National Mapping Council. Frequently in practice the form of contract is even shorter than the specifications given in this contract as both the purchaser and the contractor rely to a certain extent on good will.

TECHNICAL SPECIFICATIONS TO BE INCLUDED IN CONTRACTS FOR AERIAL PHOTOGRAPHS

(after U.S.F.S. specimen contract)

1. Area covered as shown on the flight plan and maps.
2. *Confidential nature of survey.* No information regarding the survey will be divulged except by permission of the ordering company.
3. *Ownership of photographic negatives.* These are the property of the purchasing company, regardless of where they are kept or stored. No sales of reproductions will be made without approval of the purchasing company.
4. *Responsibility for damages.* The contractor will be fully responsible for the safety and liability of his own personnel and equipment.
5. *Scale.* A stated scale calculated for a stated ground elevation will be maintained. Permissible variation, 5 per cent.

6. *Overlap*. Along the line of flight this shall average 60 per cent and shall be between 55 and 65 per cent. Between flights, it shall average 30 per cent and be between 15 and 45 per cent.

7. *Broken flight lines*. Two exposures of the new photography shall overlap the last two exposures of the old photography.

8. *Tilt*. This should average less than 2 degrees and should not exceed 5 degrees. It should not interfere with the use of the photographs in radial line triangulation.

9. *Crab* (deviation of photograph alignment from the line of flight). This should not affect more than 10 per cent of the width of the photograph.

10. *The make, model and serial number of camera and lens*. These should be specified.

11. *Film and filter combinations* (to be specified).

12. *Photograph quality*. This must conform to the submitted samples. All photographs shall be free from blemishes or cloud shadows.

13. *Season of photography* (to be specified). The desired date of completion of photography and delivery of materials should also be given.

14. *Flight log*. This should be supplied to purchaser and should give time, dates, etc., for each strip of photography.

15. *Delivery*. Standard delivery consists of two sets of contact prints and one set of index mosaics. Both sets of prints should be free from staple holes or other blemishes, unless the purchaser wishes to save money by accepting one set of prints (used in making index mosaics) with staple holes. Additional materials desired should be listed in detail.

16. *Radial line plots and mosaics*. If ordered, it should be clearly stated who will furnish ground control (i.e. survey data) and what control points will be used. The purchaser should list points such as fire towers that he would like to have correctly located.

17. *Business arrangements*. These include prices, delivery schedule, payments, cost of additional materials (extra prints, etc.) and cancellation provisions.

(f) Ground control

If the area to be covered by photographs has been previously covered by accurate topographic maps as in England and Wales, ground control can usually be provided by selecting suitable features common both to the Ordnance Survey maps and the photographs. If possible the photographs should be tied in directly with bench-marks or trigonometrical points. Accurate planimetric maps can also be used, provided the height of the datum plane can be determined from spot-heights on the map. If these spot-heights on the map are not recognizable on the photographs then it may be necessary to carry out a ground survey between recognizable ground points and the spot-heights. Tacheometers and geodemeters are now used to speed up the determination of distance between ground control points, although in forested country the trees provide a ground obstruction to their use.

GROUND CONTROL

The advantage of having accurate maps, prior to flying, is that a ground survey to establish control points is not necessary or need not be carried out until after the aerial photographs are available. A *control point* is defined as a station in a horizontal or vertical control system that is identified on the photograph and used for correlating data on that photograph. It is preferable to retain the term *ground control point* for stations actually established on the ground and to use the term *photogrammetric control point* for correlating data between photographs by selected points on the photographs. Suitable control points are provided by road intersections, railway crossings, isolated trees, change in direction of wood boundaries, the intersection of power line traces and other prominent features, hedge junctions, isolated buildings and footpath intersections. When preparing a topographic map, by a Stereotope for example, four well-distributed photogrammetric control points are required on each half of the photographs. These are points determined from existing maps, aerial triangulation or by ground survey.

If, as in many parts of Africa, there are no satisfactory maps, then it is necessary for ground control points to be established before taking the aerial photographs. The importance of these ground control points will be appreciated better after considering the section on photogrammetrical mapping. In aerial photo-interpretation, including forest inventory, the provision of horizontal ground control points is usually adequate. If the country is mountainous, however, vertical control may also be necessary.

Kelsh (1940) carried out an experiment in Maryland, covering 140 square miles at 1/20,000, to ascertain how many ground control points are required. It was found that with horizontal ground control points the average error was 18 ft and the maximum error was 90 ft. The points were well distributed so that there was one at each corner and one at the mid-points. If only four points were used, one at each corner, the errors were 43 ft and 125 ft respectively.

In Mexico (Huguet *et al.*, 1958), forest maps for inventory were required for an area of 500,000 hectares within which there were only two established geodetic survey points. By ground cruising from these and using a theodolite and luminous signals to fix triangulation points, a further twenty-four control points were established at distances of 20–25 km. These points were plotted at a scale of 1/15,000 and used as a map base for filling in detail using mechanical templets and a Multiscope. The forest map was checked against a topographical survey made separately, which confirmed the accuracy of the forest map using the twenty-four triangulation points.

In Australia, for mosaics and small and medium scale maps, extensive use has been made of astronomical fixes, triangulation and traverses and recently 'Shoran'. It is doubtful whether chain and compass surveys and altimeters should be used in mountainous country.

The *geodemeter* is now widely used for establishing distances between trigonometric points and base line measurements for aerial mapping. The principle of this instrument is as follows. A ray of light projected from one end of a line to be measured is reflected back at the far end of the line to the recorder. The time interval of the ray on its path is measured and from this the distance is calculated. This is in accordance with Fermat's principle, namely that of the several paths available for a ray of light by reason of the media of different refractive indices, the path taken will be the one requiring the least time. The instrument has been reported as having an accuracy of one part in 100,000 for measured distances of 10 miles.

A similar type of instrument is the *tellurometer*, which was developed in 1957 in South Africa by Wadley. Instead of relying on a light signal, radio waves are used to measure distance between 500 ft and about 50 miles. The receiving set changes the phase of the wave and retransmits it to the emitting or master set. The distance is again calculated from the time taken for the return signal. Accuracies as high as ± 12 in. in vertical heights and ± 6 in. in measured distances of 30 miles across open country have been achieved. Both instruments are expensive.

Targets. Frequently ground control points in the form of targets need to be placed around the perimeter of the area to be photographed (and sometimes within the area if it is large). The minimum size of target will vary with the scale of the photography and to a lesser degree with its colour contrast, surrounding vegetation and its shape. White targets on a black background are the most conspicuous. Cheap highly reflective plastic 'cloth', in orange, red or yellow, to provide a cross on the ground is effective; and so are tree logs painted white. Specular reflecting surfaces must be avoided (e.g. new tin sheeting). In Finland, targets at about 2 m above the ground have been found to be more conspicuous than targets on the ground. As a rule of thumb, target length should be about 0·001 times the denominator of the representative fraction (e.g. about 10 ft for 1/10,000 photographs) and about one-fifth the length in width (e.g. 2 ft).

8

ECONOMIC CONSIDERATIONS

INTRODUCTION

Frequently a decision on the taking of aerial photographs is made without those responsible being conversant with the more important economic considerations. Usually a decision has to be made to sub-optimize under conditions of subjective uncertainty or subjective risk, and seldom under conditions of objective certainty. If the decision is made under conditions of partial objective certainty in a vertically integrated industry then, in the writer's experience, the objective certainty applies only to the aerial photography; and top management in reaching a decision normally fails to consider opportunity cost and to include in the price ratio all relevant costs of photogrammetry and photo-interpretation to the final product (e.g. completion of a forest survey). Recently a well-known ecologist expressed concern when informed that aerial photography would cost 3*d* an acre. Yet this was a small cost in comparison with the overall cost of establishing and analysing data from a grid of quadrats on an acre without aerial photographs (20*s* to 40*s*).

Provided all fixed, variable and on-costs, including administrative, executive and technical costs, are taken into consideration up to the final product, it will normally be found that the cost per end-product is reduced by using aerial photographs. Frequently new survey costs can be reduced by re-flying an area after the lapse of a few years. In Sweden, for example, re-flying of large blocks of commercial forest, at 5 year intervals, is considered to be economically justifiable (Francis, 1957). In Gippsland (Victoria) a financial saving is being made, at management level, by re-flying areas after land clearance and planting. In New Zealand most cities are re-flown at least every 10 years (Asch, 1961). In Washington state, Kummer (1964) reported that the most economic method of assessing hurricane damage to three-quarters of a million acres was to have the area re-flown.

(a) Aerial photographic costs

Several important factors require consideration in relation to the total cost of aerial photography in a geographic region; and if a single factor is to be chosen as being overall the most important, then usually it would be scale. For example Harrison

& Spurr (1955) quoted the following relative costs of two sets of black-and-white prints and an index mosaic:

Scale	1/30,000	1/20,000	1/15,000	1/12,000
Relative cost	1·0	1·5	2·2	2·25

If costs are to be kept reasonable there should be a free market in which competing contractors can tender for the work. For low costs, aircraft and equipment need to be locally available in the season(s) favourable to aerial photography, and there should be sufficient other work in adjoining regions to enable the contractors to operate throughout most of the year. Frequently government policy favours the development of a monopoly either by deliberately using military aircraft or contracting to its own civil aircraft or by unintentionally creating an unstable market by contracting at irregular intervals.

Local climatic conditions also influence costs. For example, in high rainfall areas along the equator flying is often restricted to the mornings at certain times of the year, due to thunderstorms in the early afternoon. In Britain, with long periods of cloud cover and unpredictable periods of suitable weather, costs tend to be high. In Australia, as flying can be undertaken at most times of the year, costs are fairly low. Conditions are normally favourable to winter flying in the tropical north and summer flying in the temperate south. Frequently if a contractor is given the option of flying an area at a time convenient to his company then the contract price will be lower than if the aerial photography is required on or near a specified date. Such an option on time of flying enables the contractor to cover the area when conditions elsewhere are unsuitable for flying or in conjunction with other local contracts. To the purchaser of a small number of photographs this option can result in a saving of up to 50% of the normal cost.

In addition, the contract price will vary with the shape, size, locality and scale of the photographs. A long rectangular block, requiring fewer flight lines and a minimum of turns by the aircraft at the beginning and end of the flight lines, will cost less to fly than a square or an area of the same size with more flight lines. An area near the main operational and maintenance base of a contractor should cost less to cover than a distant area requiring the setting up of a temporary base. For example, the positioning fee within a few miles of base at Melbourne was £40, but was double that sum for an area about 100 miles away. A large area, e.g. 1,000 square miles, costs less per square mile than a small area (e.g. 10 square miles) at the same distance from base since the cost of getting the aircraft airborne and to the flying area remains the same. The minimum cost for a very small area covered on a single pair of photographs tends to equal the positioning fee.

Further, the cost of aerial photography will reflect costs related to the type of aircraft used. In the past, secondhand DC-3 Dakotas (150 m.p.h.) have been very popular. These have a service ceiling of at least 20,000 ft and a range of 8 flying hours. Recently Beechcraft (150 m.p.h.), Cessna 180s and Lockheed Aero Commanders

(225 m.p.h.) have been favoured. The cost per flying hour for the latter will probably vary between £30 and £35. This includes depreciation on the aircraft over about five years, fuel, maintenance and insurance. To this must be added the salaries of £20 to £30 per hour for 30 flying hours a month of a crew of three or four (captain, photographer, navigator). There is also depreciation on the cost and installation of the camera to be written off over, say, five years. The total cost per flying hour will then range from about £50 to £70.

Total flying range of an aircraft will vary possibly between 6 hours and 8 hours. Part of this time, as mentioned previously, is used in reaching and returning from the mission area and the remainder in taking photographs. Thus out of 6 hours flying, $3\frac{1}{2}$ hours might be spent in the mission area. Of this probably only 40%, depending on flight-line planning, will be actual productive time during which photographs are continually taken (i.e. 1·4 hours). In this time, at a ground speed of 160 m.p.h., a distance of $1·4 \times 160$ miles will have been photographed. For 9 in. by 9 in. photographs having a side-lap of $33\frac{1}{3}$%, the area covered at scales of 1/10,500, 1/15,840 and 1/21,000, will be approximately 224, 336, 448 square miles respectively. At a cost of £65 per hour for 6 flying hours this is approximately £1·7, £1·2 and £0·9 respectively per square mile. To this must be added the cost of the negative film processing, one set of proof prints and one or two sets of contact prints. Negative film will probably cost about 4s per exposure.

Actual costs that can be quoted for black-and-white aerial photography are 5d per acre in Holland for 1/20,000 photographs (Stellingwerf, 1963); 1·9d and 4·4d per acre in Victoria for 1/15,840 and 1/7,920 photographs; U.S.$3,500 for a half million acres of forest land (i.e. 0·7d per acre) in Washington state for 1/56,000 photographs and about five times this cost ($3\frac{1}{2}$d per acre) for 1/12,000 photographs (Kummer, 1964); and 0·6d to 2·2d per acre for photographs in eastern Canada (Gimbazevsky, 1964). Tolkning (1955) has pointed out that, under Swedish conditions doubling the scale increases the costs three-fold. Additional prints cost 3s to 5s each in Australia, 7s 6d in the U.K., 17s 6d in Sweden and New Zealand, about 5s each in eastern Canada and 3s 6d to 8s in the U.S.A. In New Zealand a government royalty of 14s 6d per photograph is charged as a means of recovering the initial cost of the aerial photography.

Provided the negative processed film has been retained by the purchaser of the initial photographs and he has legal rights to further printing, it is well worth while to consider setting up a laboratory.

To establish a simple photographic laboratory the basic equipment excluding an enlarger, for contact printing may cost about £75; and will comprise safelight, contact printer, electric timer, three trays, print washer, print dryer and print tongs. If the processing of aerial film is anticipated additional equipment will include film processing tanks, an aerial film drier and a mechanical timer. A room approximately 70 sq. ft to 90 sq. ft should be large enough to accommodate the basic equipment. Enlargers cost £70 to £5,000. Equipment for developing can be purchased for a minimum of £350. It should then be possible to produce prints at 3*d* to 9*d* each, excluding overheads and depreciation.

Colour aerial photography. On first considerations, the taking of colour photographs appears not to be recommended. Colour films are more expensive to purchase and more difficult to use, process and duplicate. Wastage is higher and the transparencies may fade in time and be attacked by fungi. For colour, the cost may be £50 for 75 ft of film and 10*s* to 35*s* for prints (according to contact size) as against £35 for 200 ft and 3*s* to 6*s* respectively for black-and-white. Chemicals are several times more expensive for colour than black-and-white.

The price ratio of colour transparencies to black-and-white photographs, including cost of film and processing and based on the provision of acceptable sets, may vary between 3/1 and 5/1. To make a decision on these price ratios, however, would be erroneous, as there are much larger costs to be included which may be considered as fixed costs for purposes of comparison. The conclusion will be, as pointed out by Smith (1963), that the total cost of aerial photography and provision of transparencies may not be greatly increased by using colour, as the largest cost factor, namely the taking of the photographs, is the same. For a large contract under favourable weather conditions, the price ratio should not exceed 6/5, after the fixed costs are included.

Mikhailov (1961) commented that by employing spectro-zonal photography in forest photo-interpretation in the U.S.S.R. costs were reduced considerably compared with black-and-white photography. Mikhailov considered that the reversal colour film process incurred little additional expense, but for a negative film process the cost was raised considerably due to the printing paper being much more expensive and the procedure more laborious. He estimated the cost of a photogrammetric survey to be increased by 10% to 20% when colour photographs are used. Colour prints from transparencies (i.e. reversal film) may cost twice as much as colour prints from negatives, and usually the exposure tolerance is greater with negative film.

(b) Aerial photogrammetric costs

To generalize on photogrammetric costs is impracticable as there are so many imponderables. It may, however, be safely stated that the photogrammetric mapping costs are more than the cost of the aerial photography. Ground control is usually the variable cost of greatest magnitude, being negligible when accurate maps already exist and being greatest and exceeding possibly one-third of the total cost in unmapped mountainous terrain. Quoting U.S. costs as percentages van Asch (1961) has given the following breakdown for topographic mapping from photographs:

Aerial photography	10%
Ground control	30%
Plotting	34%
Draughting	21%
Miscellaneous	5%

For topographic maps of New Zealand cities costs vary between 14s 6d per acre at a scale of 1/2,400 with 5 ft contours to 50s per acre at a scale of 1/1,200 with 2 ft contours. In the Middle East, the cost of mapping urban areas at 1/2,500 can cost as much as £4 per acre.

Fortunately, the requirements of the forester or ecologist in respect to planimetric mapping, as outlined in Part III, are less exacting and therefore the costs are less. Provided expensive ground control can be avoided, the cost may be only a fraction of those quoted in the previous paragraph. For example, in Gippsland bridging and planimetric mapping by contract at a representative fraction of 1/7,920 from photographs at 1/15,840 cost 4d an acre. A first-order machine was used by the contractor to prepare the basic map showing roads and legal boundaries and rivers. As a separate operation, forest detail was entered later on to the map sheets using a sketchmaster. In Western Australia (McNamara, 1959), 1/15,840 forest-type maps have been produced, using the dyeline process at a cost of 1·2d per acre. Sketch-maps, copied directly from the photographs, cost 0·3d to 0·4d an acre. Kummer (1964) in western U.S.A. reported that the cost of providing 'photo-maps' (Ozalid prints) at the contact scale of the enlargements (3½ times) was 0·3d an acre. This included the purchase of enlargements (£2 each) and the 'Ozalid' prints from the photographs at 8d a copy. Copycat prints (United Kingdom) and Dalcopy prints (Australia) can be obtained at about 6d a contact copy, excluding royalties, labour and depreciation on equipment. This is equivalent to less than a total cost of 0·2d an acre at 1/15,840.

(c) Photo-interpretation costs

From consideration of costs relating to photography and photogrammetry, it will be appreciated that the photo-interpreter can reduce his costs by direct use of the aerial photographs, or by relying only on simple planimetric maps or sketch-maps in conjunction with use of the photographs in the field. On economic grounds, the preparation of topographic maps prior to interpretation should be avoided unless they are required for other purposes. In Sweden, piece-rates have been used for identifying and delineating separately conifers and hardwoods on black-and-white infrared photographs (9 in. by 9 in. format). One example is given in table 8.A.

Stellingwerf (1963) working in Holland with 1/20,000 photographs gave the following break-down of costs for 5,000 hectares of forest. The percentages shown in brackets have been calculated from his data. For 10,000 ha. the total cost per hectare fell by 11%. One man was reported as being able to examine 500 photoplots a day in conjunction with stereograms.

Aerial photography (including photographs)	1s per ha. (55%)
Office work (including photo-interpretation)	5½d per ha. (25%)
Field work (measurement and checking interpretation)	4½d per ha. (20%)
Total cost	1s 10d per ha. (100%)

ECONOMIC CONSIDERATIONS

The opening comment relating to photogrammetry applies equally well to photo-interpretation. Similarly, as mapping from aerial photographs is now preferable to mapping entirely by ground survey, so it is preferable on economic grounds to carry out field studies in conjunction with photographic interpretation. During oil exploration in the Sahara Desert, the time taken to produce geological maps for the Compagnie des Recherches et d'Exploitation de Pétrole was reduced to about one-third after photographic interpretation was introduced (Richard, 1962). In Ceylon, Mott (1956) concluded that a forest survey using aerial photographs costs only about 30% of survey without aerial photographic coverage. Francis (1957b) gave the cost for 100% survey of selected valuable areas of tropical forest as U.S. $1·75 to U.S. $2·06 with the aid of photographs and $2·20 to $2·70 without photographs. In the south-eastern United States, tests have indicated that forest ground survey without photographs is much slower and gives an unfavourable cost ratio of up to 100 to 1 (Anon, U.S.F.S., 1959).

TABLE 8.A

Piece rates (1966)—1/15,000 *photographs*

| Percentage of forest/ photograph | Number of forest stands per photograph | | | |
| | 10 | 10–24 | 25–39 | >40 |
	(price paid in shillings)			
<25	8	16	24	34
25–49	10	22	36	48
50–74	12	26	48	54
75–100	14	32	50	70

Data relating to the cost of very large scale photographs, e.g. 1/1,000, have been given by Lyons (1964). Costs were separately determined for thirty-three quarter acre plots by ground measurement, photographic measurement and a combination using double sampling. Data included plot location, plot size, tree height, tree diameter and crown diameter. By ground survey only the cost per single plot was £17 10s 0d. The cost using a helicopter for the aerial photography, without ground work but including all interpretation and measurements off stereo-pairs (about 50 trees per plot) was £1 6s 0d. Actual aerial photography included in this cost was about 2s 6d a stereo-pair, and processing and printing 1s 6d. The interpretation of the individual trees on the photographs was the most expensive (35 trees per hour). The cost of a photo-plot in conjunction with ground sampling was £1 15s. The variance ratio $\frac{1,067^2}{1,246^2}$ (0·73) showed that the same standard error will be provided by 100 photographic plots as by 73 ground plots, whilst the cost ratio for the same efficiency was 50/6·48 = 7·7. That is 7·7 photo-plots are completed for the price of one ground

92

plot! Lyons also determined that, by double-sampling techniques, a 'photo-ground' forest survey for the same cost has a sampling error of 87% of that for a ground survey; and that for a specified objective the 'photo-ground' survey is only about 75% of the cost of a ground survey.

The writer is not aware of any published data relating to the cost of plant community studies carried out in conjunction with aerial photographs. There seems to be little doubt that the use of aerial photographs will be quicker and result in fewer quadrats being required for a similar degree of accuracy.

PART THREE
Elements of Photogrammetry

9

THE AERIAL PHOTOGRAPH

INTRODUCTION

So far, interest has been focused on the procurement of aerial photographs. Attention must now be given to the use of the completed product. Aerial photographs may be used singly, in pairs or as an assembly. Single photographs are frequently used in the field as sketch-maps; or can be used advantageously from a light aircraft for surveying the boundaries of insect or fire damage. Sometimes enlargements are used. An assemblage of several photographs is particularly useful as a rough map when initially planning a project. When, however, the details on the photograph are to be carefully interpreted, it is advisable to examine the area of interest as recorded in a stereo-pair of photographs. This particularly applies to forestry and ecology, as the forester is normally interested in tree and stand height and the ecologist is interested in the minutest details of community structure as recorded on the photographs.

Stereo-pairs of photographs of the same area are also needed in the preparation of accurate planimetric maps. Each photograph taken from an aircraft is a perspective view of the area of interest; and may be compared with a perspective view provided by an artist when landscape painting. Whereas the artist is required to produce a perspective drawing, the photogrammetrist needs to reproduce true measurement or an orthographic projection of the perspective view provided by the photograph. This calls for the preparation of planimetric maps and possibly topographic maps. The latter, however, are considered to be outside the needs of the present study and will only be briefly introduced via the contouring of planimetric maps.

(a) The vertical photograph

A photograph is said to be vertical when it is taken with the camera axis vertical. As the aircraft tips and tilts during flight, the photograph will no longer be truly vertical; but in practice it is usually accepted as such provided the resultant tilt does not exceed 3° to 5° in any one direction. An oblique photograph is one taken with the camera axis pointed downwards between the horizontal and the vertical. The camera is maintained in this oblique position in the aircraft. An oblique photograph is taken when a miniature camera is held in the hands and pointed downwards at an angle to the ground. For example, the amateur photographer on a flight from Paris to Rome is taking oblique photographs, as he photographs the Alps or Mont Blanc through aircraft windows.

Vertical and oblique photographs both have distinct advantages. Opinion at present in the world is strongly in favour of vertical photographs. Vertical photographs are simple to use photogrammetrically, as a minimum of mathematical correction is required. Further, vertical photographs are suitable for use with relatively inexpensive equipment and can be usefully handled by an interpreter with little photogrammetric training. In a pair of oblique photographs, the scale varies not only towards the horizon but may also vary within a single large object, e.g. hill. In favour of oblique photographs, Jonsson (1960) records that the accuracy of the elevation being mapped is greater using oblique photographs than vertical photographs. In Canada successful bridging of strips of small scale photographs, up to 200 miles, has been achieved using oblique photographs along the lines of flight (Blachut, 1957).

On the single vertical aerial photograph, the geometric centre is termed the *principal point* provided the camera is correctly adjusted. This is located at the intersection of the two diagonal lines from the corners of the photograph or at the intersection of lines between opposite fiducial marks on the photograph or by measuring and marking the mid-points on the sides of the photograph.

In marking up photographs, it is convenient to mark the centre point by pricking through the surface of the photograph with a pin and surrounding the pin-point with a faint red ink circle about one fifth of an inch in diameter. The centre point is then transferred stereoscopically to the adjoining photograph by pricking through its transferred position and inscribing it with a circle of identical diameter. This point is known as the *transferred principal point* or *conjugate principal point*.

By extending a faint red ink line from the edge of the circle of the principal point to the circle of the conjugate principal point the *flight line* of the aircraft is established on the photograph. This is also called the base line or (photographic) *air base* when applied to the principal points of two consecutive photographs in the same flight line. The air base, as recorded on the photograph, is used for lining it up in relation to similarly marked photographs, in the determination of heights and for preparing maps. If the same air base as recorded on two adjoining photographs is measured, these will usually be found to differ slightly in length. This is due to resultant tilt and changes in the flying height of the aircraft above the datum plane. The line passing through the principal point, being parallel to the flight line, is known as the *x-axis* and the line perpendicular to it passing through the principal point is the *y-axis*.

After marking the principal point and transferred principal points on each photograph, the next step is to choose points towards the edge of each photograph. These points must be common to the adjoining photographs. Thus if there is approximately 60% overlap between photographs of the same flight line and a side lap of 20% between flight lines, each point will be common to six photographs. Normally photographs are taken so that the corresponding principal points of photographs in adjoining flight lines are opposite to each other. If a single strip of photographs is being used, then each point will only be common to three photographs. These additional points are termed *wing points* or *pass points* or *photogrammetric control points*. As the last name suggests, the points are used in a manner similar to ground control points to provide orientation between photographs for mapping. Usually there are three wing points on each side of the photograph. The wing points, in combination with the transferred principal points, provide a total of eight peripheral points on a photograph. They serve a further useful purpose by delineating the centre-area on each photograph, which will contain least radial displacement, change in film density, etc. Each centre-area is known as the *effective area* of the photograph.

If there is a ground control point on a photograph, it is usual to show this clearly by encompassing the point with a distinctive symbol, e.g. triangle. A photograph marked up with the principal point, conjugate principal points, wing points and a ground control point is shown in fig. *9.1*.

Frequently the *caption* on the photograph fails to show all the data which may be required in future work. Ideally, the following should be given in the 'caption' on the bottom edge of each photograph (the data of the caption of fig. *9.1* is given in brackets as an example):

1. A geographic reference (e.g. Vic., Broadford Forest).
2. Photograph number and flight line number (e.g. 9, 1867).
3. Date of flying (e.g. 5.2.65).
4. Time of day of flight (not shown).
5. Altitude (e.g. 9,100′).
6. Focal length of lens (not shown).
7. Type of film (not shown—not necessary).

Throughout the world, where the inch is the unit of measurement, photographs are frequently 9 in. by 9 in. Occasionally, photographs 7 in. by 7 in. are taken; and the corresponding 19 cm by 19 cm are commonly used in countries with metric standards. For certain engineering work the smaller photographs are favoured. Older photographs, still available in Australia, are $5\frac{1}{2}$ in. by $5\frac{1}{2}$ in. and 7 in. by 9 in. In New Zealand, photographs 7 in. by 9 in. are still being taken. These measurements exclude the margins and the caption at the bottom of each photograph. In North America, it is customary not to have a margin around the photograph. This is to be preferred, as the margin serves no useful purpose, especially as normally only the effective area of each photograph is used. Frequently when matching photographs and assembling them as an air index map it is necessary to cut off the plain margins.

It is frequently convenient when using photographs as a mosaic for field work to assemble the photographs of each flight line correctly and join them together by 'scotch' tape ($\frac{1}{2}$ in. to $\frac{3}{4}$ in. width) along the two edges parallel to the flight line. Certain proprietary tapes have the added advantage of not adhering so strongly to the surface of the photograph that it is damaged when removing the tape. Each 'flight line' of photographs can then be separately rolled up for field work. Usually, the 'scotch' tape can be peeled off without damaging the edges of the photograph, unless the edges and the tape have been exposed to hot sunlight or left in contact for several months.

When examining a pair of photographs under a stereoscope, the photographs should be correctly orientated. This requires:

(a) that the flight lines are correctly aligned by checking with a ruler's edge that the two principal points and the two transferred principal points are in the same straight line;

(b) that the distance between the principal points is suitable for stereoscopic viewing, being at the eye base distance less about one-fifth of an inch for a hand stereoscope. Often it is convenient to fix the photographs to the table with adhesive tape. Careful scribing of circles of identical diameters around the principal point, wing points and conjugate points facilitates the orientation of the photographs.

(b) The oblique aerial photograph

An introduction to oblique aerial photographs is desirable for two reasons. By studying the oblique photograph a clearer knowledge of the geometry of the aerial photograph is attained. This knowledge may be needed for, example, when using 'vertical' photographs having excessive tilt. An oblique photograph may be considered as being a photograph with deliberate tilt of known angle. Secondly, the taking of oblique photographs has been important in the past both in peace-time and war-time and remains an important peace-time method for special purposes. For example, the coverage of Canada by aerial photographs was only made economically practical by using oblique aerial photography. Normally however, oblique aerial photography is not to be recommended for mapping, since, as mentioned earlier, the photogrammetry becomes complicated. This aspect cannot be emphasized too much. In the United Kingdom, an attempt was made to use R.A.F. obliques as an aid to accurate stock-mapping of small, intensively managed forest plantations, but had to be abandoned owing to difficulties resulting from the fact that the photographs were obliques. Obliques can be useful in the identification of the species of trees, as in the side view of the tree on the oblique photograph the branching habit is more often recognizable. Obliques are also a valuable aid to teaching regional geography in schools. For the professional interpreter, obliques have the disadvantage of providing a perspective view with many ground objects concealed.

Interest in single obliques has been stimulated through satellite photography. Oblique photography has been found useful in game census, and for locating sites of archaeological interest in the United Kingdom (St Joseph *et al.*, 1966) and in France (Goguey, 1966). In fig. *9.2*, part of an oblique and part of a vertical photograph of the same area are shown. The outline of an ancient ditch is only conspicuous on the oblique photograph. Sometimes the chance of recording an image of archaeological interest by oblique photography may be lost within 24 hours and often favourable conditions do not prevail for more than a few days. Extreme drought in temperate regions, light snowfalls and slight flooding have been recorded as providing suitable conditions.

Oblique photographs may be grouped into two classes. If the horizon is visible in the photograph, then it is termed a *high oblique* (fig. 9.4). If the horizon is not visible then the photograph is termed a *low oblique*. A further subdivision of low obliques can be made by distinguishing between low obliques properly taken by setting the camera axis at the known angle to the vertical, termed the angle of tilt, and obliques formed by accidentally tilting the camera in the aircraft from its vertical position at the time of exposure due to tilting of the aircraft.

Most extensively, oblique photography has been used in Canada and the United States. With twin oblique cameras, e.g. *Twinplex*, the cameras either point forward and backward along the line of flight or converge inward towards each other at right-angles to the line of flight. An assembly of three cameras, one vertical and two oblique, was very popular for many years in North America, e.g. *Trimetrogon*. This combination enables the entire area of the project to be covered by oblique photographs, and at the same time strips of vertical photographs are provided along the widely spaced flight lines. Andersen (1956), for example, gave an interesting account of such photography in Alaska for forest typing. The nominal scale of the oblique photographs on the isometric parallel was 1/20,000 and the vertical photographs 1/40,000.

To introduce the various technical terms, a tilted photograph will now be considered (fig. 9.3). Assume there is a photograph *ABCD*, which was taken at an angle of tilt ($t°$) to the vertical (*nON*). 'O' is the *perspective centre* through which all rays of light are assumed to pass when reflected by the object to form an image on the film; '*p*' is the *principal point* of the photograph, being at the foot of the perpendicular from the perspective centre to the photographic plane. The point at which an imaginary vertical line from the ground, passing through the perspective centre, cuts the plane of the photograph is called the *nadir point* (*n*) or photographic nadir point. This line *nON* is known as the *plumb-line*. If the photograph is truly vertical, then it is seen that the principal point and the nadir point will coincide. The *optical axis* lies about *Op*; and *Op* is the calibrated focal length (f'). The optical axis is also known as the *photograph perpendicular*. Any plane that contains the perspective centre (*O*) and a line in the plane of the photograph is known as a *perspective plane*. Thus *Ovn* is a perspective plane. The perspective plane at right-angles to the photograph (i.e. *Opn*) is the *principal plane* and cuts the photograph along *pn*.

The relation of these various points may be more easily understood from the next figure, which represents a side view of the previous figure, having the principal plane lying in the plane of the paper. From fig. 9.4, it is seen that angle *nOp* is also equal to the angle of tilt (opposite angles). The point on the line bisecting the angle of tilt, and subtended from the perspective centre to the plane of the photograph is known as

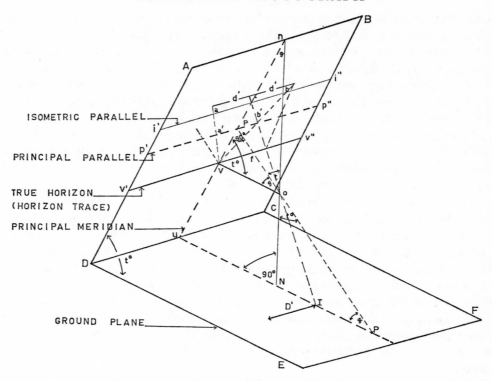

FIG. 9.3. Geometry of the tilted photograph (see text).

the *isocentre* (*i*). The isocentre (*i*) lies at the mid-point between the nadir (*n*) and the principal point (*p*) and all three coincide on a truly vertical photograph. The isocentre and nadir are very important due to their relation to tilt and radial displacement as discussed later.

The plane *CDEF* (fig. 9.3) is termed the *ground plane* or horizontal ground plane. Where the plane *ABCD* meets the plane *CDEF*, in a horizontal line, *CD*, it is known as the *horizontal trace* or *perspective axis*, along which any point coincides with its

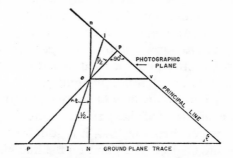

FIG. 9.4. Fig. 9.3 as seen in the principal plane (see text).

102

homologue. Points N, I, P, are the *ground nadir, ground isocentre* and *ground principal points* respectively. Sometimes the student confuses the *horizontal ground plane* with the *horizontal plane*. The latter, the *horizontal plane*, contains the perspective centre (O) and meets the photographic plane along the *horizon trace v' v''*. It is in a plane parallel to the ground plane. On a high oblique photograph, the horizon formed at the junction of the sky and the land is termed the apparent horizon (fig. 9.5). This horizon only coincides with the true horizon when viewed at sea-level. The *horizon trace* or vanishing line is an imaginary line in the plane of the photograph, being important in the study of oblique photographs. It represents the image of the true horizon on which the *vanishing point* (v) occurs, and is used in determining scale and measuring areas on oblique photographs. This point is also termed the horizon point.

The vanishing point (v) is defined as the image on the plane of the photograph of the point towards which a system of parallel lines of light in object space appear to converge. Thus in fig. 9.3, a, b, a' and b' are all points on the broken lines corresponding to the same ground distance D' from NP. A horizontal trace passing through the isocentre and continued in the photographic plane is termed the *isometric parallel, i'i''*. The isometric parallel is the only parallel in an oblique photograph along which the scale is the same as the scale of a vertical photograph taken under similar conditions. It follows that the scale along this line is represented by f/H, where f = focal length or approximately Op. H is the flying height. A measured distance on the ground (D') can be set off on the line $i'i''$ to determine d' on the photograph, i.e. $d' = D' f/H$. Similarly other distances can be calculated.

The theory of the perspective grid or Canadian grid shown in fig. 9.5 is based on the principle of perspective drawing, namely that parallel lines in an object can be shown as intersecting at a common point, i.e. the *vanishing point*. This figure should be carefully compared with the relevant parts of fig. 9.3. Details in an oblique photograph may be conveniently transferred from a superimposed Canadian grid to an orthographic grid on the map base by locating forest types, etc., in relation to their grid references.

The scale (S) at various points on an oblique photograph and distances between points are provided by the following formulae:

1. On the isometric parallel: $S = f/H = d/D$.
2. On the parallel passing through the principal point:

$$S = \frac{f \sin 90 - t}{H}.$$

3. On parallels other than in formulae 1 and 2:

$$S = \frac{d_1 \cos 90 - t}{H}$$

where d_1 is the distance from the true horizon to the point measured in the plane of the oblique photograph and parallel to the principal meridian.

4. On rays parallel to the principal meridian connecting the principal point (p) and the nadir (n):

$$S = \frac{d_1 d_2 \cos {}^2 90 - t}{f.H}$$

where d_1 is as in formula 3 and d_2 is the distance to another point near by on the same ray. This provides a mean scale.

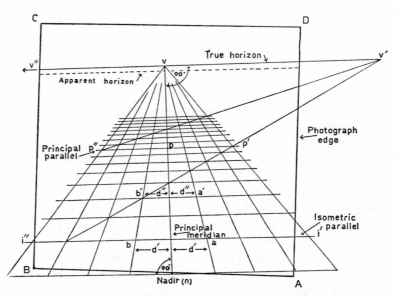

FIG. 9.5. Geometry of the perspective grid (see text).

5. The distance $v'\, v$ is given by the formula:

$$v'\, v = \frac{f}{\cos 90 - \theta}$$

where v' is the vanishing point for grid diagonals. In fig. 9.5, v' and v'' are both diagonal vanishing points.

6. For an oblique measurement on the photograph, i.e. other than along meridians and parallels, resolve the location into the respective components and use Pythagoras' theorem.

The angle complementary to the angle of tilt (i.e. $90° - t°$) is termed the *angle of depression* (θ), which represents the tilt of the camera axis from the horizontal (fig. 9.3). When this angle (θ) and the focal length of the camera lens are known, then the distance from the principal point to the vanishing point, isocentre and nadir can be calculated. These points all lie along the principal meridian, which is a straight line at right angles to the horizon on the photograph.

7. To vanishing point: $p\,v = f\tan\theta = f\tan(90 - t) = f\cot t$

8. To isocentre: $p\,i = f\tan\frac{1}{2}(90 - \theta) = f\tan(\frac{1}{2}t)$

9. To nadir: $p\,n = f\tan(90 - \theta) = f\tan t = f\cot\theta$

Rectified prints. The process of projecting an oblique photograph or a slightly tilted photograph on to a horizontal plane to produce the equivalent vertical is termed *rectification*. In the United States the term transformation is sometimes used.

To transfer detail from a slightly tilted photograph on to a horizontal plane such as a map base is relatively easy, and may be carried out on a suitable sketchmaster, such as manufactured by Zeiss in Europe or Kail in the U.S.A. (see fig. *12.2*). However, to provide an accurate rectified negative of an oblique photograph (or oblique negative) is much more complicated and expensive. Normally ground control has to be provided for each photograph; and complications occur due to the fact that the lens of the rectifier may be different from the focal length of the camera. For ground control, usually four points are required on each photograph, although three points will suffice, when the interior orientation of the camera-lens and focal plane are known.

Ratioed prints. Aerial photographs are normally used as contact prints; but sometimes they are enlarged or reduced in size for a special purpose. These photographs are then known as ratioed prints or ratioed photographs. An example is the special ratioed diapositives used in the Multiplex. If enlargements are used then it is essential to use a mirror stereoscope and not a hand stereoscope. It may be found more convenient to cut the enlargements into strips and to view the strips under a mirror stereoscope.

Nyyssonen (1955) determined the height of trees and stands from enlargements (1/5,000) of negatives at 1/10,000. He did not obtain any appreciable differences between measurements from the enlargements and from contact prints, which is understandable as the stereoscope itself provides adequate magnification. The enlargements, however, facilitated measurement and reduced the effect of technical errors connected with the adjustment of the micrometer of the parallax bar. On ratioed prints (1/5,000, 1/10,000) and contact prints of the same (1/14,000) Nyyssonen observed that crown closure in Finland is best estimated on 1/10,000 and poorest on 1/5,000. Some workers have found that in the course of making enlargements that a little of the forest micro-detail contained in the negative may be lost in the printing. This is probably due to stray light. Other detail, however, may be recorded, which would not be reproduced in a single contact paper-print. In contact printing the grain of the paper reduces the resolving power, whilst in enlargements of two to four times the resolution is improved but the tonal range is probably reduced.

Contact print. This is a print made from a negative in direct contact with the print paper. Negative and print scales will be the same.

10

RADIAL DISPLACEMENT

THE radial displacement of an image on a photograph is due to both tilt of the aircraft at the time the photograph was taken and to changes in the ground elevation (i.e. topographic displacement). It is important always to remember that the displacement due to tilt is radially outward from the isocentre; and displacement due to topography is radially outward from the nadir. The mnemonic 'ToN' will help the student remember that topographic displacement is from the nadir.

On a nearly vertical photograph, however, it is usual to assume that the principal point, isocentre and nadir coincide. This simplifies calculations; and is convenient, as

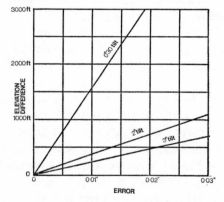

FIG. 10.1. Graph used to correct for errors in parallax measurement caused by tilt and changes in the elevation of the ground.

(Stephen H. Spurr, *Photogrammetry and Photo-Interpretation*, 2nd edition, © 1960. The Ronald Press Company, New York.)

the principal point is readily located as the geometric centre of the photograph. The combined effect of a little topographic displacement and a little tilt in a nearly vertical photograph will emphasize the displacement outward of an image from the principal point unless they are in opposite directions. Extra care must therefore be taken to ascertain the tilt in photographs of rugged terrain when measurements are to be made from the photographs. Obviously there is no topographic displacement in a photograph of level terrain, e.g. flood plain. Graphs are readily prepared from which the

106

parallax error of photographs (of known nominal scale) can be read off according to small changes in tilt and differences in the ground elevation. Such a graph for $\frac{1}{2}°$, 2° and 3° tilt at a nominal photographic scale of 1/15,840, is given in fig. 10.1. The *nominal scale* is the photographic scale at the principal point.

Example. What is the error in parallax measurement on a pair of photographs with $\frac{1}{2}°$ tilt at elevations of 750 ft. and 3000 ft above the mean ground datum? (Answer: 0·005 in. and 0·018 in.).

(a) Topographic displacement

If a truly vertical photograph of an area with considerable variation in the height of the ground above sea-level is compared with a topographic map of the same area,

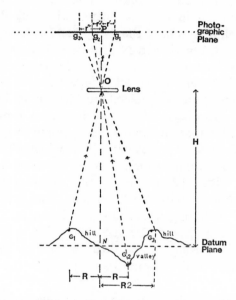

FIG. 10.2. The influence of changes in ground elevation on radial displacement in the vertical aerial photograph. (G_1, G_2, G_3) are points on the ground and G_1, G_2, G_3 their corresponding points on the photograph (see text).

it will be found that the measurements of the distances between images do not agree with measurements of the same distances on the topographic maps or on the ground. This may be due to elevation differences in the topography. As mentioned above, the topographic displacement (relief displacement) is radially outwards from the nadir point; but on vertical photographs, with less than 3° to 5° tilt, it is assumed that the nadir point and the principal point coincide.

The effect of topography on radial displacement in a truly vertical photograph is illustrated in fig. 10.2. The hill-top (G_1) is displaced radially outwards from the

107

principal point on the photograph due to the topographic relief of the hill above the datum plane being nearer to the camera. A similar point (G_2) in the bottom of a valley and exactly the same radial distance (R) from the ground nadir (N) is displaced radially inwards towards the principal point due to the topographic relief of the valley-bottom being farther from the camera. Additional consideration of the diagram will help in formulating the following generalizations: that there is no topographic displacement at the centre of a photograph; that the topographic displacement is radial from the centre and that all objects above the datum plane are displaced outwards and all objects below the datum plane are displaced inwards.

The displacement of a further hill-top (G_3) on the same datum plane as the first hill-top (G_1) and of the same height, but at a distance $2R$ from N, will be displaced on the photograph more than G_1 and proportional to its distance from the centre of the photograph. Images at the same distance and on the same datum plane, irrespective of which side of the principal point they are located, will be displaced the same distance from the centre of the photograph. If the flying height above the ground datum at which the photograph is taken is increased, then radial displacement is decreased and if the flying height is decreased then the radial displacement is increased. In other words, topographic displacement is inversely proportional to the height of the photography above the datum plane for a specified focal length. As large scale photographs are normally preferred in forestry and ecological studies, topographic displacement assumes greater importance than in small scale mapping projects.

Topographic displacement also varies with the focal length for photographs at the same scale. Thus if a photograph was taken at H ft with a 6 in. lens, and was again taken at the same scale with a 12 in. lens at $2H$ ft, the topographic displacement is twice as great with the 6 in. as with the 12 in. lens. This may be confirmed graphically as an exercise by preparing a diagram similar to fig. 10.2, but in addition providing a second lens at a distance $2H$ and subtending rays from the hill-top (G_1) and the ground nadir (N) through the optical centre. If 'r' is measured in its new position for a 12 in. lens, it will be found to be less than the measurement 'r' for a 6 in. lens.

Height displacement of objects. Basically the measurement of the height of buildings, trees, etc., is identical with the measurement of topographic height. It is convenient, however, to separate height displacement from topographic displacement, as often objects of similar height, e.g. trees, require measurement on the same photograph at different elevations. There is nothing difficult about such calculations, provided each displacement is considered separately. The displacement of the top of a tree, building or television mast from the bottom as shown on the photograph is similar to the displacement of the top of a hill above the datum plane. Height displacement is again radial from the nadir, and as tilt is slight it is assumed to be radial from the principal point. The remarks previously given concerning topographic displacement equally apply to height displacement; and will not therefore be reiterated. The greater the height of the object viewed on the photograph, the greater will be its radial displacement on the photograph. Thus in fig. 10.3 the top of the television mast is displaced radially from the principal point a greater distance than the top of the tree. It will be observed that similar triangles are formed by the lines subtended

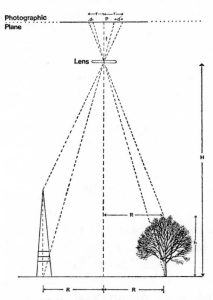

FIG. 10.3. Diagram illustrating the effect of the height of an object above the ground on the radial displacement of its image on the vertical aerial photograph (see text). If this figure is also studied in conjunction with fig. 2.4, it will be appreciated that flying height can affect the radial displacement of the recorded object on the photograph.

from the object on the ground to the optical centre of the lens. As 'H' is large compared with 'f', it is assumed 'H' and '$H+f$' are the same.

Thus $r/f = R/H-h$ or $fR = r(H-h)$
and $(r-d)/f = R/H$ or $fR = H(r-d)$
as fR is common, $r(H-h) = H(r-d)$
or $d/r = h/H$ or $h = Hd/r$.

This is the basic *displacement formula*. 'r' is the distance from the principal point to top of the image of the object and is measured radially; 'd' is the distance radially from the true base of the object to the top of the object. It is fairly easy on large scale photographs to measure from the base of a television mast to its tip or similarly an open-grown tree; but it is not practical to do so for a hill on a single photograph. This weakness, combined with the fact that true location of points cannot be determined as accurately as by radial line triangulation makes the formula primarily of academic interest. The determination of the apparent height of trees, etc., will be further considered in Chapter 15.

Example. A hill-top on a photograph is 3 in. from the principal point; and on a second photograph the same hill is 5 in. from the same locum. Will the hill be the same size on each photograph? If not, what is the percentage increase in size on the photographs? Assume the hill-top is 1,000 ft above the datum plane and that the pictures were taken at an altitude of 9,000 ft above sea-level. The datum plane is 1,000 ft above sea-level. (For formula see previous section.)

$$d/r = h/H \text{ or } d = r\,(h/H)$$
$$d' = r_1\,(h/H) = 3\,(1,000/8,000) = 0{\cdot}375 \text{ in.}$$
$$d'' = r_2\,(h/H) = 5\,(1,000/8,000) = 0{\cdot}625 \text{ in.}$$

(Answer: The hill is not the same measured size; and the percentage increase is $375/625 \times 100 \cong 60\%$.)

(b) Tilt

As mentioned earlier, this term refers to the angle (t) at the perspective centre between the photograph perpendicular and the plumb line. Tilt is the combined effect of lateral tip of the wings of the aircraft, also referred to as *tilt*, and the dipping of the aircraft fore and aft, termed *tip*. Some writers prefer to use these separately as tip and tilt, whilst others refer to both as tilt or *combined tilt* or *resultant tilt*. *Longitudinal tilt* or *y-tilt* is that which is due to the nose of the aircraft being lowered and displacing the nadir point in the $+x$ direction along the X-axis; *x-tilt* or *lateral tilt* or *list* causes the nadir point to be displaced along the Y-axis. In Europe, omega (ω) is used for lateral tilt and phi (ϕ) for longitudinal tilt.

Eliel (1939) pointed out that with an $8\frac{1}{4}$ in. focal length lens an error of 1% in scale is introduced at the edge of a 9 in. by 9 in. photograph for each $1°$ of tilt and that for a 6 in. focal length the error in scale increased by 25% for each $1°$ of tilt. Tewinkel also has estimated that in the U.S.A. under good flying conditions 50% of vertical photographs taken for mapping are tilted less than $1°$; 90% less than $2°$, a very few more than $3°$.

The effect of tilt such as would occur in a vertical negative due to side movement of the aircraft is shown in fig. 10.4. G_1 and G_2 are two identical trees of the same height, h, on the ground; and d_1 and d_2 are their image displacements on the negative when the camera was tilted $t°$. Other lettering is as in fig. 10.2. It is assumed that the principal point and the isocentre are close together, and therefore the radial displacement is from the principal point. The figure shows that the tilt of the camera has displaced the image of G_1 outwards on the side of the negative of upward tilt and the displacement of G_2 is inwards on the side of downward tilt. Also the size of the image is compressed by tilt on the upward side of the photograph and is expanded on the downward side. Further consideration of the diagram will suggest that one of the safest ways of calculating the nominal scale of a photograph is

110

to use two known or measurable ground distances about the same distance away from the centre but on opposite sides of the flight line.

The scale of a tilted photograph changes in a regular manner throughout the photograph, but does not change along a line perpendicular to the direction of tilt. Two such lines at right-angles to tilt are $p'\ p''$ and $i'\ i''$ in fig. 9.3.

In fig. 10.4, if a further identical tree, G_3, is considered with its image towards the edge of the negative, it will be observed that its image displacement (d_3) is not the same as for G_2. In fact, consideration of further identical objects recorded on the photograph will lead to the conclusion that the effect of tilt is exaggerated the farther

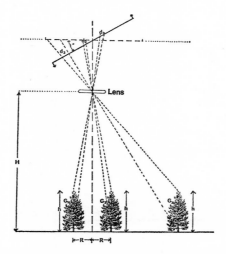

FIG. 10.4. Diagram illustrating the influence of the tilt in the camera at the time of photography on the radial displacement of the image (see text).

the images are from the principal point. This emphasizes the importance of using only the effective area of each photograph and being more cautious the farther the measurements are from the centre of the photograph.

The scale (S) of any image on a tilted photograph can be calculated by the following formula (assuming only a few degrees of tilt):

$$S = \frac{f - y \sin t}{H}$$

where y is the distance of the image from the isocentre (or when the tilt is slight from the principal point). H is the flying height above the ground datum.

Example: A camera with a 6 in. lens is used at an altitude of 23,000 ft to take a photograph of a small object at 3,000 ft datum. If the tilt is 3°, what is the photographic scale at (a) the principal point, (b) 5 in. from the principal point on the upper and lower sides of tilt?

I   111

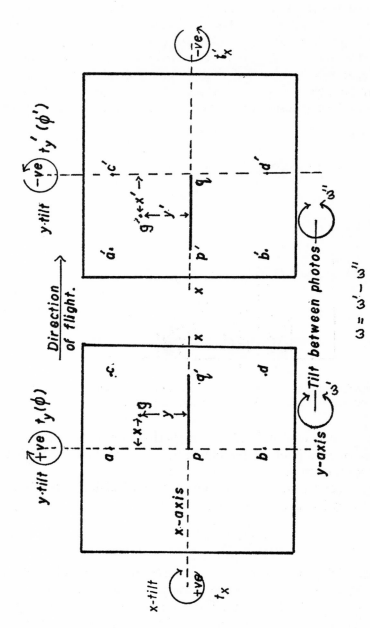

FIG. 10.5. Two photographs, centres p and q, recording on level ground the same object, G, as images g and g'. If there is no x-tilt then the distance g to pq and the distance g' to $p'q$ will be the same. If there is x-tilt, then g and/or g' will be displaced parallel to the y-axis. The distances y and y' will then not be equal and there will be y-parallax in the uncorrected stereo-model. The x-axis is the longitudinal axis of the aircraft lengthwise through its fuselage and the y-axis is at right-angles in the same plane.

(a) $\quad S = \dfrac{\frac{6}{12} - 0}{20,000} = \dfrac{1}{40,000}.$

(b) $\quad S = \dfrac{\frac{6}{12} - 5\sin 3}{20,000} \cong \dfrac{1}{42,000}.$

(c) $\quad S = \dfrac{\frac{6}{12} - (-5)\sin 3}{20,000} \cong \dfrac{1}{38,000}.$

The determination of tilt. For many purposes the determination of the tilt and the correction of it may be omitted, provided the tilt is under 3° to 5°. However, the need to correct the stereo-measurements is sometimes desirable or essential in relation to the determination of apparent height. This is particularly so when determining heights for research purposes. Two methods of determining the tilt inherent in photographs will therefore be briefly introduced; and then a method of tilt adjustment by interpolation will be described. A standard text-book on photogrammetry should be referred to for further details or alternative methods.

Of historical interest is the 'want of correspondence' method, which was described by Hotine about 1929 (Hart, 1943); and used widely and successfully for determining the tilt of photographs in relation to the air base and the lateral tilt of one photograph in relation to another. Hotine's method has, however, the disadvantage of requiring four very carefully positioned ground points towards the edges of the photographs in order to maintain a satisfactory standard of accuracy. Also the elevation of each point requires to be known.

In briefly considering his method, fig. 10.5 should be examined. The x and y co-ordinates are represented by dotted lines, passing through the principal points p and q. The direction of flight is the x-ordinate and p' and q' are the transferred principal points. $p'q$ and pq' are the flight lines. The y-ordinate is measured at right-angles to the x-axis. The point G on the ground at the base of a tree will be represented by g and g' on the photographs. If there is no tilt in the photographs, the distances g to pq' or g' to $p'q$ will be the same. If these distances are different then there is tilt between the photographs and there is said to be want of correspondence (k). Expressed algebraically $k = (y' - y)$. The difference can be measured stereoscopically using a floating-dot mechanism or by resolving the x- and y-axes separately.

The four points a, b, c, d, are chosen on the photograph on either side of the flight line to form rectangles $acq'p$ and $pq'db$. On a suitable machine the x- and y-co-ordinates can be determined. ω is the relative tilt of the left-hand photograph to the right-hand photograph. t_x is the true angle of tilt about the x-axis. ϕ is the direction of tilt to the air base. Finally the following equations are set up and solved for ω and ϕ. For further details Hart (1943) should be consulted.

$$\text{For } a, \; y_a\omega + x_{a'}\phi' = \frac{k_a f}{y_a}$$

$$\text{For } b, \; y_b\omega + x_{b'}\phi' = \frac{k_b f}{y_b}$$

$$\text{For } c, \; y_c\omega + x_c\phi = \frac{k_c f}{y_c}$$

$$\text{For } d, \; y_d\omega + x_d\phi = \frac{k_d f}{y_d}$$

The disadvantage of the 'want of correspondence' method has been overcome by several workers including Robbins (1949), Crone (1951), and Thompson (1954). A simple graphical method of determining tilt for oblique photographs is illustrated in fig. 10.6. Four map points A, B, C, D are chosen in the form of a parallelogram or rectangle. These points are located on the photograph (a, b, c, d) and the sides are

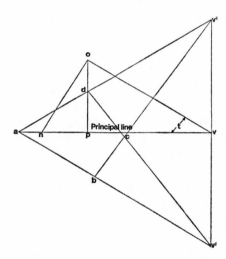

FIG. 10.6. A simple graphical solution to determine tilt of an oblique photograph is illustrated by the above diagram (see text). The line $v'v''$ is the horizon line as shown in figs. 9.3 and 9.5. On some government photographs in the United Kingdom, the principal point, isocentre and nadir are shown; and hence by measuring the distance between the principal point and nadir on a photograph, its tilt (t) in degrees is readily determined (i.e. $\tan t = pn/op = pn/f$).

extended outwards to meet at v' and v''. $v'v''$ is the horizon or vanishing line corresponding to the line at infinity in the map plane. The principal line or meridian is at right-angles to the horizon and passing through the principal point p. The focal length (f) is drawn at right-angles to p and ov is joined. Then angle ovp is the angle of tilt ($t°$).

TILT

Adjustment of height estimates for tilt by interpolation. The procedure to be described is based on a method developed by Thompson (1954), who adapted Robbins's simple method (1949) for what he terms the crude estimates of height for tilt on medium scale vertical photographs. Thompson's method requires the use of only a parallax bar, mirror stereoscope and a slide rule.

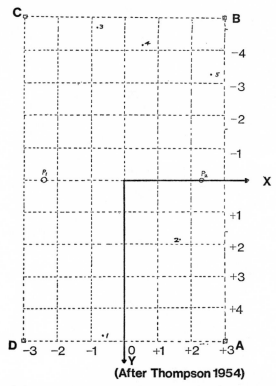

(After Thompson 1954)

FIG. 10.7. Gridded overlay for the adjustment of height estimates by interpolation.

The method requires five ground control points common to each pair of photographs. This means that the centre photograph will have ten points marked on it if a strip of three is being used. Tilt is not actually determined, being essentially a method of interpolation between points of known height. The points are described by their co-ordinates along the x- and y-axes with the x-axis being along the flight line and the origin being at the mid point between the principal point and the transferred principal point. No four points should be in line or nearly in line nor three points in line perpendicular to the flight line. Ideally four of the five points should be in the corners of the rectangle formed by half the photograph and the fifth point should be in the centre. The co-ordinates of each point are read off from a transparent gridded overlay (in ¼th millimetres). This is illustrated in fig. 10.7. Each grid-square is 2 cm × 2 cm.

115

RADIAL DISPLACEMENT

TABLE 10.A

Formation and solution of equations (after Thompson, 1954)

Line	Point	a_0	x / a_1	y / a_2	xy / a_3	x^2 / a_1	$h'-h$	Solution	
1	1	1	−0·6	+4·8	−2·9	+0·4	−62		
2			+41·9	−45·1	+6·29	−0·26	−64·8	−64·8	
3	2	1	+1·6	+1·7	+2·7	+2·6	−200		
4			−111·8	−16·0	−5·86	−1·69	−64·7	−64·7	
5	3	1	−0·8	−4·7	+3·8	+0·6	+27		
6			+55·9	+44·2	−8·24	−0·39	−64·6	−64·6	a_0
7	4	1	+0·7	−4·2	−2·9	+0·5	−68		−65
8			−48·9	+39·4	+6·29	−0·32	−64·7	−64·7	
9	5	1	+2·6	−3·3	−8·6	+6·3	−201		
10			−181·6	+31·0	+18·7	−4·09	−65·1	−65·1	
11			−3·2	+8·1	+5·7	−6·4	+139		
12				−76·1	−12·4	+4·16	+223·3	−69·8	
13			−1·0	+5·0	+11·3	−4·2	+1		
14				−46·9	−24·5	+2·73	+69·7	−69·7	a_1
15			−3·4	−1·4	+12·4	−6·2	+228		−69·8
16				+13·1	−26·9	+4·03	+237·8	−69·9	
17			−1·9	−0·9	+5·7	−6·3	+133		
18				+8·5	−12·4	+4·09	+132·8	−69·9	
19			−1·0*	+2·53**	+1·78	−2·00	+43·4	*i.e. $\dfrac{-3\cdot2}{-3\cdot2}\times-1\cdot0$	
20			−3·4	+8·61	+6·06	−6·80	+147·8	**i.e. $\dfrac{+8\cdot1}{-3\cdot2}\times-1\cdot0$	
21			−1·9	+4·81	+3·38	−3·38	+82·5		
22				+2·47*	+9·52	−2·20	+42·4	* i.e. 5·0−2·53 (eliminates a_1)	
23					−20·6	+1·43	−23·2	−9·39	
24				−10·01	+6·34	+0·60	+80·2		a_2
25					−13·7	−0·39	+94·3	−9·41	−9·41
26				−5·71	+2·32	−2·50	+50·5		
27					−5·03	+1·62	+53·9	−9·43	
28				−10·01	−38·6	+8·92	+172·2	Subtraction eliminates a_2, a_3	
29				−5·71	−22·0	+5·07	+98·1		
30					+44·9	−8·32	−92·0		
31						+5·42	−97·4	−2·17	a_3†
32					+24·3	−7·57	−47·6		−2·17
33						+4·92	−52·5	−2·16	
34					+24·3	−4·50	−49·6		
35						−3·07	+2·0	−0·65	a_4†† −0·65

†† $a_4 = \dfrac{+2}{-3\cdot07}$; † $a_3 = \dfrac{-5\cdot42 - 92\cdot0}{44\cdot9}$

TILT

If the estimated elevation of an object above datum is h, then it can be shown that:

$$h' = h + (a_0 + a_1 x + a_2 y + a_3 xy + a_4 x^2)$$

$$\text{or } h' - h = (a_0 + a_1 x + a_2 y + a_3 xy + a_4 x^2)$$

where h' is the true height above datum corrected for tilt and the terms in brackets are corrections to be added to the estimated height.

In the formula, it is assumed that h' is small compared with the height of the aircraft above the datum plane; and the tilt is sufficiently small for the powers above the first power to be neglected.

Referring to table 10.A, the details are entered as follows: If for point (1), $x = -0.6$ and $y = 4.8$, then $xy = 2.9$ and $x^2 \doteq 0.4$; and $h' - h = -62$ being the difference between the computed height and the known true height. The latter can be determined from a topographic map or by measurement on the ground. As the coefficients of a_0 are all unity, it is eliminated immediately. Similarly lines 3, 5, 7 and 9 are completed for the other four points and line 9 is subtracted from these to give lines 11, 13, 15 and 17. A process of elimination is pursued to provide

$$a_4 = \frac{+2.0}{-3.07} = -0.65$$

in line 35 and then by back substitution the other coefficients are determined. For example, line 19 is determined by dividing the leading coefficient of line 11 into each of the coefficients and multiplying these by the leading coefficient of the next equation, similarly for lines 20, 21. Lines 19, 20, 21 are subtracted from 13, 15 and 17 to eliminate a_1. The results are entered in lines 22, 24, 26.

This provides:

$$h' - h = -65 - 69.8x - 9.4y - 2.17xy$$

and therefore the true height of any other point in the area can be determined.

Lens Displacement

The effect of image displacement, due to the aberrations in the lens, is usually insignificant for agricultural, forest and ecological studies. Lens displacement is normally termed lens distortion and can be measured as radial displacement of a ray trace. It will be recalled from Part I that the focal length of the lens is calibrated to give the minimum distortion over the field of coverage at the time the exposure is made. Lens distortion may be radial or tangential. A curve of tangential distortion against distance from the principal point is nearly symmetrical in magnitude and sign and reaches a maximum at some distance from the principal point. The maximum magnitude is not normally towards the edge of the photograph. Tewinkel (1952) remarked that, although lens distortion can be ignored for graphical purposes, the effect is noticeable when elevations are determined stereoscopically. To overcome

117

this, first-order plotting machines often have a means of correcting for lens distortion. Hart (1943) observed that lens distortion does not exceed 0·025 in. for a $3\frac{1}{4}$ in. focal length lens or 0·04 in. for a 5 in. lens. On a 9 in. by 9 in. metrogon photograph the maximum distortion is about 0·005 in. about 4 in. from the principal point and 0·006 in. at the extreme corner.

FILM AND PRINT DISPLACEMENT

Both these displacements are erratic and not radial from the centre of the photograph. The greater distortion usually occurs in the print paper. Polyester as a base is remarkably stable, particularly as compared with cellulose. Film shrinkage probably varies between 0·03% and 0·1% and print shrinkage between 0·3% and 0·2%. Hart (1943) commented that a 15% change in humidity will result in a change as much as 0·2% in waterproof papers. Calhoun (1960) described a novel method of detecting local distortions by making a moiré pattern of known or measured dimensions on a strip of film at time of exposure; and after development examining the strip of film for distortion of pattern. He found that local distortions caused by waterspots on the film after processing could be 18 microns. Adelstein & Leister (1963), by using a moiré pattern to measure quantitatively dimensional changes of aerial film after processing, concluded that cellulose-base films may show random linear displacements of 30 microns. This is believed to occur during processing. Comparable errors of polyester base films were only 5 microns.

11

RADIAL LINE TRIANGULATION

THE principle of two-dimensional radial line triangulation is not only important to the trained surveyor and photogrammetrist, but also provides the agriculturist, ecologist, geologist and forester with a simple means of introducing basic control to field maps and sketches. The technique used is easy to understand and apply; and, according to the equipment used, a planimetric map of an acceptable standard of accuracy can be prepared. If, in addition, a topographic map should be required, it is often quite satisfactory to contour the planimetric map prepared by radial line triangulation. Possibly, radial line triangulation provides a satisfactory basis for 95% or more of the maps needed in land-resource studies.

In three-dimensional triangulation, the true map position of ground control points are plotted from the stereoscopic images of suitable points common to correctly orientated adjoining photographs. The mathematics of three-dimensional triangulation is complex, but impetus has been given to analytical triangulation by the development of electronic computers.

Two-dimensional radial line triangulation depends on the principle that the centre of each photograph serves as a station from which radial lines, at constant and true angles, can be subtended to ground points imaged in the periphery of the photograph, irrespective of changes in general scale. It provides a method of fixing ground points in space by using overlapping photographs on which the ground points are imaged; and is used to establish the principal control points of a planimetric map base on to which detail is later added.

The flight line between the two adjoining principal points of overlapping photographs is used similarly to a base line in plane tabling, rays being subtended from each principal point to the images of the ground points being fixed. A similar procedure (graphical resection) is used in coastal navigation to provide a 'fix' using two bearings or a 'cocked-hat' using three bearings. The principle of line plotting from aerial photographs was expanded between the two world wars by Bagley in the U.S.A. and by Hotine in the United Kingdom (e.g. Arundel Method).

Fig. 11.1 illustrates the method of two-dimensional radial line triangulation. Photographs A B C D and E F G H have centre points p and q and transferred centre points p' q'. The base lines are formed by p q' and p' q. Bearings of a suitable image (m) are taken from the two station points p and q. Thus the true position of the object

M is determined by the intersection at *m* or *m'* of the two bearings from *p* and *q*. On either photograph, this is achieved by transferring the angle of the ray from the other photograph. The transferred ray is shown as a broken line. The true location of the object can be also obtained by superimposing contiguous single-weight photographs on top of each other on a light table so that the flight lines coincide. The true position of the object will again be located at the intersection of the rays. Other points can be located similarly to the location of *m*. These may include the six wing points around the periphery of a photograph.

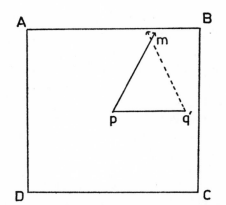
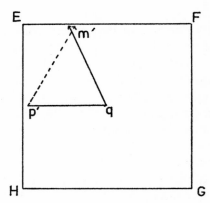

Fɪɢ. 11.1. The diagram illustrates the method of determining the true position of an object by radial line triangulation (see text).

The principle illustrated forms the basis of several simple radial line-plotting techniques. These include:

(i) the overlay method,
(ii) the hand template,
(iii) the slotted template,
(iv) the mechanical template.

(N.B. Template may also be spelt templet.)

Each of these will now be considered.

The overlay method. Each photograph is marked as described previously to include the principal point, transferred principal point and six wing points; and in addition important objects (e.g. mountain peak, road intersection, ground control point) are also marked on the photographs. Each photograph is then placed separately under a sheet of transparent or semi-transparent tracing material, (e.g. matt acetate, ethulon, Astralon or good-quality tracing paper), and the principal points and flight lines transferred. When the flight lines common to two photographs do not coincide, a mean of their lengths is used. Each photograph is then separately reinserted under the

120

tracing material and aligned on its flight line segment according to the location of the principal points. Rays are subtended on the overlay from the principal point to the wing points. The intersection of the two rays subtended to a common wing point is the correct location of that wing point. Eventually, due to end lap most wing points will lie at the intersection of three rays or in the cocked-hat formed by three rays. After preparing the tracing material by marking the correct positions of the wing points, etc., the photographs are once again reinserted and details are traced on to the map base (see Chapter 12). Normally only the details contained within the effective area of each photograph are transferred. If there is a second flight line, then the corresponding segment of its flight line is positioned on the overlay by resection of radial rays through established wing points.

The method has the advantage of not requiring separate templates nor expensive equipment, as the photographs themselves are used to provide the radial line triangulation. However, the scale of the map will correspond to the nominal scale of each photograph and cannot be controlled.

Hand template method. Instead of using the photographs directly, templates are used (fig. *11.2*). These are cut to the size of the photographs from suitable semi-transparent material, e.g. matt acetate. A template is placed on top of a photograph and the nine points and other important points transferred to it together with the base line and rays subtended from the principal point to the wing points, etc. A second template is prepared similarly. The radial line plot in its simplest form will then comprise two adjoining templates, as shown in fig. 11.1, and can be used to determine a point such as *m*.

Although seldom used in practice, being cumbersome and suitable for only a few photographs of relatively flat terrain, it has the advantage over the previous method of providing a certain amount of variation in scale by allowing an adjustment to be made to the air base. The scale, however, as controlled by the common air base of the first two photographs predetermines the general overall scale for other photographs, since the third template is adjusted to make its radial line intersect on common wing points fixed by the intersecting radial lines of the first two photographs. If photographic detail is to be copied by a sketchmaster, the control points mentioned above can be transferred via hand templates on to a firmer map base, e.g. 80 lb cartridge paper.

Slotted template method. The hand template introduces the principle involved in the slotted template. As the name suggests the wing points are represented by radial slots instead of points. Each template after preparation will be similar to the template shown in fig. *11.2*. There are eight slots. Two represent the direction of flight along the air base, and six the direction of the radial lines to the wing points. An additional point or two (e.g. ground control point) may be added, in which case these will also be represented by slots. Initially the nine points are pricked on to the template by placing the photograph on top of it and pricking through the points marked on the photograph. A special *centre point punch* and a *slotted template cutter* are used to punch out the centre point and radial slots in the template. These cost up to £500 and are shown in fig. *11.2*. The templates can be cut out of any suitable material, which is smooth and tough. Surplus used X-ray film is ideal and cheap.

Slotted template assemblies may be used for a few photographs or for large lay-downs. The templates are joined together by special hollow studs and clips which fit into the slots (fig. *11.2*). Where only a few photographs are to be used, administratively it may be convenient for the cutting of the templates to be done at a central office; and the completed templates then returned to the field camps for use locally by field staff for compilation of maps.

When a large number of templates are used in a single laydown, they are usually trimmed to the smallest size possible to facilitate the sliding movement of one over the other 'concertina-wise' to fit the map base. To provide correct location of trigono-metrical points and ground control points specially weighted spring studs are used. Longer slots permit a greater scale adjustment. Thus, if the scale of the map is to be much larger than the scale of the photographs, templates will require to be larger than the photographs. Normally, templates are about the same size as the photo-graph as this is sufficient for adjusting the template-assembly to a conventional map-scale.

Mechanical template method. The mechanical or spider templates are provided by 'meccano' type strips, which are assembled as spokes of a wheel, the hub or centre being at the principal point of the photograph (fig. *11.2*). Mechanical templates have the advantage of having parts which can be used time and time again for other lay-downs; but they are slow to assemble and are clumsy and often difficult to use as large template lay-downs. They are, however, excellent as a training medium; and suitable for use with a relatively small number of photographs. Scale is adjusted by alteration of the two 'Meccano' type arms representing the overlapping air bases of two adjoining photographs. The 'Lazy-Daisy' outfit manufactured in the U.S.A. costs about £200.

Stereotemplates. These are used in topographic mapping; and have been given the name to distinguish them from the normal type of template. On a stereotemplate the slots are cut to represent the direction of the points when viewed in a stereomodel provided by two adjoining photographs of the same flight strip. In preparation of the normal template only single photographs are used when transferring the point from the photograph to the template. Stereotemplates are assembled as a lay-down in the usual way. It was claimed initially that stereotemplates enabled a greater amount of plotting to be carried out on a suitable first-order machine before complete horizontal ground control was provided. Experience has shown the method to be tedious and sometimes inaccurate when fitting machine-plotted contours to a planimetric base provided by the stereotemplates (Ovington, J. J., 1957). Stereotemplates, however, are free from tilt and topographic errors inherent in the photograph.

12

PLANIMETRIC MAPPING

INTRODUCTION

To the forester, the importance of the map in field work does not require an introduction nor does it need emphasizing. It enables the forester to find his way about little-known country by providing information relating to the local terrain. It may provide details concerning forest boundaries (i.e. as a cadastral map); and it may give adequate data about the forest itself in the form of a stock map or type map. In the case of a forest inventory, if a stock map is not already available or not sufficiently detailed, which is usually the case, it becomes necessary to provide one or to apply a suitable sampling technique, such as point sampling, directly to the photographs. A map in conjunction with forest sampling in the field enables an estimate to be made of the volume of standing timber.

Maps fulfilling each of the functions mentioned in the previous paragraph can readily be prepared from suitable aerial photographs. Such maps could be prepared by traditional ground survey methods alone, but these days traditional methods would usually be more laborious and more expensive, and might be less accurate. It is readily appreciated that having aerial photographs eliminates much tedious ground work in the collection of data for preparation of a stock or ecological map, and permits the field worker to concentrate on the most pertinent aspects of the inventory. It is useful before proceeding with photo-interpretation to learn a little about the preparation of maps from aerial photographs. The knowledge so gained will be valuable when considering, for example, micro-areas.

A choice has to be made between planimetric maps and topographic maps. Planimetric maps of an acceptable standard can be prepared using the simplest of instruments and methods as outlined shortly in section (a); and fortunately the requirements of the photo-interpreter can normally be satisfied by studying a planimetric map in conjunction with aerial photographs of appropriate scale. Details relating to micro-areas of interest and not shown on the planimetric map can often be obtained directly from the photographs. Frequently the magnitude of errors arising from the use of a planimetric map is smaller than the magnitude of errors occurring in the field collection of pertinent data for an inventory and the overall accuracy may not be improved by using a topographic map. Several simple techniques used in the preparation of planimetric maps will be considered in conjunction with the methods of radial line triangulation previously outlined. Topographic mapping will be left to the skilled photogrammetrist using one of the first-order machines mentioned later; and will not be considered further in this text, beyond the description of an easy method of contouring a planimetric map in section (b).

A *planimetric map* is a map presenting the horizontal positions only of the ground features recorded on the photograph; and is distinguished from the topographic map by the omission of relief in measurable form, i.e. no contours. It is important to appreciate that a planimetric map is not a direct tracing of the ground features shown on the photographs, but is an accurate drawing to scale of the features in relation to each other. A carefully prepared planimetric map or an orthophotomap will have been corrected for the radial displacement of objects on the photographs. The aerial photograph, an enlargement of the photograph, a photo-map, an index mosaic and controlled and uncontrolled mosaics are therefore not strictly maps. These will be briefly discussed in section (c), under the heading 'Photographs as maps'.

A few generalizations on map scale may be helpful, although there will be exceptions. For agricultural and ecological studies, maps normally need to be at 1/10,000 or larger. For studies of intensively managed forests, maps at 1/10,000 (or 1/10,500), 1/15,840 and sometimes 1/20,000 are convenient. For less intensively managed forests, regional biological surveys, plant associations and land unit surveys, maps at 1/25,000, 1/50,000 and 1/63,360 are suitable. For geomorphic and geological studies land systems and plant formations/sub-formations, possibly maps at 1/50,000, 1/63,360, 1/100,000 and 1/125,000 will be preferred. These latter groups, when simplified, can also be shown on maps at 1/250,000 to 1/1,000,000. 1/250,000 to 1/1,000,000 is popular for mapping regional geographic data. The Directorate of Overseas Surveys, London, produces maps showing land-use, land-form and vegetation patterns at scales between 1/50,000 and 1/500,000 (Brunt, 1966). The vegetation of northern Tunisia is being mapped at 1/50,000 and finally reduced to a scale of 1/200,000 (Floret, 1966).

(a) Planimetric mapping

The overlay method. The simplest method of providing an approximate map of an area, covering only a few photographs, is to use the overlay method introduced in the previous chapter. The method is suitable for one to three flight lines and having only a few photographs in each flight line. The map scale will vary as the scale within each photograph. The map will be fairly accurate, provided the terrain is level; but as it is not corrected for radial displacement, it will be inaccurate in mountainous country and when tip and tilt exceed 3°/5°. The map provided by the overlay method is suitable for planning a forest inventory (timber cruise), for preliminary surveys of farm-land, land-form and land-use, for preliminary ecological, forest, soil and hydrological typing, and for use as a general map where an accurate map does not exist.

The wing points and transferred principal points on each photograph provide the boundaries of the photograph's effective area.

The reader will recall that each photograph is being placed under the transparent map base for a third time and aligned, so that the control points and the flight lines on the photographs and map base coincide. Important features are then transferred to the map base by pencil. It will be found that when an adjoining photograph is aligned with the map base, the outline of previous features transferred to the map base will not coincide with the same features in the periphery of the photograph. This is due to radial displacement. A compromise between the two photographic positions has therefore to be drawn on the map base. This is not the object's true position, but is close to it. The true position of major features, such as a peak or a cross-road, can be located by additional radial ray intersections from the principal points. The network of points provided by the wing points and principal points will prevent errors accumulating from photograph to photograph, but will not eliminate errors occurring within the sections formed by the principal points and wing points.

Three-photograph method. If only a strip of three photographs of the same flight line are required to cover the area to be mapped, then the map base can conveniently be prepared directly on drawing paper by firstly pricking through the wing points, transferred principal points and principal point of the centre photograph. The direction of the wing points is then extended outwards in pencil before finally locating the flight lines of photographs one and three over the pin-holes of the flight line of photograph one. Pin-holes are pricked on to the drawing paper to show the location of the wing points of photographs one and three. Rays are extended in pencil to these points on the drawing paper and their true positions are located at the intersection of the radial lines. Photographic detail is transferred by a sketchmaster or other suitable equipment.

The hand template method. Instead of using a transparent overlay, hand templates may be used as described previously to provide the map base. Details from the photographs are transferred either as outlined in the overlay method, or, if the map base is opaque, then the detail is transferred by a sketchmaster or other suitable equipment.

The slotted template method. A more efficient method by far, which has been widely used, is the template method. Much of Australia, 2,974,000 square miles, has been mapped planimetrically in the manner to be described. The method has the advantage of being simple, providing a planimetric map with hill shading, eliminating accumulative errors and not requiring elaborate equipment. For large areas and most satisfactory results, small scale photographs (e.g. 1/50,000) are to be preferred.

For large lay-downs of templates, sheets of welded aluminium (about 26 gauge in the U.S.A.) or polyethylon (30 in. by 36 in. in Australia) are spread on the floor. On this base, the map projection is marked off at the desired scale and ground control points accurately located. The commonest map projections are the orthographic, Lambert, Conformal, Polyconic and Mercator. In Australia, for example, the national

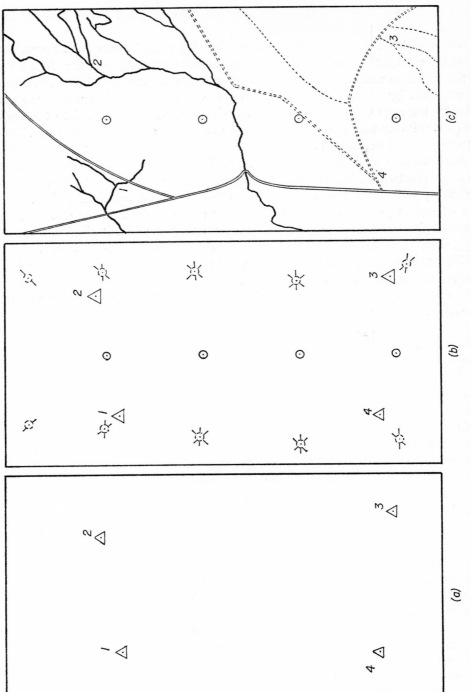

Fig. 12.1. The preparation of a planimetric map (see text). (a) Ground control points plotted. (b) Principal points and wing points plotted. (c) Physical and man-made features plotted. Finally (not shown), ecological, forest or landform detail is added.

F IG. 1. 1a. Wheatstone's reflecting stereoscope, 1838 (see page 2).

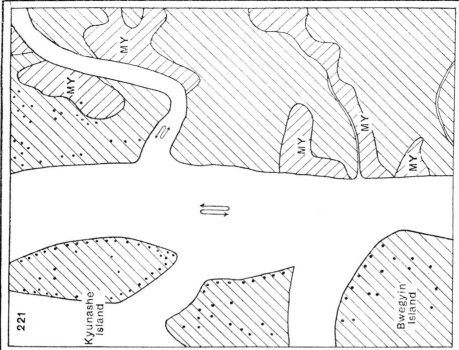

Scale about 3 Inches = 1 Mile.

221

Kyunashe
Island

Bwegyin
Island

MY

Kanazo Forest -------------

Kanbala -------------

Myinga ------------- | MY |

FIG. 1. 1b. As long ago as 1924, aerial photographs were used on an ecological basis in the tropics to prepare forest maps. The above map and part of the panchromatic photograph show associations dominated by *Heritiera minor* (Kanazo), *Sonneratia apetala* (Kanbala) and *Cynometra ramiflora* (Myinga). The photography of the Irrawaddy Delta forests was commenced in February 1924, using a 6 in. focal length lens and a flying height of 10,000 feet.

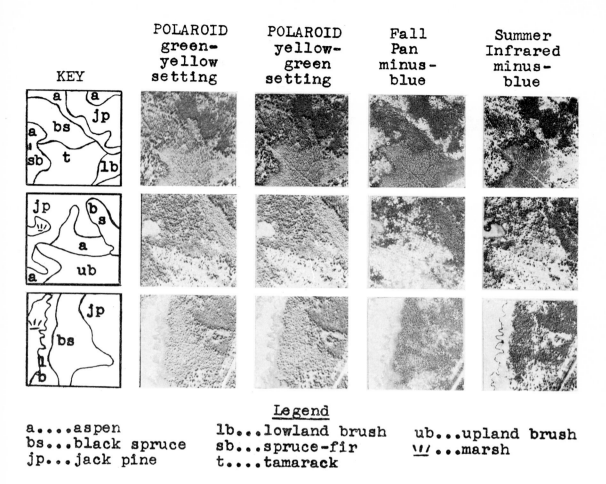

| KEY | POLAROID green-yellow setting | POLAROID yellow-green setting | Fall Pan minus-blue | Summer Infrared minus-blue |

Legend

a....aspen   lb...lowland brush   ub...upland brush

bs...black spruce   sb...spruce-fir   \//...marsh

jp...jack pine   t....tamarack

FIG. 3. 3. The film-filter combination and season of photography can have a pronounced effect on the tone and the contrast of the recorded black-and-white images, as is illustrated above. Note particularly the differences between the conifers (bs, jp, sb, t) and angiosperms (a, lb, ub).

FIG. 6. 2. The two oblique photographs of the same area were taken on the same day within a few minutes of each other. Photograph (*a*) with the camera pointing towards the sun, shows the outline of the ancient earthworks 1, 2, 3, 5 and 6; but on photograph (*b*), with the sun behind the camera, earthworks 5 and 6 are absent and the outline of 4 is conspicuous. (Dijon, France).

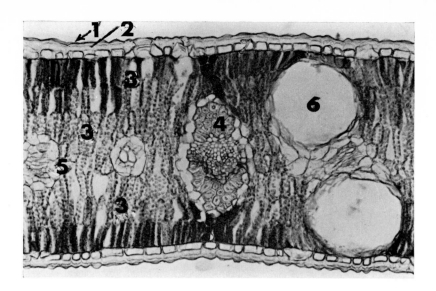

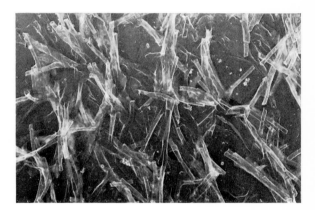

FIG. 6. 5. The leaf is a complex structure, as will be observed under high power magnification. (*a*) Transverse section of an isobilateral leaf of *Eucalyptus obliqua* (magnification: 200 times). (1) waxes when present are situated on the cuticle; (2) cuticle; (3) palisade cells containing chloroplasts; (4) vascular bundle; (5) paryenchyma bundle of mesophyll; (6) oil duct. (*b*) and (*c*) Micrographs of the distinctive leaf surface waxes of two species of the same genus (Eucalyptus). Left, *E. cincrea* (magnification: 10,000); right, *E. ovata* (magnification: 16,000).

FIG. 6. 6. A 'normal' photograph (*left*) and infrared imagery (*right*) of the same area. In the infrared imagery, a sub-soil feature, not visible on aerial photographs is quite conspicuous (1, 2).

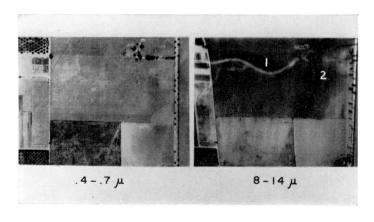

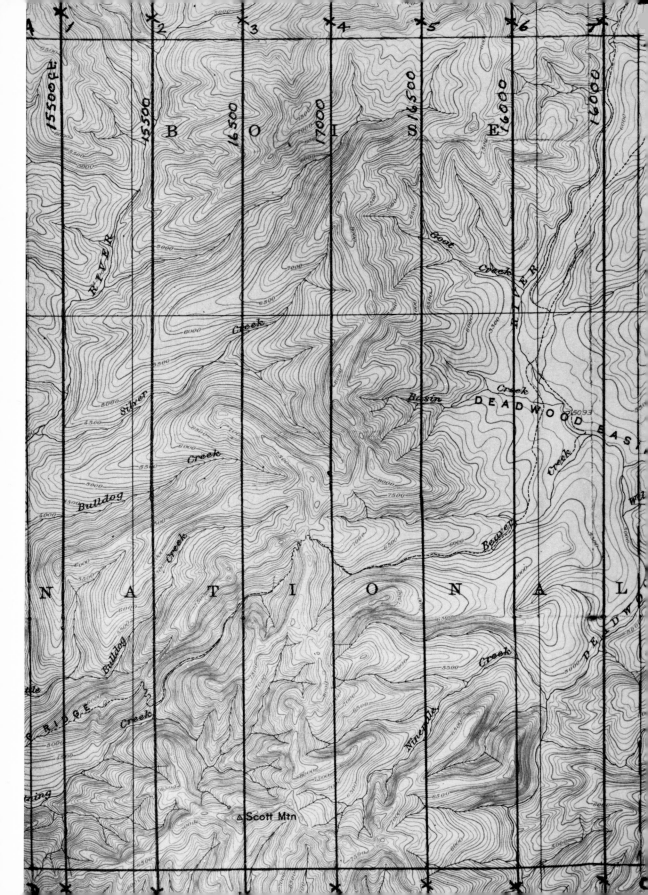

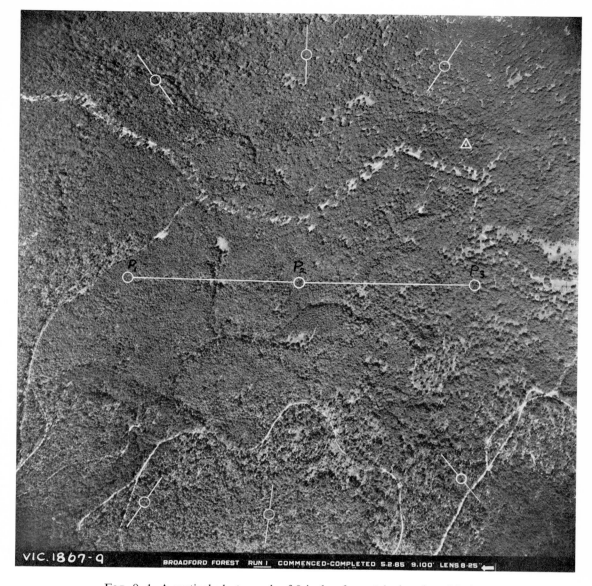

Fig. 9. 1. A vertical photograph of 9 inches format (reduced to 6 inches format) on which the principal point (p₂), the two conjugate principal points (p₁,p₃), the flight lines between these points, the six wing points and one ground control point (triangle) have been shown in white. The forest roads show up whitish against the dark forest (scale: about 1/13,500). (Sclerophyll forest on granodiorite, Victoria).

(*Opposite*)

Fig. 7. 5. Flight plan (scale 1/125,000). The area to be photographed has been lettered A B C D. Flight lines have been numbered and the crosses on the flight lines indicate the beginning and end of the photography.

FIG. 9. 2. Montagne de Bussy, Alisia, France. (*a*) Vertical panchromatic photograph (1/5,000) showing fields of cereals and shrubs on outcrops of limestone at the top of the photograph (wet summer, July 1963); (*b*) the same area on a large-scale, low oblique panchromatic photograph (July, 1963) showing the outline of ancient earthworks (indicated by arrows). These are part of a camp used by Julius Caesar in about 52 B.C. The area was photographed systematically from 1959 until 1966.

FIG. 11. 2. To the left is a *slotted template punch* (or cutter) with a template mounted on the table of the punch. To the right is shown a *centre point punch*, used for punching out the principal point in the template before transferring it to the slotted template cutter. In the foreground are three assembled slotted templates; note the studs used for joining the common slots of contiguous templates. On the extreme right, a *spider template* is shown resting on a photograph. This provides an alternative method to slotted templates. The centre stud is above the principal point of the photograph and the radiating struts conform to the directions of the wing points and flight lines. Similar spider templates are assembled for each photograph and then joined together by studs.

FIG. 12. 2a. The Zeiss Aero-Sketchmaster (German). The operator views part of the aerial photograph (3) through the eye-piece (1) and sees the photographic images superimposed on the base-map (2). (See general comments overleaf.)

(b) Nash and Thompson Sketchmaster (British)

FIGS. 12. 2 and 12. 3. *Mapping instruments*. These may be grouped either according to the operating principle of the instrument (i.e. optical, opticalmechanical or mechanical) or according to the obtainable mapping accuracy of the instrument (i.e. first-order, second-order, third-order or below third-order). Generally, photo-interpreters will be satisfied with third order or lower.

FIGS. 12. 2. (a) and (b) are vertical sketchmasters. (c) is the Hilger and Watts Stereosketch, which incorporates a mirror stereoscope (1) and an adjustable drawing table (2).

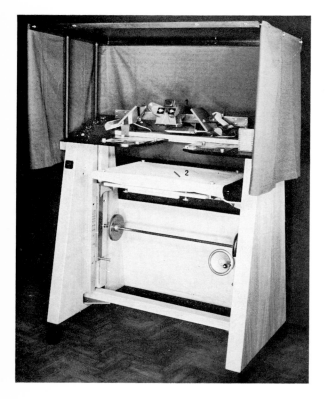

(c) Stereosketch

FIG. 12. 2d. The S.F.O.M. Orthophotoscope produces orthophotomaps from stereo-pairs of diapositives (1). The stereo-image is recorded on light sensitive paper which is exposed through an adjustable slot (2).

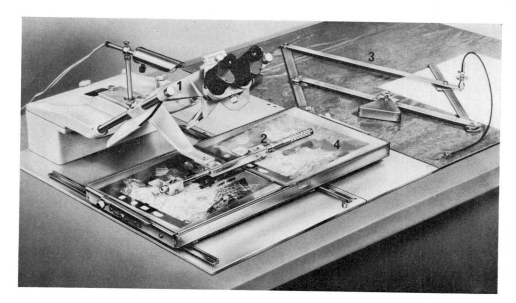

FIG. 12. 3a. The Zeiss Stereopret, which incorporates a mirror stereoscope (1), a parallax bar (2) and a pantograph (3). The stereo-pair of photographs are mounted on the sliding table (4) to which the pantograph arm is attached.

FIG. 12. 3b. The S.O.M. Stereoflex, which is a third-order optical-mechanical instrument. The operator views the stereo-model through semi-transparent mirrors (1) and the point-source of light in the small plotting table (2). The latter is attached to a pantograph for mapping.

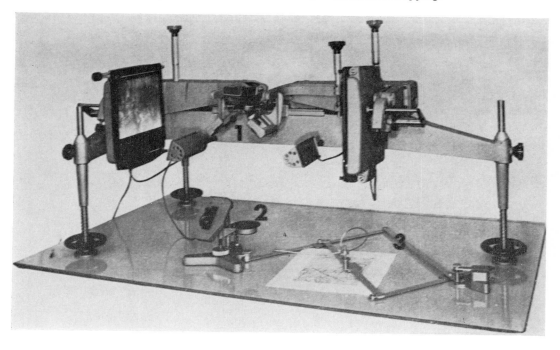

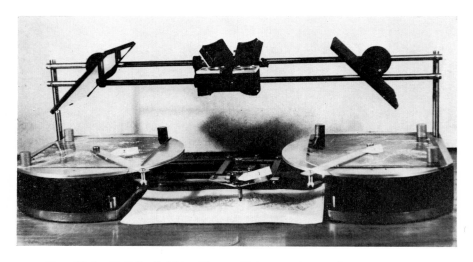

FIG. 12. 3c. Kail Radial Line Plotter. This comprises a mirror stereoscope, two tables on which the photographs are mounted and two radial arms which rotate at the principal points of the photographs and are connected to the drafting mechanism (1).

FIG. 12. 3d. The Zeiss Stereotope, a third-order instrument, provides a mechanical solution of errors due to radial displacement.

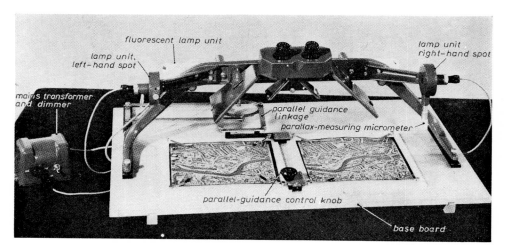

FIG. 13. 5b. The Hilger and Watts mirror stereoscope. Instead of a parallax bar having graticules in contact with each photograph (as in Fig. 12. 3a) two small spots of light are injected into the optical system of the stereoscope, when measuring parallax differences.

FIG. 13. 5c. A zoom stereoscope.

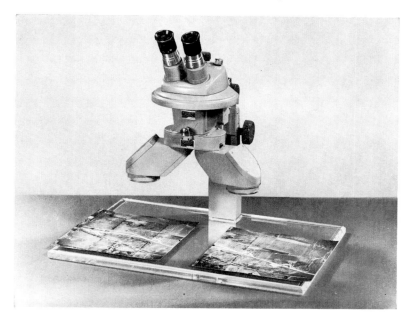

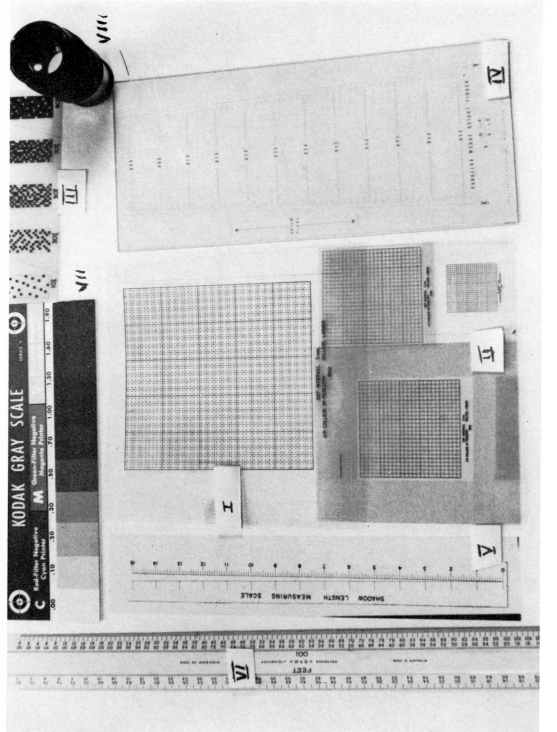

Fig. 16. 3. Measuring aids. (1) dot grid with 2mm. between dots; (2) three grids with dot intervals of 1mm., 0.5mm., and 0.25mm.; (3) crown closure (crown density) scale (see section c of Chapter 16); (4) parallax wedge (see Chapter 13); (5) shadow wedge or crown diameter wedge; (6) engineer's scale graduated in 1/50th inches so that readings can be made to nearest

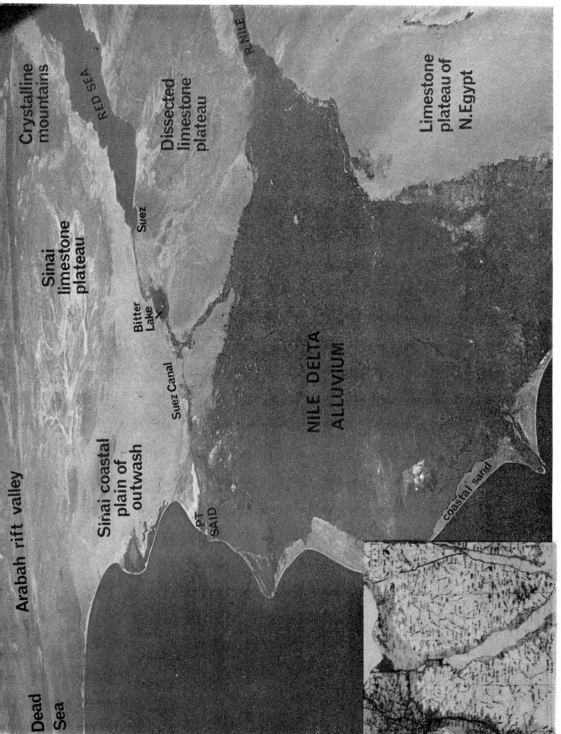

Fig. 17. 1b. This oblique photograph of the Earth's surface was taken from a Gemini-4 spacecraft in June 1964 and shows the Nile Delta, Suez Canal, Mediterranean Sea and Red Sea. The entire land area forms part of a geographical region which is divided politically by two countries (Israel and Egypt). The Nile Delta can be considered as one land system, which can be further divided into land units (e.g. coastal sand and alluvium).

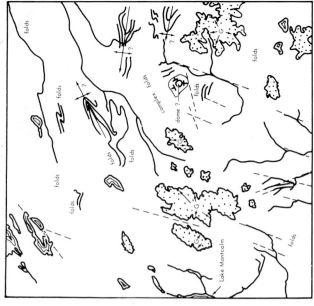
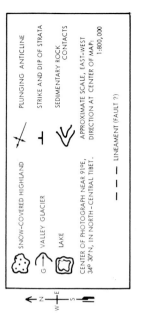
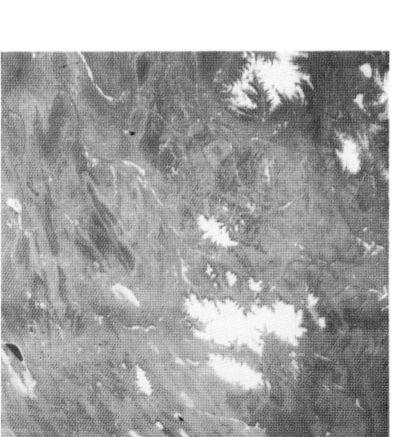

FIG. 17. 1c. A colour photograph taken with Anscochrome 200 film, of north-central Tibet, and a geological-physiographic sketch-map of the same area.

grid is a transverse Mercator with 5° parallels. Errors due to curvature of the earth and atmospheric refraction are so small that they may be ignored, when compared with other sources of error. For long flight strips, electronic computers offer a satisfactory approach to extension control and bridging.

On the aluminium sheets the slotted templates are assembled within the overall boundary provided by the map projection at the desired scale. The principal points and wing points are pricked on to the aluminium foil or polyethylon before removing the templates. If tinfoil or aluminium sheeting has been used, a semitransparent medium, e.g. matt acetate, is laid over it and the points transferred. Next the semi-transparent sheets are placed on a work bench, and by means of a vertical sketchmaster or other appropriate equipment the photographic detail is transferred to the sheets to provide the planimetric map. Steep slopes, mountains, etc., can be shown on the planimetric map by careful shading to represent the gradient and change in elevation. If a small scale map is required, the map is assembled from 'ratioed photographs' of the planimetric sheets.

If only a few templates are used, the ground control points are plotted to scale on good-quality drawing paper, e.g. 80 lb cartridge paper (fig. 12.1a) and the assembly of slotted templates is then fitted into these points. At least four carefully chosen, well-spaced control points are needed (see Chapter 7). The location of all principal points and wing points are then pricked on to the paper (fig. 12.1b). The templates are removed and at the same time the principal points on the map base are numbered. Finally, the principal man-made features and physical features are transferred on to the base map using a sketchmaster or simple stereoscopic plotting instrument (fig. 12.1c). The maximum and minimum scales of the map depend on the size of the slots in the templates and the scales to which the sketchmaster or plotter can be adjusted. Overall scale will be uniform between the triangles formed by the wing points, etc., and errors will be confined within each triangle.

Transfer of land-form, forest and ecological detail. After preparing the planimetric map by one of the methods described previously it is suggested that each stereo-pair of photographs of the map area should be carefully examined stereoscopically for the purpose of interpreting, classifying and delineating the pertinent details and transferring this detail to the planimetric map in accordance with the objective of the study. It will be appreciated that up to now a planimetric map has been prepared which shows man-made features (e.g. roads, railways) and terrain features (e.g. water courses, ridges, peaks). To this must be added the boundaries of areas occupied by each class of growing stock to provide a forest stock map or the boundaries of the photo-communities to provide an ecological map (see Chapter 22), or land units for a terrain map (see Chapter 18). The procedure to be adopted in transferring the detail may be considered as stage four.

Sketchmasters are popular, as the data already on the map provides a satisfactory network into which ecological or land units or forest stock details can be fitted. This procedure should always be treated ocautiusly and each case must be considered on its own merits. The I.T.C. reported that in Europe the Zeiss sketchmaster, as compared with other sketchmasters, is preferred for transferring detail; but for mountainous terrain a radial line plotter is the more satisfactory. In the U.S.A., the Multiscope was popular for this type of work a few years ago; but a radial line plotter is now more often used.

The importance of carrying out the forest or ecological studies in continual association with field visits cannot be emphasized too strongly. At least 10% and up to 50% of the total time normally requires to be spent in the field. The length of time allocated to field work depends considerably on the purpose of the survey.

Cadastral maps. These are planimetric maps showing the legal boundaries. When an adjoining property comes under new ownership or if it is found necessary to adjust the forest boundaries due to logging or afforestation, it may be advisable to carry out an additional survey using aerial photographs. The field worker will possibly visit the owners of all common boundaries and record on his forest map any disputes in ownership for legal verification later. Frequently, it will be found useful to have stereo-pairs of photographs at the meeting, as this will help in locating additional ground points when in discussion; and can frequently help in settling what could be a legal dispute, since a neighbour will often accept evidence from a photograph even if he cannot use a stereoscope himself. The writer (1960) used aerial photographs in a boundary dispute at Pembrey in South Wales, and Huguet (1958) reported the successful use of aerial photographs for cadastral survey in Mexico. An up-to-date cadastral map can serve as a valuable map base when preparing forest stock maps, ecological maps, etc.

INSTRUMENTS

Monocular mapping instruments. Both sketchmasters and reflecting projectors are used, although the former are the more popular. The *sketchmaster* employs the principle of the *camera lucida*, in which the photographic image is seen superimposed on the map base through a semi-transparent mirror surface of a prism. The semi-transparent surface is provided by coating one surface of the prism with a very thin layer of silver. Probably the two most popular are the vertical sketchmaster manufactured by Aero Service Corporation (U.S.A.) and the Aero-Sketchmaster by Zeiss (West Germany). The Zeiss Sketchmaster (fig. *12.2*) reduces to $\frac{1}{2}$ and enlarges to 2·5/1.

A substitute for an instrument using the principle of the camera lucida is the *reflecting projector*, resembling a photographic enlarger. As the name suggests, a light source of adequate brightness is used to reflect an image of the photograph from the back of the projector through the lens on to the map base. Reflecting projectors have the disadvantages of being bulky, and of requiring the use of a semidarkened room. These instruments cost £400 upwards; but a home-made type, described by Meyer (1961) can be assembled for £50, being suitable for transferring photographic details on to existing maps. Recently Kail has marketed a convenient compact table type of reflecting projector, not requiring a semi-darkened room for use. Transparency slides, taken with a standard 35 mm camera and projected on to a suitable copying screen, provide a cheap and frequently satisfactory method (Howard & Kosmer, 1967).

Stereoscopic mapping instruments. The vertical sketchmaster and the reflecting projector have two disadvantages. Firstly the photographic detail is transferred to the map base by means of a monocular instrument, so that within any one of the triangles formed by the principal points and the six wing points, errors caused by radial displacement will not be corrected (see Chapter 11). Secondly it is often necessary to examine each photograph separately with a stereoscope before monocularly transferring the detail to the map or map base, in order to ensure that the details seen monocularly are interpreted correctly. These disadvantages can be overcome by using an instrument in which a stereo-model is seen as the details are transferred to the map base.

Instruments for the preparation of planimetric maps from a three-dimensional model may conveniently be grouped into three classes. The first type of instrument uses a *mirror stereoscope combined with a camera lucida.* The operator sees with both eyes the stereoscopic model formed from the two photographs; and views the map base as he did when using a sketchmaster. In so doing, he is able to transfer the details of the three-dimensional model on to the map base. An instrument employing this principle is the *Stereosketch* (Hilger and Watts). Viewing is at photographic scale or in the range 1/0·45 to 1/1·25.

The Stereosketch gives satisfactory results under most conditions, provided tip and tilt is not high and relief displacement is not excessive. Some operators complain of eye-strain and unskilled operators may find that the model moves in relation to its position on the map base. This latter disadvantage occurs also with other types of instruments and is due to optical parallax between observer, stereoscopic model and map base. The drawing table below the stereoscope has a vertical movement of 12 in. and a tilt of ±5°.

A second group of instruments uses the principle of *radial line plotting* introduced via the overlay method, hand templates and slotted templates. A large number of objects on the photographs could be located in their true position by drawing radial lines as described earlier; but the method would be cumbersome and slow. It has, however, been successfully used by Desjardins (1943a), who employed black thread to provide the direction of the ray from each principal point.

Improvement, however, can be made by using two straight edges pivoted at the principal points of a stereo-pair of photographs. The true location of an object is given by the intersection of lines subtended by placing the straight edges over the image common to the two photographs. The method is slow; but can be again improved by connecting the straight edges, by a suitable linkage mechanism, to a recording pencil or other drafting device. This technique, coupled with a mirror stereoscope, is used in several radial line plotting machines, including the *Kail Plotter* (fig. *12.3*) and the *Hilger and Watts Plotter.* In the Kail Plotter, two radial arms of perspex operate from the centre of two metal tables on which are mounted a stereo-pair of photographs. A scale adjuster permits the drafting pencil to record between 1/3 and a little larger than 1/1 ratio. During viewing of the stereo-model the radial arms can be seen but not the drafting pencil. Difficulty is encountered, and resetting is necessary, when sketching close to the flight lines. Also, as in surveying, the accuracy is reduced when the rays intersect at oblique angles. The machine does not correct for tip and tilt. Some operators have difficulty in keeping the intersection of the radial lines fixed in relation to a selected point in the stereo-model. However, an operator using photographs of fairly level terrain is reasonably efficient within one to two days' training. When preparing a map, using a radial line plotter, the base map can be prepared as outlined earlier by means of slotted templates.

Finally, there are machines in which the principle of the fused floating dots (i.e. floating mark) is introduced (see Chapter 13). A stereo-pair of photographs are viewed under a mirror stereoscope as described above; but at the same time a floating mark is introduced in place of the radial lines. The mark is made to float on the apparent surface of the stereo-model; and is used in conjunction with a suitable drafting mechanism to trace the outline of the images on to the map base.

In its simplest or earliest form, the 'floating dot cum tracer' comprised a parallax bar having a centrally mounted pencil, e.g. *Tracing Stereometer* (Zeiss). The scale of the tracing is obviously the same as the scale of the photographs. With slight modification, *Fairchild's Stereocomparagraph* and *Abrams' Contour Finder* operate on this principle. These have the stereoscope, parallax bar and tracer as a single unit, and can be attached to a parallel drafting arm. In the *Stereopret* (fig. *12.2*), the parallax bar is fixed to the stereoscopic viewing head and is stationary. Movement of the floating mark and coverage of the stereo-model is achieved by a moving table on which the photographs are mounted. Some new operators, when tracing, experience difficulty in keeping the floating dot in contact with the stereo-model. Transfer of the photographic detail to the map base is carried out by a pantograph arm attached to the table. The scale of the pantograph can be adjusted from 1/5 up to 3·0/1. Errors resulting from radial displacement cannot be eliminated but adjustment can readily be made to correct for y-parallax. In the *K.E.K. Plotter* the floating mark is used to establish a datum plane and the stereo-model is introduced into this plane. The *Stereoflex* and *Stereotope* (fig. *12.2*), to be described later, may also be used for planimetric mapping.

Topographic mapping instruments. The simplest topographic plotting instruments incorporate a parallax bar and mirror stereoscope under which the three-dimensional model is viewed. Plotting, as outlined previously, and form-line tracing is by a floating dot and pantograph attachment. There is no provision for correction of the y-parallax in the *Abrams' Contour Finder* and the *Stereocomparagraph*. The *Stereopret* has an adjusting arm to correct for y-parallax only and provides perspective contours. The *Stereotope* uses built-in mechanical analogue computers (rectiputers) to correct for topographic displacement and tilt. The interpreter views the model formed from the stereo-pair of photographs as in the Stereopret; but the parallax bar is replaced by a fixed distance floating mark. The floating mark is raised or lowered by the x-parallax thimble, which alters the distance between the horizontal photo-holders. A y-parallax thimble controls the y-movement of the right-hand photo-holder. The two photo-holders and rectiputers are mounted on a carriage which can be moved freely; and are connected to the pantograph (scale range: 1/5 to 2·5/1). A relatively new instrument of increasing popularity and also using a mechanical system for rectifying errors due to tilt and relief displacement is the Cartographic Stereomicrometer (Galileo-Santoni).

In a second group of instruments, an attempt is made to completely eliminate combined tilt by tipping and tilting the photographs out of the common plane. Instruments include the *K.E.K. Plotter* and recently the *S.O.M. Stereoflex*. The delineation of contours is subsequent to the preparation of the planimetric map. In the Stereoflex, the three-dimensional model is viewed at a little less than photographic scale through semi-transparent mirrors. As the model is always subject to exaggeration of the vertical scale with respect to its horizontal scale, a small plotting table (the height of which can be adjusted), is operated above the map base to remove the optical parallax. This is an important innovation so far not commented on. The floating mark is provided by a point-source of light in the plotting table.

With all the instruments so far mentioned, bridging between photographs of the same flight line is impracticable or tedious; but the instruments are considerably less expensive and relatively portable compared with first-order machines. It is, however, sometimes convenient to have a basic map provided by a first-order machine and then to add detail by a third-order machine such as the Stereotope. For example, one first-order machine can supply map bases for up to ten Stereotopes.

In the *Manual of Photogrammetry*, the first-order instruments are classed as automatic stereoscopic plotting instruments and subdivided by projection systems. The most popular first-order machines used in mapping include the *Balplex ER-55*, *Kelsh Plotter* (C-factor: 1,000), *Multiplex* (C-factor: 750), *Kern PGI Plotter*, *Nistri Photocartograph*, *Stereomat*, *Stereoplanigraph* (C-factor: 1,200), *Stereotopograph*, *Wild Autograph* (C-factor: 1,200) and *Thompson-Watts Plotter*. The most widely used instruments vary considerably between countries, depending probably more on government policy, marketing and 'after-service' than on the instrument being the most suitable for the local conditions. All first- and second-order instruments are expensive, costing £3,000 to £10,000. Some require bedding on concrete or on a similar base and cannot be considered as even semi-portable. For photogrammetric mapping, they have the advantage of providing accurate maps with small scale photographs, e.g. 1/30,000 to 1/100,000, so that the number of ground control points is reduced per unit area of ground. Further bridging between photographs is possible and is frequently used for providing additional control points. Tilt, lens distortion and model deformation can be eliminated. The optical system of the instrument may, however, limit the choice of focal lengths to be used with the aerial camera.

PLANIMETRIC MAPPING

The *Multiplex* and the *Kelsh* employ a principle not so far described. This is *double image projection*. Diapositives of a stereo-pair of photographs are placed in the instruments and projected downwards on to the drafting table to form a double image. The double image (or anaglyphic image), when viewed through one blue and one red lens, forms a normal three-dimensional model. A light-source floating dot and a pantograph are used to plot each model on the map base at the correct scale. Other instruments involve the application of optical mechanical principles by the creation of a virtual image by the stereoscope and the human mind. A light-source floating dot is used for plotting and for contouring. Print-size diapositives are most commonly used.

There is little doubt that, as colour aerial photographs come to be more widely used, the light systems will require modifying/improving. For example, using the well-known *Stereoplanigraph E-8* and a *Stereo-plotter A-8*, Maruyasu & Nashio (1962) found that colour photographs provided no positive advantage over mono-chromatic, particularly due to the small bulbs providing reddish light through a refracting optical system.

C-factor. The reader will observe that the term 'C-factor' has been introduced after the names of a few instruments to describe the relative accuracy of the machine. The C-factor is commonly used in the U.S.A. to indicate a relationship between the least contour interval and photographic scale, being found a simple but practical method of comparing instrumental qualities, i.e.

$$\text{C-factor} = \frac{\text{Flying height above ground}}{\text{Least contour interval}}.$$

The least contour interval is defined as double the range in elevation within which 90% of the points of the measured height fall about the mean. Thus if the range for 90% of the points is 3·5 ft, the least contour interval is 7 ft.

The application of C-factor is appropriate when standards are uniform and similar aerial photographic equipment is being used, but may be unreliable when there are great variations in the quality of the photographs, type of equipment and skill of the operators. Skilled photogrammetrists may differ in assessing the C-factor according to local conditions. Colner (1962) quoted the Kelsh Plotter as having a C-factor of 512 and Tewinkel (1962) appraised the instrument with a C-factor of 1,200 to 1,700. Recently Lyon (1964) has suggested a modification of C-factor; and used the term coverage-contour factor, which takes into consideration the area of the stereo-model as a function of flying height. In continental Europe, *map error* is used instead of C-factor, being the error expressed in terms of one standard deviation under the conditions provided by the film, camera-lens, photographs, plotting machine, etc.

Bridging. A photograph mounted in a suitable stereo-plotting machine and correctly orientated in space in relation to known ground control points can be used stereo-scopically to transfer selected photographic control points on to the next photograph. The second photograph can then be held in the machine, whilst the first photograph is removed and the next photograph is introduced. The process may be continued until the end of the flight strip, when a 'tie-in' is made with similar ground control points on the last photograph. This process is termed *bridging*. If ground control points are only available on the first photographs, then the term '*extension*' would be applied in lieu of bridging. Once the photographic control points have been established on each photograph by means of a first-order machine, e.g. Wild A5, details from each stereoscopic pair of photographs can be transferred to the map base by a simpler and less expensive instrument, e.g. Stereotope. This procedure was used by the Snowy Mountain Authority, as slotted templates were found to be unsuitable due to the rugged nature of the terrain.

(b) Topographic maps

As a part of photogrammetry, topographic mapping is a complex subject about which a great amount has been written by the photogrammetrist, trained in mathematics and surveying. The *Manual of Photogrammetry* is considered about the best lucid reference on the subject.

When showing relief the easiest method is to prepare a planimetric map and then to use shading to show changes in the topography. This technique has been used effectively by the Division of National Mapping in Australia. A second method is provided by sketching *form lines*. These lines are similar to contour lines but are sketched by visual examination of the stereo-model. A true topographic map, however, provides contours. The Directorate of Overseas Surveys, London, has obtained satisfactory results by establishing stereoscopically 20 to 25 photographic control points per photograph and then sketching in form lines at 50 ft contour intervals. The map so produced is prepared in about one-third the time of a contour map. The 'styling' of form lines and contours can sometimes be improved if the photogrammetrist has a knowledge of geomorphology, e.g. valley forms (Verstappen, 1964).

The simplest method of preparing a 'topographic' map is as follows. Firstly, a planimetric map is provided by one of the methods outlined previously. Then, combined with additional ground survey to determine the elevation of the ground control points or the determination of the heights from existing maps, height control points are established on the photographs. Next, using a suitable floating dot machine, incorporating a parallax bar and pantograph, lines are traced off on to the planimetric map. The Fairchild *Stereocomparagraph* and the Zeiss *Stereopret* are both suitable instruments provided tilt is negligible and the ratio $\dfrac{\text{elevation change}}{\text{flying height}}$ remains small.

Unless the contour intervals are fairly widely spaced, the inexperienced operator tends to run contours into each other. As a check, it is also advisable to establish additional height control points on the photographs by parallax bar before starting. Frequently an additional form line can be inserted free-hand between two contours.

Mott (1956) gave an interesting account of contouring in dense jungle in Ceylon based on ability of the photogrammetrist to recognize the height of prominent trees above the ground on photographs at a contact scale of 1/20,000. A similar technique has been tried when mapping from photographs of New Guinea. Mott observed that the canopy had a flattening effect on the formation of the terrain, but it was obvious from the photographs that the main ridges stood out reasonably well and that the general shape of the terrain could be deduced from an appreciation of the height of the canopy in association with spot heights obtained in forest clearings. Lack of moisture on the top of hills and ridges was considered likely to give rise to stunted growth compared with that in the valleys where the moisture content was known to be greater. As part of the area had been contoured by ground control previously, it was possible to compare the results. In general the shape of the contours showed a remarkable consistency. The maximum discrepancy was about 20 ft and the mean deviation was ±10 ft.

(c) Photographs as planimetric maps

As a photograph is a perspective view and a map is an orthographic projection, errors will occur in using photographs directly as a map, due principally to radial displacement and sometimes to misinterpretation of shadows. However, armed with knowledge of the errors that are likely to occur, it is often convenient to use individual photographs or assemblies of photographs as a temporary 'map'. Obviously the enormous advantage of the stereoscopic view is lost, but a person who has been accustomed to viewing photographs non-stereoscopically can develop a skill in interpreting local features particularly if the shadows are favourable to viewing. Frequently, this type of skill is not acknowledged or is sometimes treated as a heresy! Photographs or photograph-assemblies used as maps have frequently an advantage over maps in that they may contain information not shown on maps made from the photographs by normal photogrammetric methods. This is particularly the case in ecological, forest and land-form studies.

If angles are required to be determined from photographs, a transparent 360° protractor is suitable. It is convenient to drill or punch a pin-hole through the protractor at its centre of rotation; so that it can be pinned in position on the photograph or mosaic and rotated as required. North may be determined from flight line data and sun shadows and should be shown. A bearing and distance towards the edge of the photograph may need adjusting due to radial displacement. This particularly applies to mountainous areas; and is difficult to check in areas of closed forest, e.g. tropical rain forest, due to lack of prominent features. If a bearing on a photograph is checked on the ground by magnetic compass, both bearings must be expressed in degrees true or degrees magnetic. Both the flight line and other radial straight lines from the principal point will also provide straight lines on the ground.

Aerial photographs, to be used as maps, may be conveniently classified under the following headings:

(i) *Rectified prints*. These have been considered previously. They are single prints made from tilted photographs and corrected for the tilt by re-creating an opposite tilt at the time the print is made.

(ii) *Ratioed prints*. When prints are enlarged or reduced in scale compared with the original negative they are said to be ratioed (see Chapter 9).

(iii) *Enlargements*. As mentioned above these are ratioed prints. Spurr (1952) recorded that modern aerial photographs can be enlarged at least two to four times and yet produce images of high pictorial quality. Studies in Japan (Maruyasu & Nishio, 1962) to compare black-and-white photographs and colour prints indicated that the sharpness was similar at about the same scales; but delicate tonal differences became noticeable in the colour enlargements (which were not observed before) and these differences were helpful in the interpretation.

An enlargement at the same scale as a contact print will not have identical qualities for stereo-interpretation and photogrammetry. Enlargements are bulky to handle and difficult to study under a stereoscope. For study of tropical forests, which often have a great complexity of pattern, enlargements have been found useful both in the office and field in New Guinea and by the Directorate of Overseas Surveys, London.

(iv) *Photo-maps*. This term has been applied loosely to enlargements of small scale photographs and to mosaics of two or more rectified prints; but is here restricted to enlargements of the effective area of a photograph. In New Zealand, by enlarging two to four times the effective areas only of photographs taken at a scale of about 1/60,000 a print is obtained which can be used satisfactorily as a 'map' for many purposes. Possibly in the near future, due to the introduction of small scale photogrammetrical photography, this type of enlargement will have a much wider appeal. In the U.S.A., photographs at 1/56,000 were enlarged $3\frac{1}{2}$ times, gridded and then copied by Ozalid printer to provide field 'maps' for timber assessment in Washington state (Kummer, 1964).

(v) *Index mosaics*. The aerial photographs are assembled so that common images on contiguous photographs overlap and the indices at the bottom of each can be seen. The assembly is then photographed and printed at whatever scale is required. This is then used as an index for easy reference to the photographs. In areas not yet mapped, an index mosaic can be helpful to field work until the photographs, map mosaics or maps are available.

(vi) *Uncontrolled map mosaics*. A map mosaic comprises the assembled effective areas of the aerial photographs. These are fitted together, similar to pieces of a jig-saw, so that the edges of each effective area are carefully matched with the adjoining edges. Normally, the edges are 'feathered' into each other to give the appearance of a single photograph. Single-weight photographs are preferred.

Careful inspection of a mosaic will often show that matching cannot be perfectly attained due to topographic displacement and tilt. There is likely to be considerable difference in scale in different parts of the mosaic, especially if the terrain is rugged. Also errors are likely to accumulate in any direction as there is no means of localizing them, as was possible by using slotted templates for planimetric maps. However, as only effective areas are used, the margins of the photograph which are likely to give the greatest errors are excluded.

(vii) *Controlled map mosaics.* These are similar to the previous mosaics, but photogrammetrical control points and ground control points are used to provide overall control of the scale of the mosaic. In preparation of controlled mosaics rectified prints are normally used to remove excess tilt and ratioed prints are used to provide photographs at approximately the same overall scale. The final scale of the mosaic is provided at time of printing as for uncontrolled mosaics; but the assembled effective areas prior to printing may comprise contact prints, ratioed prints and rectified prints. Controlled mosaics are useful in the field for initial planning and for the purpose of providing an overall idea of conditions. Often a clearer understanding of geomorphic structures and general hydrology can be obtained from the mosaics than from individual photographs. Mosaics are probably the most useful photographic aid in regional geographic studies. In Sweden, mosaics are printed with contours.

Orthophotomaps. An orthophotomap is a mosaic of uniform scale formed from an assembly of orthophotographs. It may also be termed an *orthophoto-mosaic. An orthophotograph* is a photographic copy of the area covered by an aerial photograph in which the effects of the lens aberrations and the radial displacement due to tilt and topography have been removed. It contains the same detail as would be recorded in an orthographic aerial photograph, if the latter were possible. An aerial photograph is a perpsective view of the ground (Chapter 9).

To obtain orthophotographs from aerial photographs an orthophotoscope is used. Basically, this is a photo-mechanical device, incorporating an anaglyphic projector; a movable slit (or scanning aperture) which replaces the table of a photogrammetrical plotter and a light-sensitive film below the slit, which records the anaglyphic image and is processed into orthophotographs. The slit aperture can be varied according to the topography and can be moved in the x- and y-directions. The film is adjustable in the z-direction (see fig. *12.2d*).

Orthophotographs and orthophotomaps are a new introduction and therefore their importance to the interpreter has not as yet been evaluated. An orthophotomap is a true map and may be contoured by conventional photogrammetrical techniques to provide a 'photo-contour map'. No doubt orthophotographs will have a wide application in natural resource studies not requiring a stereoscopic examination.

Thermal maps. In the last few years, remote sensing by infrared radiation in the range 0.72μ to 14μ has been used to provide 'thermal maps'. The infrared thermal mapping can be provided by either daylight or night flying. Definition of coast lines and other water boundaries are sharp. Highways under suitable conditions are clearly recorded due to differences in surface conditions and temperature. Ploughed and unploughed fields and ploughing patterns are clearly recorded in map form. This results from the compacted and grass-covered nature of one and the reverse conditions of the other (Harris & Woodbridge, 1964). Except for overall similarities the details recorded on thermal maps are different from the details recorded on photographs as the former represent infrared emissivity from the scene and the latter reflected solar radiation in the visible and near infrared spectra. Radar has also been studied and tested in the preparation of maps. Impetus is now being given to the study by side-scanning high-resolution radar.

(d) Digitized mapping of natural resource data

Parallel to the development of orthophotomapping is the development of a computer-based storage system of map information related to natural resources. Information is stored on magnetic tape in a form suited to rapid retrieval, rapid measurement and rapid comparisons of large numbers of maps and data related to areas, lines and points on the maps (i.e. location specific information). Such a system, the Geographic Information System, is now operated in Canada by the Department of Forestry and Rural Development (Tomlinson, 1968). Map boundaries and location-specific information related to land-use and the capability of land for agriculture, forestry, recreation and wild-life are stored in digitized form on magnetic tape. About three hundred conventional map sheets, showing an average of 800 distinct areas (map elements) can be stored on a single tape.

The procedure (Tomlinson, 1968), for digitizing and storing information recorded on a (source) map of land-use classes etc., is illustrated in the flow chart (fig. 12.3). The boundaries of each class (map element) are scribed in lines about 0·08 inches

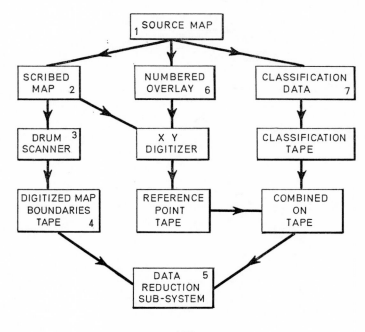

thick on a separate blank sheet. The scribed map is then placed on the surface of a drum, which rotates as it is slowly traversed by an electronic scanninghead. The scanning system is connected to a standard magnetic tape-drive and detects, within areas of 0·032 inches × 0·04 inches of scribed map, the presence or absence of a boundary line. Each time a map line or map point is detected, its X-Y-co-ordinates are recorded in digital form on magnetic tape. Finally the digitized map boundaries are combined with digitized data relating a numbered overlay of the map elements and specific information related to each map element (class).

PART FOUR
Basic Interpretation

Do not invent, observe
Auguste Rodin

13

STEREOSCOPY

STEREOSCOPY is essential to the photo-interpreter. It may be defined as the science and art of viewing two different perspectives of an object, recorded on photographs taken from nearby camera stations, so as to obtain the mental impression of a three-dimensional model of the object. Two photographs of ground objects taken in quick succession from an aircraft will provide a three-dimensional model of these objects when viewed through a stereoscope by the photo-interpreter. In the present study, it is principally the application of stereoscopy which is important. Of the facets of this subject, the ones requiring varying degrees of attention are the stereoscopic image, parallax, the stereoscope and testing for stereoscopic vision.

PROPERTIES OF THE EYE

The faculty of vision by a human eye involves the use of the eyeball, including the fovea, the retina and possibly the para-fovea, the optic nerve and the visual centres of the brain. Rays of light pass through the refracting media of the eye, similar to light passing through the camera-lens, and form an image at the back of the eyeball on the retina. The stimulus is then transmitted to the visual centres of the brain and co-ordinated with a similar perception from the other eye. The fovea is most sensitive to green light at 0.555μ whilst in subdued light the para-fovea responds most to light at 0.515μ (Middleton, 1952).

A single eye is able to determine accurately the direction of an object; but is only able to gauge distance qualitatively. This is illustrated by holding two pencils, with the points about a foot apart at nearly arm's length. and with one eye closed trying to bring the point of one in contact with the top or tip of the other. It will be found that the top of one pencil will pass usually behind or in front of the tip of the other pencil. If the exercise is repeated with both eyes open, it is easy to bring the pencils into contact. The reason is that a pair of human eyes possesses the faculty of *binocular vision* by which it is possible to obtain a conception of relief in space (i.e. a three-dimensional model). Wheatstone, in the mid-nineteenth century, showed that if two photographic images are placed in front of the eyes, so that a similar view of the image is seen by each eye, then an impression of the relief of the object is produced by the eyes in the form of a single model. To illustrate the relief effect of

141

binocular vision hold a pencil end-on a few inches in front of one eye. Then view it with both eyes. It will be found that it is only when it is viewed with the two eyes that an impression of its relief in space is obtained. It is interesting to note that visual acuity in stereoscopic work depends on a number of factors including the physiological state of the individual, which may be affected by fatigue, noise, lighting and mental depression (Nowicki, 1952).

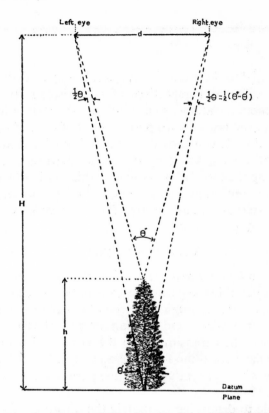

FIG. 13.1. Assume an observer in a helicopter at a flying height H above the ground sees a tree below him, then provided the disparity between the two angles of convergence (θ', θ'') is not excessive, he will experience an impression of depth for the tree (height: h) (see text). In reality the eye-base is extremely small compared with H.

An observer out of doors also obtains the impression of relief of distant objects, provided the objects are not excessively far. Let us consider an observer in a helicopter. If, in fig. 13.1, the eye base of the observer in the helicopter is d, the angle subtended at the top of the object viewed, a tree, θ'', and the angle subtended at the base of the same object θ', the difference in the angles of convergence is θ ($\theta'' - \theta'$). The angle of convergence is also known as *angular parallax* and *parallactic angle*. It is the angle subtended by the eye base at the object viewed. The *eye base* (d) or *interpupillary*

142

distance varies between about 2·0 and 2·9 in. for most Europeans. An average is 2·60 in. (65 mm) (Nowicki, 1952).

The minimum disparity in the *angles of convergence* (θ) for a person to distinguish between objects is in natural space about 20 seconds; but Pulfrich (1901) observed that for some people the angle is as small as 10 seconds. Later Fourcade concluded that the maximum is 2° for obtaining a stereoscopic image. Ogle in 1952 has shown that for quantitative estimates of depth the disparity at 6° from the fovea is up to 2°, but for qualitative estimates disparities up to $3\frac{1}{2}$° can be tolerated. Above $3\frac{1}{2}$° the eyes cannot form two photographic images into one relief model in space. Each eye behind a lens of the stereoscope observes the images recorded on the photograph from a different relative position, as it would when observing the object in natural space at a comparable relative distance. The maximum disparity varies according to the region of the retina stimulated.

The three-dimensional model using a stereo-pair of photographs is not formed where the convergent visual axes intersect, as was formerly thought; but, instead, the model is formed some distance below the table plane of a magnifying stereoscope. For example, a conventional lens stereoscope with a focal length of $4\frac{1}{2}$ in. and at $3\frac{1}{2}$ in. above the viewing table provides the image at about 16 in.

Some people, possibly 5%, do not have the faculty of stereoscopic vision and no amount of training in adulthood will give it to them. Unfortunately, there is no known physical aid to provide stereoscopic sight to a person who does not possess it naturally; but training can help those with weak stereoscopic vision. The interpreter, when beginning his day's work, will probably find that his stereoscopic vision improves for the first 20 to 25 minutes. A minimum practice period of 5 to 10 minutes is advisable before candidates are tested for stereoscopic acuity.

THE STEREOSCOPIC IMAGE

Let us now consider the same tree as recorded on a pair of stereo-photographs and examined under a stereoscope. The stereoscopic image of the tree will be formed by two photographs as shown in fig. 13.2. If dots g'_2, g''_2, g'_1, g''_1, are marked on the photographs at the top and bottom of the tree and the photographs are set up for stereoscopic viewing, then the fused dots (g'_2, g''_2) at the top of the tree will appear to float in space and above the fused dots (g'_1, g''_1) at the bottom of the tree. If the distance between the corresponding dots is measured, then it will be found that the distance between the dots at the top of the tree is less than the distance between the dots at the bottom of the tree.

The difference between these two measurements, i.e. (g'_1 to g''_1) − (g'_2 to g''_2) can be used for determining heights by parallax. The same effect of floating dots can be achieved by making two pencil spots ($\frac{1}{10}$ in. diameter) on a sheet of paper at the distance of the eye base apart or at the eye base less $\frac{1}{8}$ in. When viewed under a stereoscope these two spots will fuse into one; and if a further two spots are marked

L

about $\frac{1}{5}$ to $\frac{1}{10}$ in. closer together and about $\frac{1}{2}$ in. below the first dots as shown in 13.2b, these (g'_2, g''_2) will fuse and float above the first pair of fused spots (g'_1, g''_1). Within limits, the closer the pairs of spots are together, the higher the fused spots appear to float, and the wider the spots are apart the deeper the fused spots will appear to sink into the paper. This phenomenon provides the principle used in the construction of parallax bars, dots being permanently marked on glass slides and

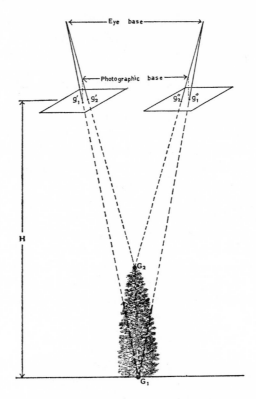

Fig. 13.2a. If instead of observing the tree from a helicopter (fig. 13.1) a pair of stereophotographs are taken and set up for stereoscopic examination (of the image of the tree recorded on each photograph under a stereoscope), the interpreter will obtain the impression of depth for the tree (in the form of a three-dimensional model).

the distance being recorded by micrometer screw-gauge. Nyyssonen (1955) commented that height of trees as measured by parallax can vary according to the colour and size of the floating marks. He found that red floating marks were difficult to use; and observed that the standard error of the estimate seems no greater when measuring by parallax wedge than by other methods.

As mentioned above, two dots viewed under a stereoscope can be made either to fuse and float above the stereo-model by moving them a little closer or to sink into the ground by moving them apart. If instead of moving the dots laterally two long rows of dots are marked on a transparent base in such a way that each two opposite dots are closer together, then by placing this V- or 'dot-wedge' on the stereo-pair of photographs a similar effect will be observed as the parallax wedge is moved at right-angles to the air base. Each pair of dots is usually 0·002 in. further apart (see fig. *16.3*—IV).

To make a reading, the transparent wedge is placed over the base of the image to be measured so that the left dot is at the base of the left-hand image and the right dot is over the base of the right-hand image. When the photographs are examined stereoscopically the dots fuse and float at the base of the image. The parallax wedge is then adjusted so that the fused dot floats at the top of the image. The parallax difference between the two readings is then determined from the side-scale on the

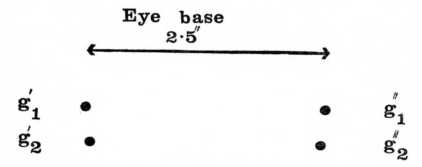

FIG. 13.2b. If the base and the tip of the tree imaged on the photographs in fig. 13.2a are replaced by spots (see text), then under the stereoscope spots g'_1 and g''_1 and g'_2 and g''_2 will be observed to fuse separately and float one above the other.

parallax wedge and substituted in the parallax formula to determine the height in feet or metres. In some wedges, lines are substituted for dots. When viewed stereoscopically, these form a wedge the apex of which provides the parallax reading.

PARALLAX

It is convenient now to introduce the parallax formula used in the determination of heights. In the United Kingdom, the term 'parallax' is often preferred to 'absolute parallax' or 'absolute stereoscopic parallax'. The latter are commonly used in the United States. The term x-parallax is also sometimes used.

Parallax is the apparent displacement of the position of an object (e.g. top of tree) in relation to a reference point. In fig. 13.3 it is assumed that the reference point is the centre of each photograph (i.e. principal point), the photograph having been taken with the camera axis in a truly vertical position as occurs when the film is parallel to the ground. On a pair of stereoscopic photographs, the position of the recorded object in relation to each principal point (p_1p_2) will vary due to a shift in the point of observation or camera station in the aircraft. The distance between the principal points $(p_1p'_2 \text{ or } p_2p'_1)$ as measured on the photographs is termed the air base; and the parallax (P) of a point common to the pair of stereoscopic photographs (i.e. *the absolute stereoscopic parallax*) is the algebraic difference, measured parallel to the air base,

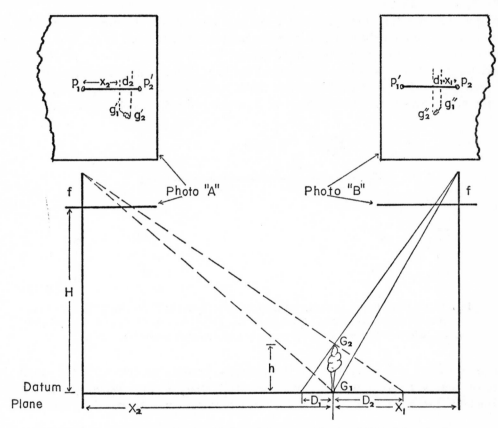

FIG. 13.3. Assume a plane flies from left to right at an altitude of H ft above the datum plane on which there is a tree G_1G_2 of height h. On the pair of contiguous photographs of the area, the tree will be recorded as $g'_1g'_2$ and $g''_1g''_2$. A formula, the parallax formula, can be derived for determining the height of the tree recorded on the stereopairs of vertical photographs (see text).

of the distance of the two images of the same object from their respective principal points. The object being measured may be a tree, as shown in fig. 13.3. If the tip of the tree is considered as point G_2. and the base of the tree point G_1 then from what has been said previously it will be expected that, in stereoscopic vision, point g_2 will float above point g_1, g'_2 to g''_2 being less than g'_1 to g''_1. Examination of two photographs stereoscopically will confirm this as being correct.

146

Instead of measuring these distances direct, the distances when using the absolute parallax formula are measured in respect of two known controls for all measurements, namely the principal points and the air base. The differences measured in this manner for the top of the tree and the bottom of the tree is termed parallax difference or differential parallax. The *parallax difference* (Δp) of an object on a pair of stereoscopic photographs is defined as the sum of the radial displacements of the top of the object from its base, measured parallel to the air base on each photograph.

In fig. 13.3 the location of a tree on the ground is shown in relation to the photographic station of the aircraft when the pair of photographs were taken. If it is assumed that the tilt of the camera is negligible and that the photographic scale is constant, then it may be shown that $h = H\Delta p/P + \Delta p$, where '$h$' is the height of the tree.

1. The parallax (P) by definition is: $P = x_1 + x_2 =$ Air base ($X_1 + X_2$) in level country, or if algebraic signs are considered $P = x_1 - (-x_2)$.

2. The parallax difference (Δp) by definition is:

$$\Delta p = d_1 + d_2.$$

3. From similar triangles:

$$h/H = \frac{D_1}{X_1 + D_1} \qquad [H + f \doteqdot H]$$

or:

$$h/H = \frac{d_1}{x_1 + d_1}$$

$$d_1 = \frac{h(x_1 + d_1)}{H} = h/H\,(x_1 + d_1).$$

4. Similarly

$$d_2 = \frac{h(x_2 + d_2)}{H} = h/H\,(x_2 + d_2).$$

5. Substituting in (2)

$$\Delta p = h/H\,(x_1 + d_1) + h/H(x_2 + d_2)$$

$$= h/H\,\{x_1 + d_1 + x_2 + d_2\}$$

$$= h/H\,\{(x_1 + x_2) + (d_1 + d_2)\}$$

or

$$h = H(\Delta p)/(x_1 + x_2) + (d_1 + d_2) = H(\Delta p)/P + \Delta p.$$

For many purposes the formula may be used as $h = \dfrac{H\,(\Delta p)}{P}$ if $h/H < 2\%$, since there are other factors affecting the accuracy which are greater than any error caused by using the adjusted formula. The parallax difference (Δp) will then be: $\dfrac{P\,(h).}{H}$

147

Sometimes the formula is written as $h = H \dfrac{(dP)}{b+dP}$ where the photographic air base (*b*) or mean of the two air bases is substituted for the absolute parallax (*P*). It will be observed that, ideal conditions excepted, $P \neq b$; and also that $P + \Delta p$ is the parallax of the top of the object. Often *h* and *H* are expressed in feet and *P* and Δp in millimetres. Although these are different units of measurement they can be applied directly in the formula. The absolute parallax is increased by longer air bases and parallax difference by shorter focal lengths. By increasing the ratio: $\dfrac{\text{flying height}}{\text{air base}}$ the intersection of the rays on a pair of stereo-photographs is at a greater angle, thus increasing the accuracy in preparation of the map base. The model is then termed a 'harder model'.

It is interesting to note that to provide a scale model for use in advanced instruments the $\dfrac{\text{eye base}}{\text{space image}}$ ratio should be the same as the $\dfrac{\text{air base}}{\text{flying height}}$ ratio (see Miller, 1958).

Johnson (1957) provided a mathematical proof that parallax difference smaller than 0·001 ft cannot be detected by average photo-interpreters. He suggested logically that instruments graduated in excess of 0·001 ft do not necessarily increase the accuracy of the measurement of height. This provides weight to the argument that simple instruments measuring to this degree of accuracy may be just as satisfactory as more complex instruments having a greater precision of accurate measurement. Thus a parallax wedge reading to 0·001 ft should be as satisfactory in practice as a parallax bar reading to 0·1 mm (i.e. 0·004 in.).

Focal lengths of the stereoscope and camera, and magnification produced by the lens of the stereoscope, influence the relief exaggeration when stereoscopically viewed. The lower the magnification and the longer the focal length of the stereoscope the greater is the stereoscopic exaggeration, but details are less conspicuous and eye-strain is increased. A camera with a short focal length will provide greater stereoscopic exaggeration than a longer focal length, e.g. 12 in. A wide-angle lens with a short focal length (e.g. $3\frac{1}{2}$ in.) and large scale photography will give for tall trees a difference in the angles of the convergence in excess of $3\frac{1}{2}°$ over much of the photograph; and thereby make it impossible for the eyes and brain to fuse the two complete images into a single three-dimensional model. For a given focal length, the greater the flying height the less will be the stereoscopic relief for the same scale photographs. The stereoscopic exaggeration will also increase towards the edge of the photographs. When examining photographs of eucalypts over 150 ft tall taken with both an $8\frac{1}{4}$ in. lens and a 6 in. lens, it was found impossible to obtain a single model of many of the tree images from photographs of the shorter focal length at a scale of 1/7,920.

Pseudoscopic views. Two items of interest involving the application of stereoscopy also require a brief description. These are pseudoscopic views and anaglyphs. In viewing terrain on aerial photographs a reversal of the relief is sometimes obtained by the eyes. Such a phenomenon is known as pseudoscopic illusion. This is illustrated in fig. **13.4**. Are these lines water courses? The answer is no. They are actually the crests of sand-dunes in the Simpson desert, Australia. Pseudoscopic illusions can also be obtained if two photographs, having been flown right to left, are placed in the opposite order, left to right, for stereoscopic viewing. A volcanic crater may give the impression of being a volcanic cone. Viewing so that the shadows fall away from the observer can also result in depressions seeming to be elevations and vice-versa. Photographs are normally viewed under the stereoscope with the shadows towards the interpreter.

Anaglyphs. An anaglyph is a picture resulting from the printing or photographic projection of two nearly identical images of a stereoscopic pair in colour, e.g. red and blue. When the images are viewed in these colours, using spectacles with one lens of each colour, a stereoscopic image is formed. The observer sees the red image through the red lens and a blue image through the blue lens. The images recorded by the eyes are then fused by the brain into a relief model in space. The principle of anaglyphs has been used in the Multiplex by projecting the black-and-white diapositives through blue and red filters on to the mapping table.

STEREOSCOPES

These may conveniently be grouped into three basic types, namely lens, mirror and prism. By far the most popular is the *lens stereoscope*, which is also called a hand stereoscope or a pocket stereoscope. It comprises two semi-convex simple lenses mounted in a frame a few inches above the photographs (fig. *13.5*). Magnification is usually 2, 4 or 6 times. In some models, the eye base is adjustable between 2·2 in. and 3·0 in. (55 mm to 75 mm). The price varies from £2 to £5. The principal disadvantages are that only a small section of the photograph can be viewed at a time and, except for a narrow strip, viewing requires flipping up the side of one photograph.

Abrams and Zeiss have developed the pocket stereoscope into a *bridging stereoscope* so that flipping is not necessary. Further modifications include the addition of a parallax bar fixed to the stereoscope as in the Abrams' model; and the replacement of the simple lenses by binoculars. Rhody (1962) in conjunction with Wild S.A. developed a *'zoom' lens stereoscope*, or as he termed it 'un stéréoscope variable'. Several manufacturers are now producing zoom stereoscopes at prices between £1,500 and £2,000. The interpreter will find that maximum magnification is limited by the photographs, being possibly ten times for prints and twenty times for diapositives.

For viewing a pair of stereoscopic photographs with a hand stereoscope, the best position as determined by the distance between the images is about ¼ in. to ⅜ in. less than the eye base, or can be determined more precisely by the optical formula for a lens stereoscope. By increasing slightly the distance between the centres of the photographs, the apparent depth of the stereoscopic image can be increased. The optical formula is: $\eta = SUD$, where

η: the separation distance between a point common to the two photographs,

S: distance between the optical centres of the lenses; and is independent of the eye base of the interpreter,

U: distance between lens and the photograph,

D: power of the lenses in diopters.

The second type of stereoscope uses mirrors to supplement the simple lens, so that a relatively large area of each pair of photographs is viewed. This avoids frequent moving of the photographs and flipping up the side of the photographs. As in the Zeiss and Wild mirror stereoscopes (fig. *13.5*), viewing is usually through 'binoculars', although viewing can be carried out without the 'binoculars'. As the silver of the mirrors is on the outer surface of the glass to avoid refraction, it can easily be damaged by careless fingering.

A separate parallax bar often forms an accessory to the mirror stereoscope. A mirror stereoscope with 'binoculars' forms part of several other instruments, e.g. the Zeiss Stereopret. In the Old Delft Scanning Stereoscope (fig. *13.5*), viewing is achieved via mirrors and 45° rhomboidal prisms. External levers control the positions of the 45° prisms and internal mirrors for scanning. The photographs are not moved during the course of scanning. In the Hilger-Watts mirror stereoscope, the parallax bar is replaced by two pin-head light sources located behind the semi-transparent silver surfaces of the mirrors. When the photographs are viewed, a single point of light is seen in the stereo-model after correcting for parallax by adjusting the micrometer-screw of one light source.

For viewing colour transparencies stereoscopically through the stereoscopes described above, it is necessary to provide illumination below the transparencies in the form of a light table. The brightness of the fluorescent lighting and the illumination table on which the transparencies are placed should be similar or the same as that used initially in processing. Often viewing is improved if a black paper mask is placed over the photographs except for the small area under close scrutiny as this reduces glare and eye-strain. Of course, colour prints are examined as black-and-white prints; but as mentioned in Part II, they are more expensive.

TESTING STEREOSCOPIC VISION

Before commencing to study contiguous pairs of photographs under a stereoscope the novice should test his sight for the faculty of stereovision. In addition, if he is intending to view colour transparencies, he should have his sight checked for colour blindness.

As a first step prior to assessing one's faculty of stereovision, it is often helpful to practise looking at a pair of stereo-photographs on which a hill or building rises abruptly above the ground datum. Monocular relief due to shadows should not be excessive, as the viewer who lacks stereovision may think that what he sees, magnified and shaded, is the stereoscopic model.

The next step is to assess the ability of the candidates to see stereoscopically. Two tests have been selected. Moessner's *floating circle test* (1954) is considered to be ideal. This consists of two banks of 160 circles (fig. 13.6a); each circle is approximately 0·03 in. in diameter. Twenty-five of the circles selected randomly in the left bank have been shifted slightly to the right by distances varying from 0·045 in. at the top to 0·0005 in. at the bottom. The test is to ascertain which circles appear to be floating. Being able to select the floating circles down to a parallax difference of 0·002 in. provides a rating of 80%, which can be considered as satisfactory. An advantage of this test for the novice is that it is relatively easy for him to concentrate his stereoscopic viewing in the small area of a circle. It is thus an exercise as well as a test.

Stereogram (Lens separation – 2·25)

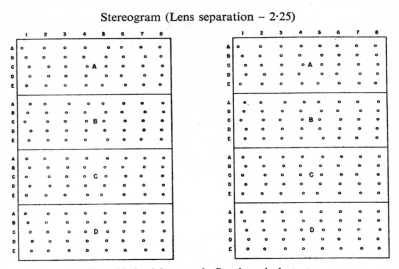

FIG. 13.6a. Moessner's floating circle test.

Finally, the Zeiss *Pulfrich test* (fig. 13.6b) should be tried, which requires the designs to be placed in sequence of depth. A suggested allocation of marks is as follows, although a fully satisfactory scoring is not possible.

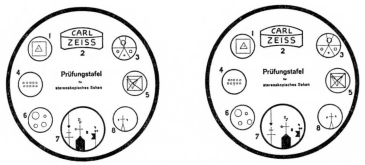

FIG. 13.6b.

  (i) a score of 90% or more is to be considered excellent,

 (ii) 85% to 90% good,

(iii) 75% to 85% satisfactory,

(iv) below 70% a 'fail'.

(Answers are provided in Appendix I).

Currently, the International Training Centre in Holland is using a modified Howard-Dolman 'three rods test' (Zorn, 1965). Basically, the instrument comprises three needles, one of which is movable; and the candidate has to decide, during a series of settings of the movable needle at about 4 m, whether it is in front or behind the two fixed needles.

After the student has been tested for stereoscopic vision and colour blindness, it may be advantageous to introduce him to photo-interpretation by his studying sets of stereograms of the type outlined by Sims & Hall (1955) as *photo-reading*. They describe tests/training in photo-interpretation, i.e. air-photo to air-photo and air-photo to air-map. The writer's own *photographic assembly exercises* at the School of Forestry, Melbourne, are useful in teaching the student to associate his training in forestry, geomorphology or ecology with what can be detected and analysed on the photograph. These may be prepared from local photographs, which are cut up into rectangular or square sections of say 2 in. by 2 in. The candidate is required to reassemble the pieces to form a photograph.

14

GENERAL PRINCIPLES

INTRODUCTION

THE aerial photograph is a permanent record of visible living objects on the ground and of their association with each other. It is also a graphic record of biomass and a tonal record of reflected solar energy; but until research into these latter attributes of the photograph are undertaken, it will not be possible to approach ecological interpretation through the broader perspective.

In the introduction, aerial photo-interpretation was defined as the art and science of studying, identifying and evaluating the significance of the images recorded on the emulsive surface of the aerial photograph. As needed, the evaluation is recorded qualitatively and quantitatively. Photo-interpretation requires keen observation, a logical mind and an advanced training in the field science for which the photographs are being used in order to analyse and to synthesize fully the available data. It is highly desirable that the chief interpreter should know the objective or objectives of the interpretation before commencing a new project. Deductive reasoning, based on training and special regional knowledge, is particularly helpful. For example, the knowledge that an area has been glacier covered may help in recognizing hillocks on the photographs as being either eskers or drumlins. If the hillock is serpent shaped, then geomorphic training will suggest to the interpreter that it is an esker and as such that it contains sand and/or gravel. The term 'special reference level' has been used as describing a specialized knowledge of processes (Vink, 1964). The interpreter should acquire a full understanding of what he observes in the images.

It is also advantageous to recognize photo-interpretation as being of two intensities, namely policy level and management level. In forest studies, small scale photographs (e.g. 1/30,000) are sometimes satisfactory for *policy interpretation*. Forest maps prepared from small scale photographs will normally provide adequate information for top-level management for such purposes as location and size of forest areas, feasibility of access, influence of topographic conditions, mill and camp locations. At management level, 'policy' maps will normally be found to be useful but inadequate; and it is necessary to replace or supplement these by new management maps prepared from specially flown photographs at a larger scale, e.g. 1/8,000 to 1/20,000. Data required in management interpretation include the identification of tree species and photo-communities, volume estimates of forest stands, stocking estimates and succession, terrain classification and site quality assessment of plantation species. In the writer's opinion, failure to recognize these two functions of forest interpretation may lead to excessive study and evaluation of detail in what should be a policy-mapping project; and in consequence overall costs per unit area will be very high for what is actually required. The high costs may then be used by those adverse to the application of photographs to a project as a reason for not extending the aerial coverage to new work.

The procedure recommended in photo-interpretation, particularly at management level, is as follows: (a) scan the photographs monocularly and/or stereoscopically, (b) select stereo-pairs of photographs representative of the types of vegetation being evaluated, (c) check these types in the field against the photographs, (d) delineate the types on all the photographs, and possibly take a representative sample of the photographic types and check these in the field to ensure typing has been carried out correctly, (e) adjust previous typing and complete all typing, (f) plan and carry out whatever field survey is needed for obtaining additional data.

Thus for a forest inventory the forest will have been broken down into its tree-types on the photographs in conjunction with ground reconnaissance, the area of each stratum assessed and a decision reached as to the number and the location on the photographs of sample plots in which field work is to be carried out. When scanning photographs as step (a), all photographs are scanned provided the number is small. If several hundreds or thousands are to be typed, then it is often convenient to inspect the photographs during a print lay-down, and later as step (b) either to select systematically a percentage after randomizing the first pair or randomly to select a percentage of stereo-pairs. From this percentage, stereo-pairs are selected which are judged to be representative of all the pertinent forest types recorded on the photographs.

As mentioned in the introductory chapter, for photo-interpretation measurements are normally a means to an end. One consequence of this is that instruments and mapping techniques are much simpler than in pure photogrammetry. There is little purpose in having an extremely accurate map if its end-use is as an aid only to other factors of a much lower standard of accuracy and if the overall accuracy cannot be improved by having a highly accurate map. For example, in forest inventory if the expected accuracy of the volume assessment is $\pm 10\%$, the same accuracy may be acceptable in preparing the map, provided the errors are not accumulative. In all quantitative studies it is necessary to separate the systematic and random errors.

Before proceeding to consider common diagnostics, a few words of warning may be needed on use of the single aerial photograph. Sometimes a person not familiar with stereoscopes will state that one is able to interpret photographs efficiently non-stereoscopically; and as an illustration may draw attention to work being done using a mosaic or single photographs. Such a statement is usually erroneous. In the first place it may be that the use of a mosaic is being confused with the use of stereo-pairs of photographs. It will be recalled that a mosaic is used as a map to formulate general impressions. Secondly, by ignoring the use of stereoscopic photographs one is failing to use a very valuable tool, namely assessment of height. In fact in certain facets of interpretation, e.g. forestry, the assessment of height by parallax is usually essential. It is interesting to note that Colwell (1954, 1960b) has judged the effectiveness of interpretation to depend primarily on three characteristics of the image, namely parallax, tone and sharpness.

Another way of viewing the significance of height and hence parallax is to consider it statistically as one major factor or variable which when combined with other major diagnostics or variables provides an increased number of permutations for photo-interpretation.

(a) Common diagnostics

The most important diagnostics are (i) shape, (ii) shadow, (iii) tone or colour contrast, (iv) pattern, (v) texture, (vi) site, (vii) association, (viii) area, (ix) height (see also Colwell, 1952; Olson, 1960; Spurr, 1960; Loetsch & Haller, 1964). The last two will be discussed separately at a later stage (Chapters 15 and 16). The others are discussed below.

(i) *Shape.* This relates to the configuration or the general outline of the object as recorded on the photograph. Some objects can be recognized on the single photograph, especially if the shadow outline can also be examined, but it is normally the three-dimensional view which is so very important. Frequently objects have unfamiliar shapes when viewed vertically downwards at the centre of the photograph and the shape may change again when viewed towards the edge of the photograph or when viewed at greatly varying scales or at different focal lengths. Until the student has familiarized himself with the characteristic shape of common objects at different viewing angles, he often finds interpretation tedious and difficult.

Shape is important in recognizing ground features and tree species. As an example of the former, refer to fig. 18.6. This shows a characteristic ox-bow lake found along the River Murray, Australia, being obviously recognized by its shape. As regards tree shape, there are numerous examples (see fig. 22.4). Two of the commonest are the cone-like crowns of young hardwoods and most conifers and the dome-like crowns of many mature hardwoods in Europe. Recent work in Canada, particularly Ontario, has high-lighted the importance of tree shape both at the centre and towards the edge of the photographs. Tree age, disease, insect infestation and crown-fires may result in a change of crown form. For example, due to the production of epicormic branches in eucalypts, the height-growth of the crown above the ground is temporarily reduced and the green crown is lengthened. The recognition of past fire damage can be important. At Mount Horror, the volume per acre was significantly higher for *E. regnans* than *E. obliqua*, this being attributed to less serious incidence of fire damage and better site in the moister localities (Lawrence, 1957). According to species, the shape of the crown may vary greatly with age. Thus Scots pine (*Pinus sylvestris*) in the New Forest, Hampshire, has a conical crown in early life and a flattened crown towards maturity.

(ii) *Shadow*. This is closely associated with shape. For the study of open grown forests, deciduous forests when leafless and some conifer forests, particularly with snow on the ground, shadow is frequently an important tool (fig. 22.5). Narrow crowns, including the cone-shaped crowns of conifers, do not resolve themselves fully on photographs; and measurement of the tree height may be more accurately determined from the shadow. This subject is further discussed in Chapter 15.

Normally fairly short shadows are preferred. When examining shadows, account must be taken of the time of the year, time of the day, latitude and degree of slope of the ground. Shadow is helpful in identifying many small objects. As early as 1947, Krinov concluded from reflectance studies at ground-level that the reflectance of soils and sands are very complex, depending not only on their structure but on the shadows produced by the micro-relief and the orientation of the relief in relation to the direction of the sun's rays. He observed that the reflectance was 200% or 300% greater at 270° to the zenith than at 0° and 90°, this being partly due to shadow. Francis (1957a) drew attention to the fact that in northern Scandinavia winter photography is impracticable due to the long shadows thrown by the hills and trees. In East Africa, the identification of the almost crownless *Acacia pseudofistula* (gall acacia) and of the palm-like *Borassus palmifera* was possible by their characteristic shadows (Howard, 1965).

(iii) *Tone and colour contrast*. Tone refers to the various shades of grey observed on black-and-white photographs (fig. *16.3*, tonal scale), being often used in delineating photo-communities/forest types at smaller scales. For example, in eastern Canada, *Sphagnum* sp. is characteristically a darker tone of grey than the tamarack trees growing in it and spruce is a darker tone again than the bog association. Colour contrast refers to differences in hue, value and chroma and provides up to 200,000 observable differences (Becking, 1959).

The causes of tonal difference were discussed in Part I. Of the factors influencing tone, light reflectance and haze cannot be controlled by the photographer; but reasonable control is provided by selecting films of the correct sensitivity and filters to control the quality of light entering the camera. Closely associated with tone are tonal contrast and edge gradient. These were also discussed earlier in Part I.

Consensus of opinion varies concerning the faculty of perceiving tonal differences. Some workers consider a 2% difference in tone can normally be distinguished. Kodak market a tonal scale with 10% differences in tone. Losee (1951) used two observers, one an experienced interpreter, to see what monochromatic tonal differences they were able to observe. Using a circular aperture, 0·2 in. diameter, he concluded that, in order to obtain 95% success in identifying tonal differences, the experienced observer required a difference in reflection density of 0·21 or more and the inexperienced observer a difference of 0·31. It was assumed that an unexposed Velox number 2 paper had a reflection density of zero. Using a tonal scale for comparison, the inexperienced observer was able to reduce his requirement for a tonal difference to nearer that of an experienced operator at 0·21.

In autumn photography tonal differences are often accentuated. For example, in Canada (Losee, 1951), the following differences were recorded in a single set of photographs: 1·25 (black spruce), 0·84 (white spruce), 0·80 (aspen), 0·52 (red and white pines), 0·4 (tamarack) and 0·32 (birch). For tropical evergreen species examples can also be quoted. For instance, using panchromatic film and a minus blue filter, betelnut (whitish) can be distinguished from coconut (dark) on photographs taken in the Pacific area. Mango trees are always blackish (Howard, 1959).

The following general observations apply to tonal differences. In northern temperate regions, using an infrared film and a deep red filter, hardwoods (angiosperms) are recorded in light tones on the photographs and conifers (gymnosperms) in dark tones. Tones perceived on the photographs frequently do not correspond with tones observed visually either on the ground or in the air. Wet surfaces normally appear darker than dry surfaces on infrared and panchromatic photographs. Krinov (1947) concluded that the reflectivity of wet surfaces and *Sphagnum* was about 2·7 times less than that of dry material. A dark smooth bitumen road may appear light on panchromatic photographs due to the fairly high reflectivity of the surface. Dirt roads and tracks appear very much darker when wet than when dry. Sub-soil when exposed usually appears light on panchromatic photographs compared with the surrounding living vegetation; but will probably record darker on infrared photographs. Dry grass, snow and sandy soils record whitish. With a suitable film–filter combination insect damage and recently burnt ground vegetation appear darker than undamaged areas. In areas with sparse vegetation, and even in closed forest (e.g. Australian sclerophyll), the tone provided by the fire pattern on the ground is often conspicuous for several years. The photographic tones and textures provided by fire may be misleading when interpreting photo-communities. Location of an image relative to the principal point can also influence tone. Trees of the same species may appear progressively darker in tone the farther they are from the principal point. Schulte (1951) noted that trees in Canada recorded progressively darker from spring to summer; and that certain species on windy days, due to the exposure upwards of the dorsal surface, may record lighter than on a calm day. It is also known from optics that a small grey object appears much darker on a white background than on a dark background. In interior Alaska, infrared photographs at 1/5,000 were used to separate the one commercial softwood from the three commercial hardwoods by tone, but the species of pole and saw-size trees could also be identified by their crown shapes (Haack, 1962). Separation of the species by tone was also possible on panchromatic photographs. Losee (1953) in Ontario, using 1/1,120 photographs, found crown characteristics better than tone.

Colour contrast in colour photography is equivalent to tone in black-and-white photography. A slight change in colour is usually more readily recognized on colour transparencies than the difference in tone on achromatic photographs. Colour photography is helpful in species identification; and colour contrast in the infrared is important in separating dead and living matter, in studying plant vigour and detecting disease and insect damage.

(iv) *Pattern*. This is associated with local geology, topography, soil, climate, plant formations and associations and human activity. Pattern is a macro-characteristic used to describe the general spatial arrangement of the images. Particularly in studying land-use, it is important to understand and to recognize patterns produced by man's activities (fig. **14.1**). In tropical areas, a difference in the vegetative pattern often records the past activity of man long after he has abandoned the area. Thus, a village site abandoned 30 years ago near Tabora in Tanganyika was identifiable by the pattern provided by the absence of tree species, Borassus palm excepted, and the presence nearby of tree regeneration still in the 'thicket stage'. Frequently the mangrove species can be identified by pattern, provided the interpreter is familiar with the local conditions. Pattern often shows up well on mosaics, being strongly influenced by geology. In fig. **17.5**, the poor drainage on basalt provides a distinctive pattern; and should be compared with the well-developed drainage pattern on the Ordovician sediments.

(v) *Texture*. Texture may be considered to be a micro-characteristic corresponding to pattern as a macro-characteristic. Texture was described by Colwell (1952) as the frequency of tone change within the images on the photograph. It is produced by an aggregate of unit features too small to be clearly discerned individually. It is the product of tone, size, shape, pattern, shadow and reflective qualities of the object and varies with the photographic scale. As the scale is reduced, so the texture becomes finer; and what may be considered as pattern on larger scale photographs provides texture on very small scale photographs. On abnormally large scale photographs, e.g. 1/2,000, individual groups of leaves contribute to texture. At 1/5,000 leaves and branches contribute to the texture; at 1/10,000 tree crowns provide the texture; at 1/30,000, it is the trees and groups of trees and at 1/80,000 groups of trees and entire woods and plantations contribute to texture.

A fifty-fifty division of black and white into extremely small units provides a fine texture. The larger and more irregular the units, the coarser the texture will be. Several examples will be quoted. At normal scales, willows, as they have small narrow numerous leaves, provide a smooth texture when photographed. Immature oaks, having bunched leaves, provide a coarse texture and old stag-headed oaks, having only a few main branches, give a very coarse texture. Eucalypts, as the leaves are commonly bunched towards the ends of branches on older trees, also provide a coarse texture. Mature *Eucalyptus delegatensis* provides a texture probably best described as woolly. Mangoes (*Mangifera indica*) provide black dome-shaped spots which tend to break up or distort the texture provided by other plants. *Eucalyptus regnans*, spruces, firs and young pines, having pointed crowns, provide a distinctive texture on small scale photographs. The texture of the ground surface can be greatly altered by ant-hills in the tropics (fig. **22.1**). In Australia gil-gais at small scale provide a characteristic texture. At certain scales, texture is also provided by shrubs and saplings, maturing grass species, soil cultivation and agricultural crops.

Density scales (fig. *16.3*), as described later, are closely associated with texture. They are commonly used in forestry to estimate the stocking of trees per unit area by crown closure. Density scales are also used in social studies and wild-life surveys (e.g. water fowl on a lake surface).

(vi) *Site*. This term has two distinct meanings. Firstly, it is widely used in the study of aerial photographs to describe the location of the image under observation in relation to its surrounding features. Its more restricted meaning connotes the sum of the factors of the environment influencing plant growth. By studying site from the quality of the forest stands on the aerial photographs, some key factors of the environment can often be appraised; and conversely, by recognizing site quality, the interpreter may be able to eliminate the occurrence of certain species. Tree growth, conveniently measured as height in relation to age, is a function of local climate, soil conditions and topography; and is often used to express site quality. Harrison and Spurr (1955) observed that site classification by topographical differences has proved valuable in some areas.

Referring back to site as implying the location or position of the image, it is important always to consider the image in relation to its surroundings. For example, an open space by itself is probably impossible to identify, but the presence of certain urban development in its vicinity, and possibly the association of cars on the space, immediately suggests a car park. Similarly, in the country but near a town, a large open space which is T-shaped or X-shaped would suggest an airfield. A detailed examination of the area for other objects, e.g. aircraft on ground, helps to confirm this suggestion. Similarly site is an essential characteristic for the ecologist and forester. For example, in fig. 18.8 (Tabora, Tanganyika) several types of vegetation are represented from the water course at A to the highest ground at B. A person familiar with the ecotypes will know that the site of the *Brachystegia–Julbernardia* woodland is above the herb-line. Similarly in Victoria the interpreter is able to locate the boundary between *E. pauciflora* (var. *alpina*) and *E. delegatensis* by knowing that the former is sited at elevations above the latter and by the former being the common short-stemmed tree species immediately below the tree-line and associated with elevations between 4,500 and 6,000 ft.

Aspect, topography, geology, soils and the characteristic natural vegetation are all important factors when examining site. The first three may be classed as macro-characteristics, and the last two, with soil moisture and tree size, may be considered as micro-characteristics. The relative importance of each varies with local conditions. Thus at high latitudes, and in many parts of Australia, aspect is probably the most important. In tropical Africa elevation and rainfall probably take precedence. In maritime western Europe, geology, aspect and elevation are important. In photo-interpretation, aspect, slope, land-form and the natural vegetation provide the key factors in the preparation of a site map from photographs.

(vii) *Association*. This is another term with two distinct meanings. In its ecological context it refers to a plant community of definite floristic composition, presenting a uniform physiognomy and growing in uniform habitat conditions (B.C.F. terminology, 1953). In general aerial photo-interpretation, the term has a much wider meaning (e.g. Churchill & Stitt, 1955).

Continuing from the examples and discussion on site, it will be appreciated that some objects are so closely associated with others that each helps to confirm the presence of the other. The term correlation is sometimes used instead of association. Frequently only one class of object is discernible on the photograph; but its shape, tone, pattern, texture, area, height and/or site are associated with another class of object not recorded or not clearly recorded. By studying one or more of these characteristics which have been observed to be associated with the object not clearly recorded, characteristics of the latter can be evaluated. Thus the forest types near Tabora (Tanzania) are so closely associated with the soil types that, with the aid of a catena diagram, the soil types can be delineated by the boundaries of the forest types as seen in stereovision. On photographs of the *Eucalyptus regnans* formation with an understorey of temperate rain forest in Tasmania, the rain forest was discernible partly by its association with open-grown mature *E. regnans*. Under these conditions the individual *E. regnans* could readily be counted at a scale of 1/24,500. Near Tabora, identification of forest types was used to delineate the best tobacco-growing soils. In eastern Canada, black spruce may be often identified and delineated by knowing that the species is associated with *Sphagnum* bog, which is readily located on photographs.

In Germany, Totel (1960) recorded that site quality can be interpreted from aerial photographs using crown-diameter sizes of even-aged stands as the association or criterion. As early as 1928, Zieger observed in central Germany that *Pinus sylvestris* (Scots pine) with shorter boles and wider crowns was associated with poor sites. Tests in Sweden have indicated that the estimation of site quality from tree height and crown diameter provides about twice the standard error as when height-age tables can be used (Spurr, 1960). J. R. Dilworth (1959) observed that for Douglas fir, site affects the relationship between crown diameter and diameter at breast height; but for long-leaf pine E. W. Johnson (1958) considered that there was no correlation between site and crown diameter/height. As the latter is possibly a function of the species in relation to its varying environment, each locality should be considered separately, until more research has been carried out on these relationships.

In Victoria, *Eucalyptus camaldulensis* of higher site quality is associated with the flood plain of the Murray. This association can be inverted for use by the hydrologist to give him a useful tool, namely that where the higher quality red gum occurs near the Murray one is in an area liable to flooding either annually or every two or three years.

In the red gum forests, the following features on photographs were listed by Davies (1954) as being associated with site quality:

160

PRELIMINARY AIDS TO PHOTO-INTERPRETATION

(a) Locality—high quality in seasonal swamps and in river bends.

(b) Crown cover—usually the greater the crown cover the higher the quality.

(c) Ground tone—this reflects the ground flora, which is related to water conditions and site quality.

(d) Crown form—more regular, denser crowns on better sites.

(e) Stand height as expressed by the ratio of $\dfrac{\text{diameter}}{\text{stand top height}}$.

In Tasmania, the height classes of old trees in regrowth areas were accurately determined from stereo-pairs of photographs at 1/24,000 scale; and were used to provide an index of the forest potential of the site.

(b) Preliminary aids to photo-interpretation

A number of aids will be briefly commented on. However, before doing so it may be worth while to comment on boundary delineation. It is often convenient to assume that forest types and other vegetative groupings have recognizable boundaries on the stereo-pairs of photographs. The boundaries so delineated will be found not to coincide exactly with the boundaries as observed on the ground but only to approximate to them. It is the process of drawing a line which is ideal. Vink (1964) has referred to the technique as 'idealization'. Frequently the process is used when mapping planimetrically broad or indeterminate man-made boundaries, e.g. hedges, ditches.

Some preliminary aids of use to the photo-interpreter include geographical maps, forest stock maps, ecological keys, stereograms, classification codes and catena diagrams. Aids used in actual measurement are considered separately in chapters 15 and 16.

Geographical distribution maps showing the normal range of plant species, of forest types, of crops, etc., are helpful in equipping the interpreter with local knowledge, enabling him to eliminate certain types or species and also to substantiate his interpretation as being correct. Thus in Malaysia, it is possible from local maps to locate areas of mangroves, lowland forest, rubber and palm-oil areas, rice-growing areas, etc. Ecologists, foresters and soil scientists are normally concerned with the observation of and mapping of detail beyond the information contained in a geographic map. *Terrain maps*, if available, are useful in showing physiographic units which by subsequent field work may yield some soil and ecological associates.

Forest stock maps. These show the distribution of different forest or stand types which have a bearing on forest management, with information on species, age, etc. (B.C.F. Terminology, 1953). For example, the U.K. Forestry Commission provides stock maps by species of compartments planted. When preparing a working plan of a typical area in West Wales, it was expedient to check the photographs against the stock map. After a few minutes' study of the photographs, it was usually possible to identify the trees on the photographs by species, by using texture, tone, size and the stock map data, and to delineate areas where the trees had failed to grow and to correct the boundaries for errors in the existing maps.

To attempt to associate the forest tree species with the ground flora could be misleading as sometimes the planting of two or more species does not follow the boundaries of the plant communities. For example, Japanese larch might have been planted both on *Deschampsia flexuosa* dominated slopes and part of the *Festuca rubra, F. ovina* topland. In the Spanish national inventory (1966), 1/30,000 photographs were

TABLE 14.A

East African vegetation key

1a. Upland sites (hills and hillsides and other well-drained areas), 2.
1b. Lowland sites (valleys and flood plains), 7.
2a. Trees, shrubs or bushes present, 3.
2b. Trees, shrubs and bushes absent or very widely spaced. GRASSLAND.
3a. Ground completely obscured by trees, shrubs or bushes 20–200 ft high, 4.
3b. Ground or grass visible between trees, shrubs and bushes up to 50 ft high, 6.
4a. Open or barely continuous cover of trees which form a single layer not more than 50 ft high. WOODLAND.
4b. Continuous cover of trees and shrubs 20–200 ft high, 5.
5a. Continuous, uneven cover of trees 40–200 ft high with crowns touching, intermingling and overlapping. Layers of different heights often visible, individual tall trees ('emergents') common. FOREST.
5b. Continuous, even, interlocking cover of trees and shrubs seldom more than 40 ft high. THICKET.
6a. Tree–shrub–grass mixture, trees and shrubs from 20–50 ft high cover more than half the area. WOODLAND.
6b. Tree–shrubs–grass mixture, trees and shrubs up to 50 ft high cover less than half the area including low bushy growths of no apparent height. SAVANNA AND SCRUB.
7a. Trees absent or widely spaced, 8.
7b. Trees present, usually abundant, 9.
8a. Herbaceous vegetation of no apparent height on seasonally waterlogged sites. SEASONAL SWAMP.
8b. Herbaceous vegetation up to 15 ft high on permanently waterlogged sites. HERBACEOUS SWAMP (e.g. Papyrus).
9a. Scattered or continuous tree or palm cover 10–150 ft high. TREE SWAMP.
9b. Even-banded tree cover 10–100 ft high in tidal areas, water often visible. MANGROVE SWAMP.

(Table prepared by I. Langdale-Brown)

successfully used in conjunction with old stock maps to identify species and to amend existing boundaries.

Ecological keys. An ecological key is a device designed to aid the photo-interpreter in the rapid and accurate identification of the images recorded on the photographs (e.g. Liang *et al.*, 1951). Both selective and dichotomous keys are used, although the latter are preferred in ecological work. A *selective key* gives a brief written description

162

with or without accompanying photographs of the principal ecological types likely to be encountered, and from these the photo-interpreter chooses the most appropriate. Table 20.A is an example of this type of key. A photo-interpreter's *dichotomous key* (table 14.A) or elimination key (Losee, 1942) (table 14.A) fulfils a similar function to a taxonomic key as used by the botanist. It may start with a separation into two groups, for example wild vegetation and cultivation, and then proceeds to divide these into regional groups before dividing them further into what might be termed their specific groups.

The writer considers it helpful to group keys into two overall types, namely *policy keys* and *management keys*. The former may be defined as a key suitable to use without field visits, to help in a general evaluation of the overall vegetation and to aid in the delineation of these general types on the maps prepared from the aerial photographs. Maps of East Africa, prepared by the Directorate of Overseas Surveys, London, fall into a category prepared with this class of key. Many keys used and maps prepared by military intelligence are also of this type. Management keys on the other hand are prepared for a specific local function, and usually require to be used in conjunction with frequent field visits and/or local ground survey.

Keys fulfil three important functions. They serve as a reference file to the trained interpreter, so that he can carry out the interpretation with a minimum of field visits. They provide a means of standardizing the classification by different interpreters at different places and at different times. Finally keys serve as a useful training device for students by helping them to recognize and to classify objects correctly.

To obtain the maximum value from keys, they should provide (a) a brief description in words of important recognition features; (b) a stereogram illustrating the features and (c) when possible one or more terrestrial photographs of the type of features covered by the stereogram. It is recommended that the location of the ground photographs should be tied in with the key. Ground photographs are not widely used, which probably results from the fact that ecological keys were developed during World War II, when ground photographs were not readily obtainable. *Stereograms*, associated with keys, are important in identifying ecological types and forest types suitable for logging, etc. Experience indicates that stereograms should have been taken in the same season as the photographs and be within 1/2,000 to 1/5,000 of the photographs to be interpreted in order to achieve their full usefulness. This is applicable to photographs at scales of about 1/10,000 to 1/30,000. Stereograms for forest studies are normally classified according to species and possibly divided into volume classes or height and/or basal area classes. If there is a choice of scales for providing the stereograms, then a second set at 1/20,000 to 1/50,000 is useful to provide information relating to geology, soils and hydrology; and similarly very large scale stereograms, e.g. 1/5,000, may prove helpful in breaking down the types at a smaller scale on the photographs. A development or alternative to the large scale stereogram is to have stereo-pairs of vertical photographs, taken at large scale, of related areas. This technique was successfully used in Western Australia for jarrah (*E. marginata*) and karri (*E. diversicolor*). Stereograms taken on the ground of vegetation can be useful in ecological studies and certainly are quicker to prepare than profile maps (fig. **14.2**).

Classification codes. A letter or number is assigned to each variable being measured or estimated on the photographs. Conifer, hardwood and mixed stands are given separate identification symbols, as also the species. A number or letter indicates the height class, density and estimated quality of the timber. Separate classification codes are used in the principal timber regions of both the United States and Canada for annotating the photographs. Thus, in the Lake states, PH-II-C is the classification code for a stand of pines (P) and hardwoods (H), of height class 21/40 ft (II) and crown closure class (C) (i.e. over 65% crown cover). A classification code enables the summary of the details being observed to be entered directly on the photograph in the immediate vicinity of the observations.

Catena diagrams to show the change of soil and/or vegetative types from, say, hill top to water's edge, are also helpful in the interpretation of communities which exist between two easily recognizable types. Catena diagrams can be built up from local knowledge and tied in with the stereograms and ground photographs. A catena diagram near Tabora, Tanganyika, is shown in fig. 18.8.

In the text, emphasis has been given to the continual use of stereo-pairs of photographs; but prior to planning the field work *mosaics* can be useful, especially in soil studies and geomorphology. Assemblies of separate photographs are clumsy, being difficult to handle in the field, whilst a mosaic can often provide important information needed prior to a detailed field study. If mosaics are not available, it may be necessary to lay down one or more flight strips by overlapping and matching the photographs correctly; and then to view these in sequence in order to formulate general impressions.

In ecological studies and forest work relating to individual trees growing close together, it is useful to have photographs of special areas of interest enlarged to serve as a field map or sketch on which detailed notes can be written in code both on the photographic surface and on the back. It often helps to pin-point selected trees, etc., on the photograph and to record the code data on the back over the pin-prick and surrounded by a free-hand circle.

(c) Computer indexing and retrieval

Recently several computer techniques have been found applicable to the handling of alpha-numerical photographic data. It is now possible to handle the following groups of data related to aerial photographs:

  (i) Photographic indexing.
  (ii) Retrieval of photographs and of data from 'format' files.
  (iii) Image identification.

Up to the present, indexing and retrieving has normally been carried out on a geographic basis, requiring a manual search by map reference. The programme techniques developed on the computer maintain the spatial relation of the photographs and maps by geographic co-ordinates, retains the flight data of each aerial coverage and adds additional information important to the interpreter. The information is stored in digital form and all parameters are handled simultaneously by the computer. The method is suitable for libraries of large numbers of photographs. Three types of logic are applied to searching and comparing search data:

(a) *equal logic* (or direct match) to match descriptions, e.g. cameras, photographs.
(b) *boundary logic* to obtain matches against date, quality, scale, etc.,
(c) *geo-coordinate logic* used by the interpreter to describe the shape of the area.

The success of the system depends on the important criteria: index storage and retrieval. The index provides data about the photography (*source data*); data in the photograph (*context data*) and data related to the context data (*content data*). The storage must provide means of handling two related information files, one being the index data and the other the photographs. The retrieval criterion must provide a means of obtaining the photograph and the related information index by source index (e.g. location, time, camera), content index (i.e. by image descriptions or keywords), context index (i.e. by a major criteria of keywords related to the pictorial content) and by a combination of the above.

For actual interpretation, the interpreter is concerned with characteristics of the images and requires to rely on keywords of context and content. The application in interpretation, at least for the present, offers considerable scope for military purposes, and should give impetus to the storage of photographs in manner suitable for retrieval of information relating to natural resources. The image has to have unique definable signatures and object groupings, recognizable to the computer 'memory'. Thus by measurements of the image of an aircraft on the ground or a ship in harbour, the computer can provide the required data and appropriate photographs. The techniques could be used for assessing agriculture crops and national production according to season. In forestry there is considerable scope, especially in volume analysis, using present definable parameters.

15

THE DETERMINATION OF
APPARENT HEIGHT

INTRODUCTION

ONE may ask why the adjective 'apparent' is used. The explanation is that there are a number of factors influencing the determination of the height of objects from the photographs which result in the measured or calculated photographic height being different from the actual height as measured on the ground. Provided these sources of error are recognized, then the true height, or a reasonable approximation of the true height, can usually be determined from the apparent height by adding a correction factor. Apparent heights are calculated from the measurements made on the aerial photographs; and the correction factor determined by comparing the photographic measurements with field measurements. The determination of systematic errors is important.

Spurr (1948) suggested that, by increasing the scale, the height measurement of objects by parallax difference should be improved, since resolving power is increased and the parallax of the same object is greater. Although theoretically correct, studies since then have indicated that for medium to very large scale photographs there is little improvement. Worley & Landis (1954) attributed to instrument error, shadow, understorey and tree crowns the offsetting of any advantage gained in scales 1/2,500 to 1/20,000. Pope (1957) examined photographs at 1/2,500 and concluded that the individual interpreter and tilt were mainly responsible for errors and that scale was not important. Johnson (1958) associated errors in height measurement more with crown shape, tree size and the interpreter than scale. Kippen & Sayn-Wittgenstein (1964), using photographs at 1/1,200 and a 70 mm format, found error was reduced with the very large scale but the results were still disappointing. Data collected from 290 sample plots on the Slave and Peace rivers in Canada showed that tree height measured at a nominal scale of 1/2,040 is no more accurate than tree height measured from medium scale photographs, due to the base of the tree being obscured in all cases by vegetation. Aldred (1964) drew attention to the importance of wind in very large scale aerial photography, as wind at 15–25 m.p.h. could cause a heighting error in parallax measurements in excess of 15% for a tree of 75 ft.

THE DETERMINATION OF APPARENT HEIGHT

Heavy shadow may cause over-estimates of height, when using a parallax bar, due to the floating dot sinking into the ground surface in the shadow on the stereo-model (Welander, 1953). In Sweden, errors in height measurement by parallax were attributed to difficulties in identifying the ground-level and tree tip (Eriksson, 1960). The total absence of shadow on photographs may result in under-estimating tree height (Eriksson, 1960).

The ground vegetation may be of considerable height in tropical forests and even in temperate forests. For example, in *Eucalyptus regnans* forest in Tasmania, having a stand height of 200 ft to 250 ft, the apparent ground level may be an understorey of silver wattle (*Acacia dealbata*) of 60 to 90 ft. In Victoria in sclerophyllous forest the apparent ground level is commonly formed by herbs with a height of 4 to 6 ft

TABLE 15.A

Results of the measurements of the dominant height—Finland

| | | | Standard error of estimate (metres) | |
| | | Systematic error (metres) | Systematic error included | Systematic error excluded |
Photo scale	Film			
1/10,000	Panchromatic	−0·1	±1·9	±1·9
1/5,000	Panchromatic	−1·8	±2·1	±1·1
1/5,000	Panchromatic	+0·2	±1·9	±1·9
1/10,000	Infrared	+0·7	±2·6	±2·5
1/10,000	Infrared	−0·1	±1·6	±1·6
1/10,000	Infrared	+0·1	±1·9	±1·9

or shrubs and small trees of 10 to 15 ft, e.g. *Pomaderis apetelata*. It is useful to record, during measurement of sample plots in a forest inventory or an ecological quadrat survey, the mean height of the vegetation in each plot; and then to relate these estimates and measurements to their respective forest types on the photographs.

In determination of height in Finland, Nyyssonen (1955) determined two standard errors of the estimate. The first standard error was determined in the normal way by dividing the sum of squared deviations by the number of observations, and taking the square root of this quotient. He then determined the standard error, after the systematic error had been subtracted from each of the observations. This provided in certain series of measurements a lower standard error. At Ruotsinkyla (Flight 1, 1/5,000 enlargements) the systematic error was 1·5 m, due to low but dense seedling stands and luxuriant heather covering the ground and the fact that the crown tops may not have been fully resolved. The mean height of the dominants was about 65 ft. Nyyssonen also considered that due to blending of the crown into the surroundings, the contrast is weak, which is particularly so for sparse stands; and so the systematic

167

error is increased. In 1966, Nyyssonen *et al.* in a further study obtained a systematic error of about 6·5 ft (2 m).

The results of a study of the dominant height of trees by Nyyssonen is given in Table 15.A. It was found that the results obtained from infrared and panchromatic films were similar. Each of the six measurement series is based on the dominant height of approximately thirty stands. Frequently for each measurement series up to 1,000 readings were taken. The average standard error of the estimate was no more than about 10% of the dominant height which is of the same order as in certain experiments in Sweden (Flygbilden, 1951; Welander, 1952) and in the central states of the U.S.A. (Moessner *et al.*, 1951). As early as 1925 in Germany (Krutzsch), a standard error of under 5% was obtained with large scale photographs of even-aged well-managed stands (cf. Zieger, 1928).

On photographs of *Pinus radiata* at Isandula in Tasmania, it was observed that the crowns under about 3 ft diameter did not resolve on the photographs. Crossley (1966) reported infrared colour film as preferable to normal film for recording crown diameters in new plantations. Along the Murray river, stands of *Eucalyptus camaldulensis* of about 90 ft top height repeatedly were under-estimated by 20 ft, using a parallax bar. This was presumably due to both resolution and the ground vegetation. The latter comprised mainly tall grasses. Both ground and photographic measurements were rechecked.

Several times previously it has been emphasized that photo-interpretation is both an art and a science. As an art one may develop the skill of assessing the height of stands of trees, or even individual trees, by examining the stereoscopic model visually without a measuring aid. The skill is built up initially by repeated reference of the photographic appearance of the stands to the actual stands visited in the field. A similar skill is often achieved by a timber buyer, who estimates correctly by eye the standing volume of timber for purchase. At the beginning of his day's work, or occasionally in the course of it, the timber buyer may make an accurate measurement as a check on his visual estimate. A similar technique is used by the interpreter, who occasionally makes a check with a parallax bar or parallax wedge. Visual estimation of height can be particularly helpful when using small scale photographs of 1/30,000 or less, as at these scales only broad height classes can be determined by parallax-bar measurements.

As a science, precise measurements of height can be made on the photographs using either single or stereo-pairs. The height on single photographs may be determined roughly by radial displacement and more accurately by measurement of shadow length. Each of these methods will be considered briefly; and parallax measurement using stereo-pairs will be considered in relation to individual trees, stands, point elevations, small areas, slopes and geologic structures.

It is important to appreciate that height of an object is exaggerated in the stereoscopic image. This is both disadvantageous and useful. To the unwary, it may lead to an incorrect visual assessment. To others, it is a more sensitive means for certain purposes of comparing differences in height. For example, topographic differences in height are more conspicuous from photographs than from a vantage point on the ground or from an aircraft. This may be helpful when studying the influence of site factors. Stand height above a shrub layer and isolated trees and coppice above a herb layer are advantageously exaggerated, and dominant trees are often conspicuously taller than co-dominant trees of the main canopy layer. Under the stereoscope a slope may appear to be two or three times greater than it is on the ground. For further details concerning stereoscopic exaggeration, reference should be made to Hurault (1949), Stone (1951), Miller (1953, 1961), Aschenbrenner (1952), Thurrell (1953), Treece (1955), Jackson (1960), Mekel *et al.* (1964).

Stone (1951) suggested the use of what he termed the *appearance ratio* as a method for expressing stereoscopic exaggeration. This is a general expression of the ratio of the height of objects through a stereoscope to the way the objects ought to look. Using a hand stereoscope and 9 in. by 9 in. photographs having a 60% end-lap, the height of the objects will be exaggerated. For example, an appearance ratio of 1·8:1 indicates that a hill 100 ft above a plain appears in stereovision the same as a hill of about 180 ft high would appear when viewed from an aircraft at the same scale. The exaggeration ratio (E.R.) is given by the formula:

$$\text{E.R.} = \frac{B}{H} \cdot \frac{(f)^2}{(f_s)(E)}$$

where B, H and f are as used elsewhere; and f_s = focal length of the stereoscope and E = interpupillary distance.

A simpler formula for the exaggeration ratio, which can frequently be used, was given by Jackson (1960):

$$\text{E.R.} = \frac{B}{H} \cdot \frac{250}{E}$$

(where E is in millimetres).

Several authors (e.g. McCurdy, 1950) have pointed out the limitations of aerial photographs in determining the approximate depth of underwater features even in clear sea-water; and this facet of vertical measurement, with the exception of one example, will not be pursued further. In small lakes at Mountain West, Utah, Moessner (1963a) determined successfully the location of the 15 ft underwater contour by first estimating the slope of the shore around the lake on 1/20,000 photographs and then extending the slope outwards into the lake. A high correlation coefficient of photo to field slope was obtained.

(a) Measurement: Apparent height on single photographs

(i) BY RADIAL DISPLACEMENT

As mentioned earlier this method is of minor interest, and therefore warrants but a brief description. Two measurements are made on the single photograph.

THE DETERMINATION OF APPARENT HEIGHT

These are the radial distance from the principal point to the top of the object (*r*) and the distance between the top and the bottom of the same object (*d*). The distance '*r*' can be measured to the nearest 1/100 of an inch by an engineer's scale graduated in fiftieths of an inch; and '*d*' is measured by a wedge or a micrometer screw-gauge.

The height of the object (*h*) in feet, is then given by the formula:

$$h = \frac{d \times H}{r}$$

where *H* as usual is the flying height above the ground datum.

(ii) By Length of Shadow

This is a very useful method for determining the height of trees and similar-shaped objects, wherever full-length shadows can be clearly seen on the photographs. It is a satisfactory method to use in photography of tree formations as diverse as winter photographs in arctic regions, when the snow provides an excellent background, and dry-season photographs in savanna woodlands of Africa and Australia, when the herbaceous growth is dry and sparse and has a high reflectivity in the visible spectrum. Obviously the determination of tree height from shadow length is impossible in tropical and temperate rain forest, closed temperate forest, and in unthinned plantations of conifers.

In determining height from shadow length, it is normally necessary to know the angle of the sun to the zenith at the time the photographs were taken. This can be ascertained by reference to nautical or air navigation tables showing the sun's declination (cos *b*), latitude (cos *a*) and time of the day (cos *c*). The angle of the sun's elevation (*x*) is then calculated:

$$\sin x = \cos a \times \cos b \times \cos c \pm \sin a \times \sin b.$$

As suggested by Spurr (1952), the elevation of the sun may also be determined by first locating the zone of no shadow (N.S.) which may or may not be on the photograph, and then applying the following formula to determine the elevation of the sun above the horizon. The altitude of the camera is above ground datum.

$$\tan x = \frac{\text{altitude of camera at N.S. (in feet)}}{\text{distance from N.S. to nadir (in feet)}}.$$

Usually the nadir is assumed to coincide with the principal point, making the divisor equal to distance in feet from the centre of the no-shadow zone to the principal point of the photograph.

The angle of the sun's elevation (tan *x*) can also be determined approximately by determining the height (*h*) of a recognizable object on level ground either by field

170

APPARENT HEIGHT ON SINGLE PHOTOGRAPHS

measurement or by parallax; and then using this height in conjunction with the photographic shadow length (h_1) to calculate the elevation angle:

$$\tan x = \frac{h}{h_1} \cdot \frac{f'}{H'}$$

As mentioned in Part III, the time of the day and date should be shown in the caption of each photograph. Otherwise this information can be obtained from the flight chart, if held, or from the company who took the photographs. If only the date of flying is known, the time of day can be ascertained from the angle between the shadow direction and the true North on the photograph.

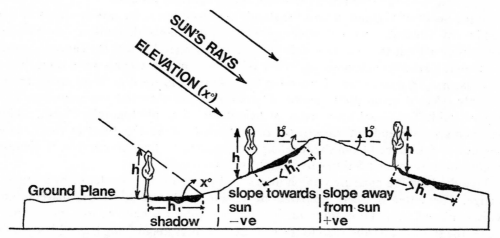

FIG. 15.1. The shadow length of objects as recorded in the aerial photograph may be used to estimate the height of the object, but the elevation of the sun must be known ($x°$) and a correction needs to be made to the photographic measurement to allow for slope ($b°$).

When the elevation of the sun (x) is known, then the height of a tree or similar object from its shadow length (h_1) on the photograph is given by the formula:

$$h = h_1 \tan x$$

(see fig. 15.1), provided it is corrected for scale by multiplying by H'/f'.

This formula is applicable to level or nearly level ground. When the ground has a noticeable slope then a correction requires to be added if the slope rises in the direction of the shadow and subtracted if the slope falls away in the direction of the shadow. As illustrated in fig. 15.1 'h' can then be determined by using the formula:

$$h = h'_1 \cos b \,(\tan x - \tan b)$$

or
$$h = h''_1 \cos b \,(\tan x + \tan b)$$

where $b°$ is the slope of the ground along the axis of the shadow.

171

If a large number of heights are to be measured from shadows, time will be saved by preparing shadow conversion graphs (see Seely, 1942, 1948). Graphs are also used by the U.S. Forest Service (Rogers, 1949).

Several writers have commented on suitable shadow lengths. According to Welander (1952) the shadow length at the time of taking the photographs should be between 1·0 and 1·5 times the height of the tree. Seely (1942) in Canada observed that the potential error in measurement of height from shadows becomes significant when the angle of the sun's rays to the horizontal is less than 35°. At 35° the length of the shadow is 1·4 times the height of the tree on level ground. Nyyssonen (1955) in Finland obtained unfavourable results with an angle of 27°. At 27° the length of the shadow is about 1·96 times the tree's height on level ground.

In the north of Finland, where the stands are more open, the measurement of tree heights by shadow length was comparable with heights determined by parallax measurement; but this was not so in the south of Finland where the canopy closure is greater. Earlier conclusions in favour of shadow measurements were often based on measuring the shadows of the most suitable trees in open areas or along tracks on only selected aerial photographs. The information obtained was used then as a key or guide for visual assessment of height on other photographs (Nash, 1949). In open-grown stands of *Eucalyptus camaldulensis* along the Murray river, stand height was determined from shadow lengths measured on the photographs. A shadow length not exceeding 1·5 times tree height was favoured.

Tree shadows are measured on photographs by either a micrometer or by a shadow wedge. The latter is easily prepared by scribing a V-wedge on transparent film. This is placed over the shadow on one photograph of a stereoscopic pair, viewed under the stereoscope and read to the nearest 0·01 in. A magnifying lens with attached scale can be used on single photographs to read off measurements to about 0·002 in. (fig. *16.3*).

(b) Apparent height by parallax on stereo-pairs of photographs

This introduces the normal method of determining height by the parallax formula which was derived earlier when discussing stereoscopy in Chapter 13, i.e.

$$h = \frac{H(\Delta p)}{P + \Delta p}.$$

As the formula suggests, parallax is influenced by the height of the camera above the ground at the time of exposure, the height of the object and the air base between two successive exposures. The accuracy of the results can also be influenced by tilt, the scale of the photographs and the focal length of the lens. Fortunately the heights of small natural objects measured by parallax difference do not require adjustment for small amounts of tilt; but adjustment is necessary when contouring, determining point elevations and possibly in research studies. Flemming (1960) estimated that for tilt of 3°, the error in the determination of a point elevation by assuming the photographs were truly vertical ranged from -3.69% to $+3.2\%$. A number of authors have suggested approximate parallax-elevation relationships for improved accuracy (Moffit, 1961; Chittenden, 1959; Robbins, 1949; Desjardins, 1943b).

Height measurements are made for several different purposes using a stereoscope and a parallax bar or parallax wedge. These may be grouped as point heights on the ground for the scale of micro-areas and for contouring, point heights for slope determination, point heights for small objects such as trees and point heights for large objects such as hills when determining the angles of strike and dip in geology. Omitting contours (Chapter 12) and micro-areas (Chapter 16) each of these will now be considered.

(i) TREE AND STAND HEIGHTS

For the height of a single tree, the parallax difference is the algebraic difference, parallel to the air base or x-axis, of the distances of the two images of the top of the tree and of the two images of the bottom of the tree from their respective principal points.

Example. The parallax difference for a tree is 1·37 mm and the air base is 92·3 mm. What is the height of the tree, if the flying height above the datum plane is 12,000 ft?

$$h = \frac{H(\Delta p)}{b + \Delta p} = \frac{12,000 \times 1·37}{92·3 + 1·37} = 175 \text{ ft.}$$

In most ecological and forest work, the above formula can be simplified, since the error so produced will be small or insignificant compared with other errors, i.e.

$$h = \frac{H(\Delta p)}{b + \Delta p} \cong \frac{H(\Delta p)}{b}.$$

For example, if the lowest flying height is 7,000 ft and the maximum tree height 100ft, the error introduced through using the simplified formula is 1·4 ft. Nyyssonen (1955) observed that the error in height produced by the approximation is in proportion to h^2/H. When the photographic scale, however, is very large, e.g. 1/1,000, and the tree height, e.g. 100 ft, is large in relation to the flying height, e.g. 250 ft, then for satisfactory determination of height by parallax, Lyons recommended using the formula as

$$h = \frac{H^2 \Delta p}{H \Delta p + fB} \text{ (Lyons, 1964).}$$

Both parallax bars and parallax wedges are used. With both instruments, the error in the apparent height of the tree at a photographic scale between 1/10,000 and 1/15,840 should not exceed 5% (Spurr, 1948; Avery, 1957). In feet, errors of 5 to

10 can be expected (Worley & Meyer, 1955); and errors of 8 to 10 ft, two times in three on photographs at 1/12,000 (Worley & Landis, 1954).

In Scandinavia, parallax bars are preferred to parallax wedges (Francis, 1957). Worley & Landis (1954) concluded that the parallax wedge gave a slightly greater systematic error. In stands of even-aged, heavily thinned *Eucalyptus obliqua* at Creswick, a dot parallax wedge was used in preference to a parallax bar (May, 1960). Johnson (1958b) in Alabama used both wedges and parallax bars to measure trees on photographs at scales 1/5,000, 1/10,000, 1/15,000 and 1/20,000 and concluded that

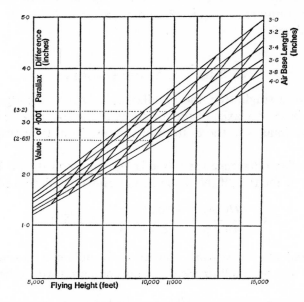

FIG. 15.2. An example of a graph used to correct measurements in parallax difference for changes in ground elevation (see text), when determining the heights of objects from stereo-pairs of photographs.

(Stephen H. Spurr, *Photogrammetry and Photo-Interpretation*, 2nd edition, © 1960. The Ronald Press Company, New York).

errors in the measurement of tree height at these scales was not associated with the scale itself but with characteristics of the tree including height.

It is interesting to note that Lyons (1960), working in British Columbia and using low-elevation photographs taken from a helicopter, recorded errors in tree height of between 2·8 ft and 3·9 ft on contact prints (5½ in. format) at scales of 1/850, 1/1,700 and 1/2,450 as compared with 1·2 ft and 2·1 ft on enlargements (9 in. format).

Various graphs or tables may be prepared, based on a large number of observations, to correct for major changes in ground elevation at the points of parallax measurements on the photographs. One such graph is shown in fig. 15.2.

Example. If the base of the tree is 1000 ft below the mean elevation of the principal points and the photographs were taken at a flying height of 10,000 ft above the mean ground datum between the principal points, what is the true height of the tree? The air base on the photographs is 3·80 in. and the parallax difference is 0·015 in. between the base and top of the tree in the stereo-model.

1. Read off graph point for 10,000 ft and base length of 3·80 in., i.e. 2·65 ft for 0·001 in. of parallax.
2. Determine flying height above datum at the tree. 10,000 ft + 1,000 ft = 11,000 ft.
3. From the point of the first intersection follow the line up to the point where an imaginary perpendicular from the 11,000 ft mark on the x-axis intersects this line.
4. Read off the number of feet at this point for 0·001 in. of parallax, i.e. 3·2 ft.
5. This is the correct value for reading off the height of the tree from its parallax difference, e.g. 0·015 in.
6. Height of tree $= \dfrac{0\cdot015}{0\cdot001} \times 3\cdot2$ ft $= 48$ ft. As an additional exercise determine the tree height directly by parallax formula. (Answer 44 feet, i.e. an under-estimate of height by 9%.)

Stand height. So far attention has been given to the measurement of individual trees as if this were the end in itself. More often, however, the height of individual trees is measured by the forester as a means of determining the mean height of a stand, for site quality assessment or volume determination. For these purposes the mean height of the stand is determined in the field from either the measurement or visual assessment of a reasonable number of dominant trees, co-dominants, a mixture of both or all trees including sub-dominants.

Unfortunately, some writers on the photographic determination of stand height do not specify clearly which trees are being measured, whether the trees are in re-latively dense stands or open-grown or heavily thinned stands and whether the stands are mixed even-aged, or uneven-aged. Normally in the determination of stand height from photographs, the sub-dominants cannot be seen or cannot be clearly distinguished; and therefore stand height is measured using either dominants, co-dominants or both. In temperate even-aged forests, ground measurements have shown a remarkable closeness of stand means based on dominant and co-dominant heights. For example, in forestry practice in the United Kingdom, the top height of conifers, based on the dominant height, of the tallest forty trees per acre, can often be provided by adding about 3 ft to the mean height. In the Australian Capital Territory, for even-aged *Pinus radiata*, the heights of dominants were obtained from sample tree height by adding 6·1 ft ±2·6 ft. It seems reasonable, therefore, in photo-interpretation prac-tice, at least as a temporary expedient, to use a constant of about 3 to 6 ft for conifers to provide the stand height of the dominants. In practice this can be very useful, as it is usually easier to measure the heights of convenient trees of denser stands on the photographs than dominants; but the mean dominant height is frequently needed when using yield tables or volume tables. The stand height of *E. obliqua* has been determined by subtracting 3 ft from the mean height of the dominants.

In tropical forest, particularly rain forest, the ground is completely covered, or nearly so, by the crowns of the trees, which form several canopy layers. However, even under these conditions it is sometimes possible to estimate subjectively, or by indirect measurement, the stand height and the height of emergent trees. For example, in Surinam, it was possible to distinguish three broad classes: 49 to 65 ft, 66 to 98 ft and over 98 ft (Heinsdijk, 1952). A prior knowledge of the communities being studied provided an estimate of the height of the top closed canopy above the ground; and then by parallax the height of the emergent crowns above this layer was determined and added to the closed canopy height to give the heights of the emergents.

Crown diameter to height ratio. As the crown diameter of an individual tree is commonly correlated with its diameter at breast height, over bark, crown measurement can be used to determine a diameter (b.h.o.b.) to stand height ratio (see Chapter 21). As pointed out by Spurr (1960), the crown diameter to height ratio is important in both site and volume studies. In *Eucalyptus camaldulensis* forest, it was found to be a suitable index of site quality which the interpreter could assess subjectively from stereo-pairs of photographs.

(ii) THE DETERMINATION OF SLOPE

Slope is of interest when determining the true size of micro-areas, site quality, sometimes the boundaries or location of plant communities and access for planting or for logging. In rugged terrain, slope may introduce a considerable error in relation to micro-areas, including sample plots (Eliel, 1939). For large areas and when many plots are established, slopes will occur in many varying directions with the result that errors usually compensate each other. Areas sloping away from the principal point on the photograph are frequently under-estimated and areas sloping towards the principal point are over-estimated.

The slope on a stereo-pair of photographs can be determined by either a parallax bar, a parallax wedge or a slope wedge. Using a parallax bar or wedge, two spot heights are separately calculated. One is at the top of the slope and one at the bottom of the slope, as observed in the stereo-model. Their difference in feet provides the difference in vertical height. The horizontal distance between the points is also determined either by direct map measurement or by photographic measurements using radial line intersections (Chapter 11) or trigonometry and the x and y co-ordinates of each point.

A simple graphical method is as follows. A template (e.g. tracing paper) is placed on the first annotated photograph and the principal point, conjugate principal point and flight line are transferred to the template. Next, radial lines are subtended on the template from the principal point to the two points on the photograph located at the top and bottom of the slope. The template is then removed and placed on top of the second photograph of the stereo-pair so that the flight lines coincide. A correction may be required for differences in length of the two photographic air bases (see Chapter 13). Radial lines are again subtended on the template from the principal point of this second photograph to the two 'slope' points. The loci of intersections of the four radial lines on the template now provide the true positions of the points in their horizontal or map positions. The distance is measured between the points on the template in inches or millimetres, converted to feet and then used in the following formula to calculate the slope in degrees (tan b):

$$\tan b = \frac{\text{vertical distance}}{\text{horizontal distance between points}} = \frac{\Delta h}{S_h}$$

A quicker but less accurate method of determining slope is provided by using sin b. In the formula the horizontal distance is replaced by the slope distance measured on one photograph. The U.S. Forest Service slope per cent scale has been designed using this principle. A transparent scale is used to convert elevation differences into slope per cent; and requires the determination of the two heights by parallax wedge. The elevation change is then located on the side of the scale and the zero point of the scale corresponding to this change is placed over one point and the actual slope per cent is read off on the scale at the point where the second photographic point cuts the scale.

Slope wedges. A parallax slope wedge comprises a series of V-shaped lines radiating out from a common centre point. A single pair of these lines resembles the outer pair of lines of a parallax height measuring wedge. When the slope wedge is placed on a pair of photographs and viewed stereoscopically, each pair of lines fuse and appear to float. The pair providing the best match to the slope of the ground gives the approximate angle of the slope in the stereo-model. A disadvantage is that each wedge is designed for a fixed photo-base and a known focal length. The Stereo Slope Comparator provides artificial slopes which can be altered (Hackman, 1956). Mekel *et al.* (1964) have described a simple method of determining slope in conjunction with calculations of exaggeration ratios.

(iii) THE PARALLAX FORMULA APPLIED TO GEOLOGICAL STUDIES

The formula is identical with that derived under 'Stereoscopy' and applied in sections (i) and (ii). The formula must, however, normally be used in full, i.e.

$$\Delta h = \frac{H \Delta p}{P \pm \Delta p} \cong \frac{H \Delta p}{b \pm \Delta p}.$$

When the datum plane is above the second reading the formula is

$$\Delta h = \frac{H \Delta p}{P - \Delta p}.$$

THE DETERMINATION OF APPARENT HEIGHT

It is usual to take all measurements from a predetermined point of known height (E_0). Hence the formula for a first elevation point (a) is

$$E_0 \pm \frac{Ha.\Delta p}{P \pm \Delta p}.$$

It is assumed that the y components cancel.

In geological studies, with both the map position determined by radial line triangulation on the photographs and height determined by the parallax formula, the true map position both vertically and horizontally of any point can be determined. The *strike* and *dip* may be obtained for beds which are well exposed in hogbacks and cuestas. The measurements are best made at least at 2 or 3 in. from the centres of the photographs so that there is adequate relief displacement. A gap, for example, in a hogback or cuesta provides an excellent location. Measurements are made of the average air base, at the base of the dip slope and at the crest of the hogback. The difference in elevation, Δ_h, perpendicular to the trend of the hogback; and the tangent of dip is Δ_h divided by the map distance (S_h), as given in section (ii) for slope.

16

MEASUREMENT OF MICRO-AREAS

HEIGHT, a single characteristic of size, was discussed in the preceding chapter. The other two parameters requiring to be considered are length and breadth. In photographic studies, these are normally combined and expressed as area. As more and more areas are rephotographed, the fourth dimension, time, is increasing in photographic importance when studying the growth of forest stands and the succession of vegetation.

The graphic representation of large areas was briefly considered in Part III, 'Elements of Photogrammetry'. It is necessary now to examine the problem of small areas or micro-areas and to consider techniques used in measuring them, as the measurement of small areas presents the photo-interpreter with a problem sometimes ignored by field workers. Crown-closure studies may be considered to be a facet of the study of small areas.

It is desirable to begin by defining the term 'micro-area', since what is small to one person may appear large to another, and vice versa. On an aerial photograph, an area can be accepted as conforming to the term if it represents only a very minor part of the effective area of the photograph. It thus depends on the scale of the photograph and by repetition it may cover a considerable portion of the total ground area being examined. For example, few would disagree that sample plots of up to 1 acre are each a small area; but, *in toto*, they may represent 10% or more of the effective area of the photographs. Again, consider the problem of determining the total acreage of patches of regeneration up to 5 acres in size in what is nominally a mature forest. These may constitute a considerable percentage of the total area of forest. In Tasmania, on photographs at a scale of 1/23,000, areas of 5 acres or less of unusual forest types are judged as small and omitted from planimetric type-maps. Another kind of small area is the continuous strip as provided by tropical riverain forest, valueless thicket along a wet drainage line in newly established coniferous plantations or the herb-line in the tropical woodlands of East Africa.

For each of these areas, it is essential to recognize the need or desirability of determining the local scale. The necessity of determining a point height in the vicinity of the object being measured was considered under tree height; and it was observed that a change in elevation could result in a 9% error in height estimation of the tree. In many field studies, this would be within the limits of normal sampling errors; and provided it did not result in an accumulative error it would, usually, be acceptable in a forest inventory. The question now arises whether a similar error will occur in relation to small areas and if the error will be smaller or larger. The answer is that the error will usually be much larger. In fact, in studies relating to small areas, it is always advisable and often essential to determine the local scale.

Two factors contribute to making the determination of local scale important. Firstly, in determining tree height, the point height of the ground is a constant in relation to the two parallax measurements taken at the top and base of the tree;

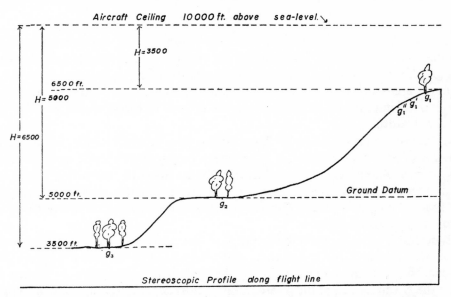

FIG. 16.1. Diagram illustrating the influence of changes of scale within the photograph on area measurements produced by changes in ground elevation. If a photographic sampling grid is placed on the single photograph, one tree would be counted per grid square at g_1, two trees at g_2 and three at g_3 (see text).

and as the flying height is considerable in relation to tree height, moderate changes in elevation will not affect too greatly the height estimate of the tree. In the case of micro-areas, under similar conditions, the point height is a single variable used to determine the scale at each plot. Secondly, as height is a linear measurement compared with area, which is two-dimensional, errors will be in proportion to the square of the linear measurements.

Consider for example three photo-plots (g_1, g_2, g_3) of identical size along the flight line common to two photographs at an average scale of 1/10,000. In stereo-profile, the mountainous area is shown in fig. 16.1 It will be observed that the mean ground datum is 5,000 ft, that only at point g_2 is a plot at the nominal photographic scale of 1/10,000 and therefore covering the correct calculated ground area. At g_1 and g_3, the representative fractions will be as follows, assuming a focal length of 6 in.:

180

MEASUREMENT OF MICRO-AREAS

At g_1

$$\text{R.F.} = \frac{f}{H} = \frac{\frac{1}{2}}{3,500} = \frac{1}{7,000}$$

At g_3

$$\text{R.F.} = \frac{\frac{1}{2}}{6,500} = \frac{1}{13,000}$$

At g_2

$$\text{R.F.} = \frac{f}{H} = \frac{\frac{1}{2}}{5,000} = \frac{1}{10,000}.$$

An acre plot at 1/10,000 at g_2 will have an area of $\frac{1}{16}$ sq. in.; and by using this same sample area of $\frac{1}{16}$ sq. in. for the plots at g_1 and g_3 the sample areas in acres of the latter will be approximately 50% less and 70% greater. Thus if two large trees are counted in the plot at g_2 then, assuming identical conditions, only one tree will be recorded at g_1 but three to four trees will be counted at g_3. A little further thought will prompt the suggestion that photographs at very small scales (e.g. 1/60,000), as commonly used in photogrammetrical mapping, will not be influenced nearly as much by changes in elevation. In contrast, the error produced by a change in elevation of only 500 ft will be $12\frac{1}{2}\%$, assuming the flying height to be 8,000 ft above ground datum with a focal length of 6 in.

It therefore becomes necessary to determine the point scale accurately for each sample plot, unless the country is undulating and/or there is strong evidence that errors will be compensating. Often when sampling large numbers of photographs it is assumed that a large number of plots are used and the errors accruing from plots above and below the datum plane will offset each other. Spurr (1960) cautioned that the scale should be taken into consideration when the relief exceeds 3 or 4% of the flying height. Provided this percentage is not exceeded then in areas of low relief the sum errors due to scale, tilt and slope will probable not exceed 5% of the true ground area. The need to consider an adjustment for tilt was forcibly brought home by Thompson (1954) who commented that, with a few hundred feet difference in elevation on photographs taken at 16,000 ft with a wide-angle lens, it is as accurate to assume that terrain is flat as to trust parallax readings of point elevations without adjustment for tilt.

A further factor influencing tree counts and average tree spacing in small areas is slope. Referring again to fig. 16.1, consider two other plots, g'_1 and g''_1, at approximately the same elevation as plot g_1 and of the same size but on different slopes. Suppose the slopes are 15°, 30° and 45° respectively for the plots at g_1, g'_1 and g''_1. Then $\frac{1}{16}$ sq. in. photo-plots on a photograph at 1/10,000 will actually cover 1·03 acres, 1·2 acres and 1·4 acres of the ground. That is plots g'_1 and g''_1 will cover approximately 20% and 40% greater ground area than plot g_1. As pointed out by Loetsch & Hallert (1964) a plot size of less than about 2 mm on a photograph makes measurement difficult; and as a rule of thumb for photo-plots of forest areas, each plot should contain 10 to 30 trees.

181

MEASUREMENT OF MICRO-AREAS

(a) Measuring techniques

If only a few photo-plots are required of a uniform area or if a very large area of country is to be sampled systematically, the simplest technique is to place the sample plots at the principal point on each photograph or even on alternative photographs. The plots will be almost evenly spaced and the scale will conform to the scale of the photograph as determined at the principal points. No correction has to be made for normal tilt or topographic displacement. When plots are located close to the centre of the photographs, there will be less displacement within the area sampled (Wilson, 1949).

If accurate small scale planimetric maps are available, the photographic nominal scale at the principal points may be determined as outlined from a survey in Iran

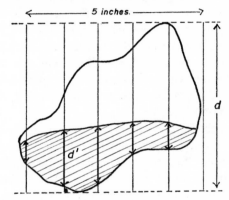

FIG. 16.2. A simple method of estimating the area of a photo-community or forest type on the single photograph is provided by a transparent grid, scribed with parallel equally spaced lines (see text).

(Rogers, 1961). The method of radial line triangulation of points common to the map and the photographs is used to determine the location of the principal points on the map. From this, the ground distances between principal points are measured. Then by dividing the ground distance between the two principal points of contiguous photographs (B) by the parallax difference (Δp) at the photo-plot, the photographic scale at the plot is provided, i.e. reciprocal of photographic scale as a representative fraction $= B/\Delta p$.

Two methods of sampling photographs described by Spurr (1948) and Loetsch (1957) use a print laydown. Parallel lines at 1 in. spacing on celluloid were used by Spurr for sampling the mosaics. In fig. 16.2, if d' is the length of the line across a photo-community or forest type and d is the length of an entire line, then the percentage of any one type is: $\Sigma d'/\Sigma d \times 100$.

Moessner (1960a) suggested using dot sampling, whenever overall volume estimates are quickly needed and the timber value is low. Each dot is classified by species, composition, forest type, stand size class, broad volume class and topographic site. Six inexperienced persons were able to estimate the volume by this method with only a 5% to 10% error.

Dot counts may be used for determining the project area from the total area, provided the total area is known. The method is equally applicable to a few photographs and to sampling from maps in a large project. In northern Iran Rogers (1961) used a dot count to determine the project area for the forest inventory from map sheets at 1/250,000. The ratio of the number of dots in the project area to the number of dots covering the map sheets multiplied by the area covered by the map sheets gave the project area. This was followed by dot counts on alternative photographs to provide the area of forest within the project area.

MONOCULAR AIDS

Grids. These are normally made of transparent materials which do not deteriorate when exposed to light; and have a grid superimposed of 1 in. squares or in metric units or acre/hectare squares at a specific photographic scale. Hectare/acres to grid scale are quicker to use, but the photographs require to be at a constant scale and the grid will not suit areas of high ground relief. On a grid, each square is normally subdivided by fine lines or dots. The dots or lines are usually black or red in colour. The latter are preferably for monochromatic photographs. Abell (1939) found that dots spaced one-quarter as intensively as small rectangles gave an error of 1·8% compared with 4·5%. Five operators were used to test the grid.

Dot grids at several scales are shown in fig. *16.3.* A combined line transect and plot grid was used in woodland surveys in Tanzania. There is no advantage to be gained by having the plots on the grid randomized in relation to each other, since after the first photograph is sampled, the plots form a cluster at fixed distance. Random plots can only be obtained by dropping the grid randomly on the photograph for each plot. This would be time consuming, and therefore if random plots are required it is usually more easily carried out using random numbers in relation to the $x + y$-axes of each photograph. Alternatively a random choice of one of several random grids may be used for each photograph to enable confidence limits to be calculated.

The most suitable number of dots per unit area of grid depends on several factors, including object of the survey, scale of photographs and required accuracy. Spurr (1948) suggested grids with 25 dots and 36 dots to the inch and (1960) for photographs at 1/15,840, 40 dots to an inch. Young, Tryon & Hale (1955) recommended 4 or 9 dots to the inch due to the number of topographic sheets involved in determining the acreage of large areas. Nine dots to the inch were used in a 25,000,000 acre survey. in Maine. Micro-dot grids with 1,096 dots per square inch have been used in determining the area of small lakes and the underwater area to the 15 ft contour line in trout surveys (Moessner, 1963).

Wilson (1949) provided the following formula for estimating the number of dots required for a specified degree of accuracy in a large scale inventory:

$$N = \frac{P(1-P)\,t^2}{(E)^2}$$

where N = number of dots; E = allowable error as a decimal (e.g. 0·02); P = assumed proportion (e.g. 30%) of the total area as a decimal (e.g. 0·30) and t = constant for the probability of an infinite population (e.g. 1·96 at 95% probability level).

Assuming these figures then

$$N = \frac{0·30\,(1-0·30)\,(1·96)^2.}{(0·02)^2}$$

Planimeters. These are frequently used for calculating the acreage of larger areas directly from single photographs, mosaics, ratioed prints, photo-tracings and planimetric maps. The planimeter arm is guided around the perimeter of the plot to obtain the dial reading which is subtracted from the initial dial reading and converted into area. Errors in measurement will occur as in other techniques when using single photographs.

Linear scales. In combination with a simple formula (see p. 182) linear scales are sometimes used for determining areas. Linear scales (fig. *16.3*) can be scribed on clear transparent material at the same scale as the photography (e.g. 1/15,840); but if the photographic scales vary, then a surveyor's scale graduated in 1/50 in. units will probably be the most convenient, and is easily read to 1/100 of an inch. The scale corresponds to a metric scale in half millimetres (i.e. 1 in. = 25·4 mm).

Magnifying scales. The transparent scale is mounted at the base of a simple 'tube' magnifying lens. The magnification is usually two to four times; and the scale may be at 1/1,000 of a foot or in 1/50 to 1/250 in. or in 1/10 mm. Such a scale is useful for reading shadow lengths, road widths, building dimensions and the apparent diameter of tree crowns on single photographs. A magnifying scale is shown in fig. *16.3*.

Scale wedges. A wedge scribed on transparent material is convenient to use for small linear measurements and crown diameters; and like the linear scale can be easily made by the photo-interpreter. A wedge is shown in fig. *16.3*.

Weighing. Naylor (1956) in New Zealand used cut-outs of photostatic copies of photographs. The cut-outs were classified according to type, then weighed and from the results the area of each type calculated. Buckingham (1959) has described the application of weighing for area measurement in Canada.

STEREOSCOPIC AIDS

By far the most efficient method would seem to be that of providing a map of the boundaries using a radial line plotting machine or a floating dot machine with an attached pantograph; and then determining the areas from the map by means of a planimeter. Such a method has certain disadvantages, but is used sometimes in forest surveys when there is considerable tilt and topographic displacement. It is, however, expensive; and it is often difficult or impossible to trace a very small area by a pantograph-arm attached to a radial line plotter. Wilson (1949) recorded that only three out of five hundred plots fell in a different stand class when located systematically on individual photographs as against being located by radial line intersection.

A simple stereoscopic technique for locating equally spaced sample plots by the radial line principle was described by Hartman (1947). The method is suitable for use with a few photographs. A transparent fan of radiating lines from a centre point, is prepared and then placed on top of the photograph so that the centre point of the fan coincides with the principal point of the photograph and the centre line of the fan lies along the flight line marked on the photograph. A similar fan is placed over the adjoining photograph. Under a stereoscope, the radial lines will appear to intersect on the stereo-model. Hackman (1957) used a stereoscopic grid which fitted on the 'stereo-pair' photographs. When viewed stereoscopically, it provided a grid for locating and transferring surface features. It was also used for flora and fauna census.

If the operator both traces the areas on to the map base from the stereoscopic image and determines the areas of the forest types, he familiarizes himself with details which might not otherwise be observed. Tomasegovic (1961) carried out measurements directly on stereo-pairs of photographs using what he describes as a 'Koolinta' or 'Quadrat Co-ordinator'. This is placed on the table of a Stereotope; and comprises a circular turntable with a freely movable pointer for covering any position of the stereo-image. Co-ordinates are read from x and y axes set at right-angles and coupled to a pantograph. It may be used for plotting points determined by rectangular co-ordinates, e.g. linear inventory on the photographs. For height and length measurements, the error was reported not to exceed that of the Stereotope. During measurement along an inventory line, quantitative and qualitative observations can be made to include quality of the trees, number of trees, dominant trees and tree heights.

(b) Crown closure

All estimates of crown closure should be made using a stereoscope so as to minimize the effect of shadow and possibly the low reflectance of the understorey or ground cover; and viewing should be within the effective area of each photograph and as near to the principal point as possible. As compared with a ground estimate of crown cover a photo-cover estimate tends normally to over-estimate crown cover. For example, Spurr (1960) quoted a study of sixteen stands of even-aged eastern white pine in which the crown closure from photographs was 89% and on the ground 77%; and commented that on photographs at scales between 1/7,000 and 1/20,000 crown closure can be estimated with a standard error not exceeding 10%. Worley and Meyer (1955) also recorded standard errors of individual measurements as about ±10%. In New South Wales, crown closure per cent was over-estimated if a stereoscope was not used, due to inability to separate the crown from the shadow (Davies, 1954). Several workers have reported a systematic error between 5 and 10% (e.g. Welander, 1952; Nyyssonen, 1955; Worley & Meyer, 1955). Welander, using

185

crown closures between 0·00 (no forest) and 1·00 (fully stocked), obtained differences in volume between individual photo-plots and ground plots of ±26%, but the overall difference in volume was small (Totals: 5,210m³ and 5,240 m³ for photo-plots and ground plots respectively).

Smaller scale photographs lead to over-estimation of the crown closure as small gaps are not visible. On a small scale photograph it is often difficult to decide whether a dot is on the crown or not. Results obtained by Nyyssonen (1955) and Welander (1952) indicated that there is no significant difference between types of film. Under existing conditions of photography, probably the smallest scale for providing reliable results is about 1/20,000.

In estimating crown closure from aerial photographs, three methods have been used: (a) ocular, (b) by comparison of the stereo-model with a crown-closure density scale (fig. *16.3*), (c) by superimposing a micro-dot grid on one of the stereo-pair of photographs and recording whether the dots (points) fall on a tree or in a space. Usually ocular estimates are too approximate and subjective to be of much scientific use. Point sampling by microdot grids are suitable for large scale photographs, e.g. 1/5,000, but are slower to use than crown-closure scales and it is doubtful if the precision is improved. In Finland, a crown-closure scale has been designed in which circles are used to represent three different crown classes (Ilvessalo, 1950). It is claimed that these give a more natural appearance to the scale than densities represented by dots or squares of equal size (Francis, 1957).

Besides determining crown closure from the aerial photographs directly, it may be determined by point-sampling, in the field, areas selected from the photographs. Robinson (1947) used a grid-viewer, described as a moosehorn, for estimating crown closure from the ground. Plane table survey has also been used. All these methods, however, are much slower than using aerial photographs directly as mentioned above and it is doubtful if there is usually any improvement in the precision. In the Abitibi inventory in Canada, key stereograms of crown cover were successfully used in relating crown cover to basal area (Losee, 1955).

Crown diameters. For individual trees, crown diameters measured from aerial photographs do not usually agree with crown diameters of the same trees measured individually on the ground. Also results may vary according to the film–filter combination. Meyer (1961) working on photographs taken with both infrared and panchromatic film at the scale of 1/15,840 obtained highly significant differences between the two.

Usually crown diameters for a stand obtained from aerial photographs show a better correlation with stand volume and mean tree volumes by crown diameter classes or other classes than similar measurements of crown made on the ground. From aerial photographs, crown diameter measurements are made more consistently, and provide a better evaluation of the functional growing space occupied by the trees (Spurr, 1960). Obviously, crown diameters obtained from aerial photographs will not include branches hidden by neighbouring trees and branches which are too thin or too poorly leafed to be resolved. In many eucalypt forests, the stereoscopic

186

measurement of the crowns on the photographs by crown diameter wedges is made difficult due to the species' poor apical dominance which results in irregular and divided crowns, so that a large tree may appear to be two trees; and due to the species' tolerance, a clump of small trees may appear to be one tree. In open stands, a tree's shadow is of great assistance in detecting the anomalies, both through its length in relation to tree height and by its shape. Worley & Meyer (1955) obtained standard errors of 3 to 4 ft on 1/12,000 photographs of oak in Pennsylvania. Nyyssonen (1955) obtained a standard error of 2 ft on 1/10,000 and 1/15,000 photographs in Finland. Both used photographs taken in summer. Nash (1948) obtained good correlation between the product of the total height of the tree and its crown diameter to the squared diameter of the tree at breast height and tree volume.

Four devices for measuring crown diameter are the crown wedge, the spot gauge, micro-dot template and magnifying scale. The former is an elongated 'V', which is scribed on transparent material. The 'V' is fitted over the crown of a tree in the stereomodel at right-angles to shadow; and at the best fit, the width of the space between the arms of the 'V' is read off in inches or millimetres. The spot or crown diameter gauge comprises spots of various known diameters. From the spot nearest the crown sizes on the gauge, the diameter of the tree is calculated. Swellengrebel (1959) obtained the best results by comparing the crown's stereoscopic image on the photographs (2·8 magnification) with a crown diameter gauge having diameters at 0·0005 ft intervals. The micro-dot template comprises several clusters of equally spaced fine dots scribed on transparent material.

Comparison of results by three methods (wedge, micrometer and dot template), for dominant Scots pine in Finland, is shown in table 16.A (after Nyyssonen). According to ground measurements the mean diameter of the crown was 4·25 m ± 1·25 m. Nyyssonen found the dot template method was the easiest to use and the micrometer the slowest.

In Victoria a Wild magnifying scale (8×) was used for determining the crown widths of *E. obliqua* at Creswick (May, 1960). Two measurements at right-angles (1/10 mm) were made, using the photograph with the sharpest crown outline. For a photographic mean of 37 ft and a ground mean of 37 ft, the correlation co-efficient was 0·6709. For the field measurements, a crownometer was used (Weir, 1959).

Davies (1954) working in Murray red gum forest (*E. camaldulensis*) recorded that an experienced interpreter could more quickly assess crown diameter classes visually than by a crown diameter wedge, and as accurately; but for quantitative measurements the wedge was preferred to other instruments, as the tree was seen stereoscopically. The 4 in. to 8 in. diameter breast height classes were determined on aerial photographs through their correlation with crown diameter.

On very large scale aerial photographs, e.g. 1/750, it is possible to measure and to consider scales separately at crown height and ground datum. Crown diameter (*cd*) measured on photographs at these scales can be corrected to ground datum scale (*CD*) by the following formula (Lyons, 1964):

$$CD \text{ (in feet)} = \frac{(H - \frac{1}{2} h_0) \times cd}{f}$$

where h_0 = tree height.

Stocking. This is the number of trees per unit area. In American forests at scales over 1/12,000, photographic under-estimates vary between 20% and 50% (Spurr, 1960), but numbers of dominants and co-dominants can usually be counted accurately (up to 5% error). In even-aged stands of *E. obliqua* in Victoria a count of the number

TABLE 16.A

Results of the diameter measurements of the dominant crown (after Nyyssonen, 1955)

Instrument	Photo scale	Systematic error (m)	Standard error of estimate (m)	
			Systematic error included	Systematic error excluded
Wedge	1/10,000	+1·10	±1·15	±0·45
Wedge	1/5,000	+0·39	±0·58	±0·42
Micrometer	1/5,000	−0·78	±0·94	±0·47
Dot template	1/10,000	−0·89	+1·11	±0·66
Dot template	1/5,000	−1·06	±1·16	±0·47

TABLE 16.B

Tree counts from aerial photographs—Holland

Scale	Optical enlargement	Working scale	Standard deviation per plot (%)
1/10,000	8×	1/1,250	10·6
1/10,000	3×	1/3,300	13·5
1/20,000	8×	1/2,500	15·6

of stems of co-dominant and dominant trees on $\frac{1}{5}$ acre plots gave a photographic mean of 6·0 and a field mean of 6·4 (correlation coefficient 0·8319). Stellingwerf (1963) in Holland obtained satisfactory results for the number of stems counted in sample plots on diapositives at 1/10,000 and 1/20,000, as shown in table 16.B. It is worth emphasizing that it is the dominants and co-dominants which are important economically and as a class are probably providing up to 90% of the current annual volume increment.

PART FIVE
Integrated Interpretation

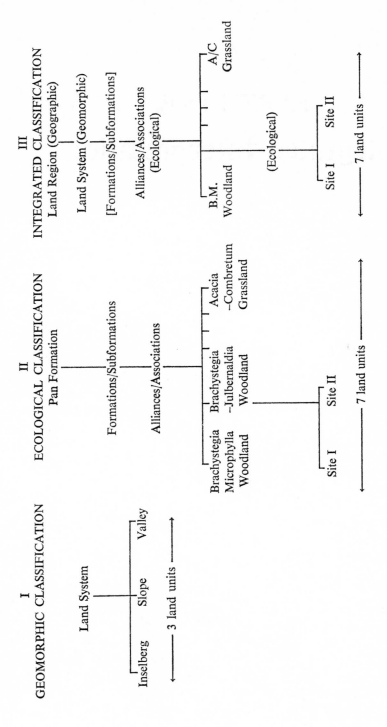

FIG. 17.1a. Three photographic approaches to terrain classification near Tabora (Tanzania). The geomorphic approach provides three units (land facets), the ecological classification seven units and the integrated classification seven units. The integrated classification in this region is mainly ecological; but when used by geomorphologists on the basis of conspicuous patterns only it provides five units. This could result in the grouping of two important soils and the failure to recognise one unit which facilitates the movement of vehicles.

17

GEOGRAPHIC AND GEOLOGIC CONSIDERATIONS

IN applying photo-interpretation to the study of the natural environment, it seems logical to begin by identifying and delineating the geographic regions in the area which is being examined. Frequently, the examination will show that the study is confined to a single geographic region; but where there is sufficient non-homogeneity in the area to justify delineating parts of two or more regions, this should be done before proceeding with the study. This procedure may draw attention to factors, which would otherwise be overlooked; and was first pursued by Bourne in central Africa and south-east England (1931). Bourne recognized geographic regions from the photographs and within these identified ecological units of similar site qualities. It is important to appreciate that the magnitude of the natural geographic units will be much greater than homogeneous ecological units. For example, Schneider (1966) mentions geographic units of a few square miles (cf. Chapter 22).

As the pattern of the regional geographic unit is principally influenced by geological structure and geomorphic development, these should be considered in sequence after photo-geography. This approach to an ecological study may be termed *vertically integrated photo-interpretation*. Alternatively, if an area is investigated by a team of experts, they will be concerned with the inter-connection of their separate studies and this would be termed *horizontally integrated photo-interpretation*. Where the ground on the aerial photographs is inconspicuous due to the presence of tall vegetation, as often happens in the tropics, it may be desirable or necessary to study the plant formations/sub-formations portrayed on the photographs in conjunction with or in place of the geomorphic examination. Even when the objective of the study is terrain classification and these ground conditions exist, it may be necessary to complete an examination of the larger photo-communities prior to the photo-geomorphic study; and it seems logical that this earlier work should be undertaken by an ecologist trained in photo-interpretation. As the land units decrease in size, the importance of photo-ecology increases. By examining the photo-geography, photo-geology and photo-geomorphology within the area of study, the photo-ecologist is fully using the photograph as an aid in studying the eco-system. As land-form classification becomes more exacting, in the future, greater attention will need to be given to the morphometric, morphographic, morphogenetic and morpho-chronological aspects (e.g. see Verstappen, 1966). Three approaches to terrain classi-fication are illustrated in fig. 17.1a.

(a) Photo-geographic considerations

INTRODUCTION

A study of photo-geography can be rewarding to the land-resource scientist as providing a basis at 'policy level' from which to work. In medieval times geography was concerned with all phenomena on the earth's surface; but by the end of the nine-teenth century, as a result of the development of distinct systematic disciplines in-cluding geology and descriptive botany, geography was for a time mainly concerned with exploration. Today it is reassuming its importance as a subject directed towards the examination of the information provided by other disciplines in relation to the many features of the landscape. Consequently, much of what will be now discussed in Part IV will be of interest to the photo-geographer.

Prior to World War II, a number of papers were published on photo-geography in English and German (e.g. Troll, 1939); and included especially English-language publications relating to work carried out by the Tennessee Valley Authority and the U.S. Department of Agriculture. In the United Kingdom, Walker (1953) published a general text; and Stone in the United States published an informative bibliography in 1954. Roscoe (1954, 1960) has drawn attention to the slower development of photo-geography as compared with photo-geology. Schneider (1966) has stressed the import-ance of aerial photographs in developing landscape ecology.

Of immediate interest is the photo-geographer's general approach to studies in a region, the recognition of a region and subdivisions of the region on aerial photo-graphs and a cognizance and application of pertinent systematic geographic disci-plines. In passing, it is interesting to note that in the *Manual of Photo-interpretation* (pp. 752–3, 1960) over thirty systematic geographic disciplines are listed.

REGIONAL GEOGRAPHY

A region is homogeneous in accordance with the criteria which are used to define it. If these criteria (e.g. land-form, plant formation) are observable on the aerial photographs, the boundaries of the regions can be delineated; and these then provide a useful basis for regional inventories. Plant formations can sometimes be distin-guished even on photographs taken by a satellite at a hundred miles or more above the earth.

On some satellite photographs of the United Kingdom afforested areas can be identified. Many of the more recent 'Mercury' space photographs (70 mm format) show details of geographic and geomorphic interest (fig. *17.1c*). Also as pointed out by Lowman (1965), there is little danger of confusing natural and artificial features, and linear features may be conspicuous (e.g. roads, railways, lineaments), but urban areas may not be recognizable. In fig. *17.1b,* the city of Port Said cannot be delineated.

The study of a region may either proceed from the hypothesis that it is homogeneous before analysing its characteristics, or the patterns of each systematic discipline can be developed separately prior to comparison and synthesis. In each process, it is necessary to learn to ignore objects recorded on the photographs which are exceedingly conspicuous, but not related to the current study. Often regions can be quickly delineated on photographs by map analysis. An important first step is to study all available maps and to determine the scale ratio between the maps and the photographs. Following this, information of interest is plotted from the maps on the photographs (or vice versa).

The photo-interpreter may be interested in a single class of features in the region, such as hill slope, consociations, forest stand height classes or soil types. A region so developed is termed a *single-feature region.* More often, however, interest is focused on two or more classes; and the region so formed is termed a *multiple-feature region.* Examples include slope and drainage, forest types and utilization classes, agricultural crops and soil erosion, and land-form and phytosociological units. If a region is formed from a study of all relevant ground features both natural and social, the geographer terms it a *compage,* as distinct from a *total* region which covers both ground features and atmosphere. A region defined only by its natural features is sometimes termed a *land region.*

Of military importance and of interest at 'policy level' is the development of *analogue geography.* In this, a technique is developed by which land units or land-form systems of one multiple-feature or one single-feature region are evaluated in a selected area and identified with similar units in another region. Small areas mapped at large scale can be compared with similar tracts elsewhere or larger regions mapped at much smaller scales. Van Lopik & Kolb (1959) described the mapping of a desert area in Egypt by terrain comparison with a selected desert area in the United States in terms of general terrain, geometry, ground features, vegetation and terrain-factor maps. In terms of military training, analogue geography enables troops to experience, in mock-battle, conditions likely to be encountered elsewhere. Possibly due to the dearth of trained photo-ecologists, insufficient emphasis has been placed on the ecological factor in analogue mapping.

SYSTEMATIC GEOGRAPHY

Several systematic geographic disciplines are pertinent, and can prove helpful in evaluating details contained in the photographs. These include physical geography, soil geography, plant geography and historical geography.

Maps based on climatic data are particularly helpful when initiating a study, due to the close relationship existing between plant associations and local climate. When seeking information from climatic maps, it is necessary to appreciate that many maps, particularly in the tropics, have been prepared from data derived by the simplest of instrumentation, e.g. weekly rain-gauge readings. Often the boundaries or iso-lines provided on the climatic maps have been drawn from this type of data in con-junction with observations on the climate as indicated by the distribution of plant formations and as observed and interpreted by the most experienced regional field scientists at the time. By examining these maps, the photo-interpreter will obtain data useful in indicating transitional zones in plant formations and possibly what species not to expect in the plant associations of the region under examination.

A knowledge of plant geography can be useful to the geologist, soil scientist or engineer in helping him to recognize drainage conditions, surface material and rock types in a climatic region. For example, Tomlinson & Brown (1962) observed that within the three geographic regions of eastern Canada (tundra, boreal zone and hard-wood zone), the common tree species are directly associated with the six principal soil types.

Historical geography has a three-fold use to the photo-ecologist. It may provide information relating to normal plant succession, as disclosed by comparison of current photographs of an area with detailed maps or older photographs of the same area. It may help in understanding the change in the sub-climax vegetation; and it may help in identifying at least some of the existing vegetation on the photographs. One example will be given of the way in which an appreciation of recent history can be useful.

In western Tanzania, information relating to the historical occurrence of sleeping sickness (*Trypanosomiasis*) helped in recognizing the regeneration of *Brachystegia* and *Julbernardia* species, and in distinguishing the association from the economically useless Itigi thicket formation on photographs at a scale of 1/30,000. Field survey had shown that the 'Itigi' thicket and the regeneration could not normally be separated on aerial photographs. The former, however, had never been cleared for settlement, whilst large tracts of *Julbernardia–Brachystegia* woodland had been cleared and settled prior to the advent of sleeping sickness in the 1930s. The coming of epidemic *Tryp-anosomiasis* resulted in the wholesale abandonment of the settlements by the Was-amba tribe; and later the areas were never resettled. As a result, the areas are now covered with regeneration. Thirty-year-old medical records were obtained, which indicated the location of the old settlements, and sometimes included sketch-maps of the areas at the time of the epidemic. Field checks of the photo-interpretation using the old medical records confirmed that the technique was satisfactory.

(b) Geologic considerations

Outside the field of forest photo-interpretation, photo-geology is the most widely used and developed facet of photo-interpretation. Recently published works of interest include *Aerial Photographic Interpretation* (Lueder, 1960), the *Manual of*

GEOLOGIC CONSIDERATIONS

Photo-interpretation (Chapter 4, 1960) and *Photo-geology* (Miller, 1961). Although photo-geology will be discussed prior to geomorphic considerations, it will be appreciated that geological photo-interpretation depends largely on landscape interpretation.

In the present chapter, the subject is introduced with two objectives in mind: firstly, to draw attention to photo-geology as providing a useful tool in initial field studies; and secondly, to provide examples, at an elementary level, of rock types recognizable on aerial photographs. In fig. 17.2, selected rock types (as modified by weathering) have been brought together in a single block diagram.

Frequently in ecological and forest studies it is desirable to recognize the principal rock formations on the aerial photographs, and then to delineate regional boundaries from an examination of the geologic features and local boundaries from an examination of the land-form. It is preferable to begin by studying existing geologic maps and references and by examining the rock types as portrayed on photographic mosaics or on the smallest scale aerial photographs available. On photographs, it is possible to recognize rocks as being igneous, sedimentary or metamorphic; and to subdivide these into several important groups. Further subdivision normally requires specialist training in geology and local experience in the field. When identifying rock types from photographs, it is common practice to add the geologic age to the identified rock (e.g. Silurian shales, Devonian sandstone), provided this information for the region is known and can be correctly associated with the rock type.

The appearance of rocks recorded on aerial photographs depends on a number of important factors, including vegetation cover, soil, absolute rate of erosion, relative rate of erosion of one rock compared with another, mineral constituents, physical characteristics, depth of weathering, structure, texture and factors influencing the aerial photograph as discussed in Part I. Obviously, many of these are interrelated, and a knowledge of the interrelationships may be helpful. For example, vegetation cover influences soil cover, rate of erosion, depth of weathering; but vegetation, in its turn, is affected by soil cover, rock, depth of weathering and rate of erosion. It is important to appreciate that, on aerial photographs, the same rock type may record quite differently, according to the longevity of weathering (fig. 17.2) and climatic conditions. For example, limestones in arid or arctic climates are often rugged, whilst in a humid climate they may be inconspicuous and gently rounded. As pointed out by Allum (1962), even two granites occurring within 8 miles of each other (Tegina, Nigeria) may differ widely in photographic appearance. An experienced photo-geologist, however, would identify both as granite by their light tone, sparseness of vegetation and soil, slow erosion, exfoliate massive appearance, clearly visible joint system, few textural features and 'homogeneity'. Where two different adjoining rock types have developed similar weathering patterns, it may be impossible to separate them on the photographs.

On rare occasions, photographs have been used successfully to locate areas in which economic concentrations of important minerals occur; and possibly, with the rapid

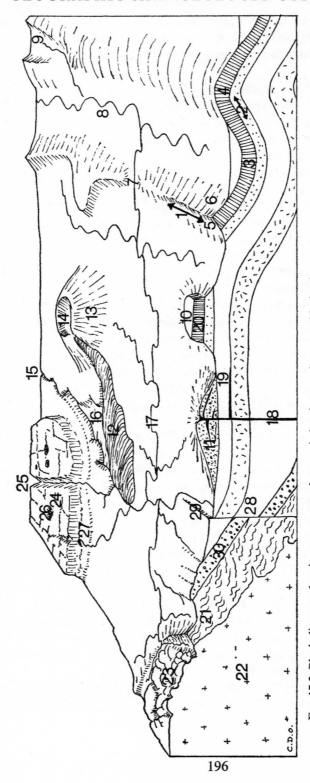

Fig. 17.2. Block diagram showing common rock types, in brackets, and geomorphic development.
1: Direction of strike. 2: Direction of dip. 3: Syncline (*sedimentary*). 4: Anticline (*sedimentary*). 5: Scarp slope (escarpment) (*sediment-ary*). 4: Anticline (*sedimentary*). 5: Scarp slope (escarpment) (*sedimentary*). 6: Dip slope. 7: Water gap. 8: Strike stream. 9: Breached anticline. 10: Mesa (*sedimentary*). 11: Eroded, extinct volcano. 12: Lava flow (*extrusive*). 13: Volcano (*pyroclastic*). 14: Volcanic crater. 15: Youthful drainage (*cf*. 17). 16: Lava-dammed lake. 17: 'Old age' stream on floodplain (*cf*. 15). 18: Volcanic neck (*intrusive*). 19: Sill (*intrusive*). 20: Cap rock (*sedimentary—sandstone*). 21: *Metamorphic aureole* (*metamorphosed sediments*). 22: *Granite* (*igneous—plutonic*). 23: Tor landscape (*igneous*). 24: *Limestone plateau* (*sedimentary*). 25: Dry valley (*limestone*). 26: Sink hole (*limestone*). 27: Spring and cave. 28: Fault (*deformation structure*). 29: Fault scarp (*deformation structure*). 30: Unconformity.

196

development of colour photography, even greater success will be achieved in the future. According to the 'Goldschmidt enrichment principle' (Lueder, 1960), economic minerals dissolved in the soil water may be absorbed by the plants, concentrated in the leaves, and then returned to the surface layers of the soil. This results in an accumulation of the minerals in the soil and influences the distribution of the plant species. As a result, so-called plant–mineral associations occur, which are recognizable on aerial photographs by tone, hue, value, chroma and texture. For example, zinc deposits have been located by luxuriant growth of ragweed and consociations of the Canadian 'zinc' pansy; and copper deposits in North America have been indicated by red campion, which is abundant in some soils with over 14 lb of copper per ton. Douglas fir, larch and lodgepole pine have all been recorded as localizing concentrations of zinc and copper in their foliage.

IGNEOUS ROCKS

Igneous rocks are formed from the magma or molten matter as the earth solidifies. The eventual characteristics of the newly formed igneous rocks, as recorded on the aerial photographs, will depend on the conditions existing at formation, the ensuing weathering, the duration of these processes, and chemical composition. Unfortunately, although there are a large number of igneous rock types, it is not usually possible on the photographs to subdivide them further than (i) intrusive, (ii) extrusive and (iii) pyroclastic (efflata); and sometimes even these broad subdivisions are not possible. On other occasions, however, it may be possible to separate the subdivisions further into acidic and basic, particularly by reference to the vegetation. In fig. 17.2, several rock groups are illustrated diagrammatically; and in fig. 17.3 the spectral reflectance curves of igneous, metamorphic and sedimentary rock samples are shown.

(i) *Intrusive rocks.* These have been formed as magma solidifies at varying depths below the earth's surface. Then, through erosion processes over millions of years, the rocks are gradually exposed. Some may have solidified as veins close to the surface and are exposed now as dykes and sills (fig. 17.2) between slightly metamorphosed rocks; whilst others, plutonic rocks, were formed at considerable depth and are only exposed due to extensive erosion. If an intrusive vein is more or less parallel to the stratified rocks, it is termed a sill; and if at the time of hardening, the magma was rising in the cracks, it is termed a dyke. Frequently as dykes (e.g. diabase) are more resistant to erosion than the surrounding metamorphosed sediments, they are conspicuous on the photographs.

Intrusives include such well-known acidic rocks as granites, syenites, diorites, phorphyries and the basic rock gabbro. With the exception possibly of granite, it is not feasible to identify the intrusive rock type, although the photo-geologist with local knowledge may be able to infer that the rock is a diorite or granodiorite.

Probably the best description of the photographic characteristics of the granitic group (fig. **17.4**) has been given by Belcher *et al.* (in Lueder, 1960). Granitic rocks are generally characterized by their massive size, occasional pendants and distinct smoothly curving boundaries (unless faulted). Frequently, the granite is conspicuous by its dome-like hills and light tones of grey on panchromatic photographs; and obviously there should be no evidence of stratification. Where the bare granite can be observed, or even when there is a thin mantle of soil, characteristic irregular patterns of curvilinear fractures can often be identified. These modify considerably the otherwise smooth surfaces. There is normally little or no evidence of recent erosion.

Surface drainage patterns are usually well integrated, non-orientated (fig. **17.4**); and, within a climatic region, the drainage system is normally poorly developed as

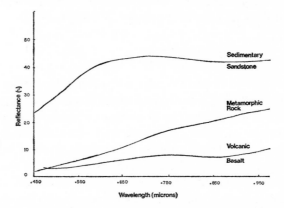

FIG. 17.3. Spectral reflectance curves of rock samples collected near Mount Disappointment, Victoria. As indicated by the curves, the sedimentary rock recorded in light tones on panchromatic photographs, the volcanic in very dark tones and the metamorphic in darkish tones. Possibly maximum photographic contrast between the volcanic and metamorphic rocks would be obtained by using pure infrared photography.

compared with systems on sediments and metamorphosed sediments. Tributaries normally join the main streams at right-angles, but at higher reaches an obtuse angle is common. On large scale photographs, the water courses will frequently be observed as being roundish rather than v-shaped.

As granites are associated with shallow soils in many parts of the world, it is not surprising to observe on photographs an abrupt change in the texture and pattern of the vegetation (fig. **17.4**). This may be due to a change in species, or to a sudden change in site quality as reflected by tree height and canopy closure. Often the texture and pattern of the vegetation is more variable and less luxuriant on granite and associated rock types than the vegetation on adjoining rock groups. Often deciduous trees will be observed to shed their leaves earlier in the season on soils over granite; and, normally, the vegetative cover is denser along fault lines with intrusive outflows.

GEOLOGIC CONSIDERATIONS

Caution, however, must be exercised when generalizing, as there will surely always be exceptions, and particularly where other factors are more limiting than, say, shallow soil. A conspicuous exception was observed at Mount Disappointment (Victoria), where the most luxuriant growth was associated with the granodiorite at the highest elevations (2,400 ft). Rainfall increases from about 30 in. on the sediments to possibly 60 in. or more on the granodiorite.

(ii) *Extrusive rocks.* These are formed when the magma flows out on the surface of the earth and loses its volatile constituents. Rock types include conspicuous volcanic lavas, dark basaltic lavas, intermediate andesites and felsitic types (e.g. rhyolite). After a little experience, the photo-interpreter will often readily be able to appreciate from a cursory examination of the photographs that the region was one of volcanic activity in recent or geologic times, and that the soils have been derived from extrusive rocks. Possibly one or more old volcanic cones (fig. **17.6**) will be observed; and the lava flow pattern will be recognized by its characteristic blister-like wrinkles. Although the terrain is frequently flattish or undulating, it may be rugged, particularly if there is, or has been in recent times, considerable volcanic activity.

The basalts and basaltic lavas are usually recognizable on the aerial photographs by tone and texture. In tone, basaltic lavas are very much darker than granite and sometimes provide a mottled soil surface on panchromatic photographs (fig. **17.6**). Sometimes due to gas-bearing flow, the lava-flow pattern is vesicular, 'ropy' or scalloped. A surface drainage pattern may be absent and results from post-flow shrinkage of the lava or the vesicular character of the lighter lava. Drainage may also be parallel, reflecting the original flows. Erosion details are not usually conspicuous, but include back-weathering and rock slides. The latter are typical. Rhyolite, highly silicic, can often be separated from the basalts due to the characteristic sickle-shaped drainage pattern, and due to its lighter tone. Rhyolites weather to kaolin.

On photographs of extrusive rocks, the image of the vegetation is very variable, being greatly influenced by aridity and age of the volcanic material. It is often sparse on very young material, but may be luxuriant, especially in the tropics, on older material.

(iii) *Pyroclastic rocks.* These are materials ejected explosively from volcanic openings; and vary in size from large pyroclastic bombs or efflata rock through breccia to the finer cinders and ashes. Evidence of volcanic activity, including extinct or active volcanoes, will immediately suggest to the photo-interpreter that consolidated and unconsolidated pyroclastic material may be present. Ash deposits may be conspicuous on the downwind side of a volcano, and the photographic boundaries can sometimes be delineated by tone, texture and possibly vegetative pattern. Usually the terrain is rugged, the topography dissected, side-slopes steep and the erosion pattern conspicuous. The drainage pattern can usually be described as dense, conspicuous, deeply channelled, and probably non-orientated. Tones vary considerably. Colour contrast may be conspicuous, as for example near Arusha (Tanzania), where the efflata soil is blackish grey and contrasts with the reddish plain soils. Provided there is adequate rainfall, the vegetation should be luxuriant and agricultural activity should appear to be intensive. Depending on the part of the world, terrace cultivation may be present. Possibly luxuriant, closed forest will remain only on the steepest of slopes and in inaccessible valleys. On pumice, however, due to its porous nature and siliceous origin, the reverse may occur. In North Island, New Zealand, the indigenous vegetative cover on the pumice is often poorly developed, but the introduced exotic conifer, *Pinus radiata*, has a current annual volume increment at 20 years in excess of 500 cu. ft per acre.

SEDIMENTARY ROCKS

These have been derived from the weathered material of older rocks. During the process of weathering, the material was transported, sorted according to size and deposited. Finer materials were carried in suspension, and the coarsest materials were rolled along the stream beds. Gravels were the first to be deposited, followed in sequence by sands, silts and clays and finally calcareous materials. Sometimes non-marine calcareous material, sands and clay have been deposited in mixture to provide a marl. After deposition, the sorted materials were consolidated in layers (fig. 17.2); and by cementation have beome coherent to provide the readily recognizable conglomerates, sandstone, shale and limestone.

On the aerial photographs, five or six groups of sedimentary rocks can frequently be recognized. That the rock formation is sedimentary in origin is relatively easy to ascertain unless the strata are horizontal or only gently inclined and covered with dense vegetation. As a major group sedimentary rocks are conspicuous on photographs due to their stratifications (fig. **17.7**) and are easier to identify than igneous rocks. In arid areas (e.g. North Africa, central Australia), typical sedimentary land-forms include buttes, mesas and hogbacks (fig. **17.7**). Frequently, belts of sedimentary rocks are roughly parallel to present-day shore lines. In arctic regions, most rock types excluding shales are resistant to erosion (Fletcher, 1964).

(i) *Sandstones*. These comprise mainly quartz cemented together into a fairly durable mass; and provide many landscape features resistant to weathering in the form of ridges and cap rocks, particularly in arid zones. Frequently sandstones form thin layers, being interbedded with shales and limestones. In humid climates the sandstones tend to form massive, prominent, flat-topped hills, which are mesa-like and unusual in a humid climate. The edges of the mesa-like tops are normally rounded; and often slope away gently. On small scale photographs sandstones sometimes give an impression of a terrain which is 'rough' and 'jagged'.

Surface drainage is poorly developed, being of low density, fairly uniform and rectangularly orientated. Fluvial erosion features are conspicuously absent; but wind erosion is prominent in arid areas. In arctic Canada, the gulleys in sandstone and conglomerates are wide v-shaped, short and not numerous (Fletcher, 1964).

The land-use pattern is often provided by forestry in humid areas. Arable farming is conspicuously absent; and, within forested land, the stand height is frequently less than in forest on adjoining rock types. In arid and semi-arid areas, the land is often only scrub-covered and not used for grazing.

(ii) *Shales.* These are the most widely distributed sedimentary rocks; the term refers to those laminated rocks consolidated from both silts and clays. Usually the topography is distinctive (fig. **17.5**). The shales are impervious and relatively easily weathered and eroded. In a humid climate, they may provide distinctive, softly rounded hills and mountains with considerable variation in the elevation between the summits and valleys; but much depends on the geologic age of the rocks. Slopes are gentle to moderate, but often very steep along the banks of perennial streams. In west Wales, where both Ordovician and Silurian adjoin, the older shales were observed in the stereo-model to provide gentler top slopes, more rounded toplands and wider valleys. Elevations between valleys and hill-tops were 600 to 800 ft on Ordovician shales, and 900 to 1,100 ft on Silurian shales. The Ordovician hills also appeared to be associated with more intensive agriculture, including ley farming and an absence of rough (heather) grazing.

On aerial photographs in general, the drainage pattern on shales is conspicuous, fine, highly integrated, fairly uniform, and commonly described as dendritic (see fig. 18.3). In arid climates, with seasonal floods, dissection of the shales is greater, hill slopes are steeper, and the drainage pattern on very small scale photographs is coarser. The tone of shales in humid areas is fairly dark and may be mottled. Both the natural vegetation and land-use on shales vary greatly.

(iii) *Limestones.* These are fairly easy to identify in aerial photographs, but vary greatly between arid and humid conditions. Limestones are characterized by a sparsity of water courses, even in high-rainfall areas, and record in light grey tones on the photographs. Valleys will usually be flat-bottomed, steep sided and devoid or almost devoid of stream tributaries. Frequently, a surface drainage pattern is very poorly developed or absent except for sink-holes. Sink-holes are photographically characteristic of many limestone areas in humid regions by providing a 'pock-marked' pattern (fig. **17.8**). The density of the sink-holes is related to the age and solubility of the limestone; and they may be round or oval, may coalesce, and in cross-section may form deep or shallow troughs, saucers or funnels. On youthful land-forms, they are sparse or absent. On tilted limestones the sink-holes are oval. Occasionally the sink-holes become flooded and develop into small lakes. In the more arid areas there is often an accumulation of talus at the foot of limestone escarpments. In limestone areas vegetation and land-use vary greatly; and possibly the photo-interpreter will be impressed by the extreme contrasts between the limestone areas and adjoining rock groups. In wet tropical regions 'tower karst' scenery, consisting of steep hills rising very abruptly from a flat plain, is commonly developed on limestone.

Sometimes *inter-bedded rocks* are formed by strata of limestones, shales and sand stones. Each will retain its own characteristic pattern and tone. If the bed-folding is prominent, it may be worth while to study the sequence of rock types where the vegetation is sparse on a steeply tilted bed; and then to relate these observations to other areas by following the sequence of folding across the photographs.

Chalk, and other soft porous limestone. Chalk is extremely porous and therefore even more conspicuous than limestone by the absence of streams. As chalk is not readily soluble in rain-water, sink-holes are not present. The ground is usually undulating with gentle slopes in the direction of dip. Dry valleys are common.

Coral. This is limestone of organic origin formed in the vicinity of old shore-lines. The line of the uplifted coral is usually parallel to the present coastline and not more than a few hundred feet above it, although it may be a considerable distance from the sea. There is usually a gentle landward slope and a more or less vertical bluff facing seaward. Streams and sink-holes are absent, and there is usually little or no evidence of erosion.

METAMORPHIC ROCKS

Metamorphic rocks were formed when either igneous rocks or sedimentary rocks were exposed to considerable heat and/or pressure. During the process, the original

TABLE 17.A
Common sedimentary, igneous and metamorphosed rocks

Source	Sedimentary rock	Metamorphosed rock
Gravel	Conglomerate	Conglomerate schist
Sand	Sandstone	Quartzite schist
Mud, clay, silt	Shale	Slate
Calcareous ooze	Limestone	Marble, schist
Peat	Bituminous coal	Anthracite
Granite		Granite gneiss
Diorite		Diorite gneiss
Rhyolite		Mica schist
Basalt		Hornblende schist, etc.

rocks were profoundly altered, new minerals have been produced, and new structures imposed. A selection of common sedimentary and igneous rocks and the corresponding metamorphosed rocks is given in table 17.A.

The geologist usually divides metamorphic rocks into two broad groups, foliated and non-foliated; but on aerial photographs, it is frequently impossible to do more than recognize the rock as being metamorphic. However, with a good knowledge of local geology, the photo-interpreter can sometimes recognize foliated rock by the characteristic stream pattern, by its tone, and by its curved and constricted drainage pattern. Some photo-geologists have been successful in separating slates, quartzite and serpentine.

As a major group, metamorphic rocks are characterized in the photographs by their greater bulk in the stereo-models than non-metamorphosed sedimentaries, by a greater resistance to weathering and by evidence of diastrophic movement. For example, metamorphosed sediments are frequently lightly folded and steeply dipping; and evidence of strains and stresses is provided by frequent fissures. When metamorphic and igneous rocks occur together in the photograph, they can frequently be separated by differences produced in weathering, by tone (e.g. granites whitish, slates dark) and by the texture provided by rock fissures (e.g. many joints in granites). Metamorphosed sediments often provide a more rounded form as compared with igneous rocks. If the igneous material is basic in origin, and the metamorphic rock siliceous (or vice versa), conspicuous differences in the vegetative pattern may be present and help to separate the rock groups.

DEFORMATION STRUCTURES

As mentioned in the previous section, the recognition of lineations is helpful in identifying and classifying rocks from the aerial photographs. In studying the lineament in one part of the photographs, where the vegetation is sparse, it is often possible to follow the bedding and folding through to areas of dense vegetation and use the information in typing vegetation boundaries.

Tectonic forces within the earth cause or have caused movements on or near the earth's crust. Such movements may be sudden in the form of an earthquake. For example, in the catastrophic Wellington earthquakes of about a hundred years ago, the beach was raised within a few hours by 8 ft or more. In contrast, in northern Norway and Sweden, the process of raising the coastline has been gradual since the last Glacial Period. In this time, the coasts have been raised by about 900 ft (Emmons *et al.*, 1939).

Features characteristic of diastrophic movement include faults, folds and raised river terraces. The latter are recognizable on photographs by tone, step-like surface, steep edges and recently developed stream gulleys. *Faulting* results from a fracture in the earth's surface accompanied by movement. Most frequently, it is present on photographs as a well-defined lineament (fig. **17.9**). Minor faults produce lineaments for short distances and are often parallel to major faults, and the presence of one helps to confirm the other. Even more frequently, faults are observed on the photographs due to the offsetting of bedding lineaments. Sometimes a fault line follows a scarp.

As distinct from a fault, a *joint* is a fracture along which there has been no noticeable movement. It is often difficult to separate faults from joints, but a detailed study of fractures has shown that these can be accurately located on aerial photographs, and their location is enhanced and extended by vegetation and topography (Boyer & McQueen, 1964). Normally the joints are much shorter, less conspicuous individually, more frequent in the same direction and often provide an etchlike appearance to granites.

GEOGRAPHIC AND GEOLOGIC CONSIDERATIONS

Particularly in arid and semi-arid regions it is possible to determine the strike, dip and the characteristics of folding in the rock and the structure geologically of the earth's surface as recorded on the photographs. Fig. 17.2 illustrates strike and dip of an anticline; and a simple method of determining the strike and dip in degrees was given in Chapter 15.

18

GEOMORPHIC CONSIDERATIONS

THE relief of the earth's surface can be placed in three orders (von Engeln, 1942). The first order is provided by the continents, oceans, basins, etc. The second order covers the geological conditions described in the last chapter, including extensive plains, plateaux and mountain ranges; and the third order encompasses land-systems including valleys, ridges, and aeolian, glacial and fluvial deposits. This third order using a biological term, could be described as 'specific' geomorphology whilst the second order would be 'generic' geomorphology. Sometimes areas of the second order on the earth's surface are termed land-regions or complex land-forms; and areas of the third order are described as land-systems, unit land-forms, simple land-form units or land units. There is, however, no general agreement on defining a land-system and its subdivisions. A subdivision of the land unit may be termed a land component but in ecological and forest studies it is normally desirable, once the unit land-form has been recognized on the photographs, to identify, or attempt to identify, the plant associations or photo-communities or forest types. Sometimes the land unit is best identified by the photo-community encompassing it, and on other occasions the photo-community can be best identified by the land unit and land component (Howard, 1966).

Lueder (1960) described a *unit land-form* as 'a terrain feature or terrain habit, usually of the third order, created by natural processes in such a way that it may be described and recognized in terms of typical features, wherever it may occur, and which, when identified, provides dependable information concerning its own structure and either composition and texture or uniformity'. The generality of this statement avoids identifying the geomorphic unit as the unit of land-form, but permits it to be so if needed. As such, the land-form unit can be helpful not only in the interpretation of the photo-community but also in soil studies, erosion classification, forest typing and land-type classification, as it provides a unit within which to work, if so desired, and can be delineated without a separate and detailed geomorphic study.

Christian and Stewart in Australia have expressed the terrain unit in terms of topography, soil and vegetation. Christian (1958) in defining the land unit describes it as part of the land surface having a similar genesis and having 'major inherent features of consequence to land use, namely, topography, soils, vegetation and climate'. At the same time he refers to the *land-system* as 'an assembly of land units

205

which are geographically and genetically related'; and previously Christian & Stewart (1953) had described a *compound land system* as a group of land systems enclosed within the one boundary for convenience of mapping. Becket & Webster (1962) have called their unit of subdivision a *facet*. This has distinct form, rock, soil and water regime. They view the physical region as a pattern of large units (i.e. recurrent landscape patterns). As mentioned previously, other workers have recognized the need to divide the landscape into natural units (e.g. site: Bourne, 1931; facet: Wooldrige, 1932; physiographic unit: Milne, 1936; ecotope: Tansley, 1939; physiotop: Troll, 1939). An example of land classification, developed by Becket and Webster, is shown in fig. **18.1**. Brink *et al.* (1966) have suggested that land unit be used as a general term and land facet and land system as the particular terms.

When beginning to delineate or to examine the landscape, it is convenient to identify the land-form as originating by deposition or degradation; and to classify the units according to the weathering agencies (i.e. aeolian, glacial, lacustrine, marinal, fluvial). Often land-systems indicative of both deposition and degradation will be present on the same photograph. A knowledge of the bedrock and recent geological history can be helpful in suggesting the composition of the materials. In some cases when it is possible to identify the parent material from the photographs, it is possible also to deduce whether the land unit is basic or acidic. Measurement or estimation of slopes can also be useful in classifying or delineating land units. Land unit terminology is summarised in Appendix iv.

(a) Degradational land-forms

The recognition of land surfaces on the aerial photographs as being degradational in origin is important. The photo-interpretation of erosion features, erosion agents and drainage patterns can be helpful in recognizing the rock types, soil types and vegetation types. Normally the erosion pattern for the same bedrock at different stages in the erosion cycle will provide a different land-form on the photograph. To become fully efficient in photo-interpretation, it is important to familiarize oneself with the local appearance of the erosion cycle on photographs. Frequently with degradational surfaces, the land unit is delineated by parent material (e.g. granite), erosion agent (e.g. glaciers), development of erosion cycle (e.g. youthful) and topography (e.g. plateau).

Slope is often a useful or sensitive photographic criterion to use in breaking down the land unit into components which are likely to be influenced by similar micro-environmental factors. On chernozem soils, extremely small changes in slope have been recorded to greatly influence the erosion process (Platonenko, 1963). Semenova (1959) reported similar soils with 1° to 2° slope to be slightly eroded and with 3° to 4° to be strongly eroded. The features of the erosion surface will vary greatly according to whether they have developed in consolidated material or unconsolidated material. Sheet erosion, blow-outs and land-slides are often associated with unconsolidated materials. Under these conditions changes in slope may define not only the land component but also the land unit.

Both wind and glacial erosion are far less common than water erosion. Glacial erosion often spectacularly transforms the entire features of the landscape. During the process of glacier movement, the valleys will have been scoured, bedrock may be exposed, hanging valleys may have formed and lower down the valleys glacial deposits may occur.

Wind erosion features are often observed on photographs of arid areas. In these areas, the impact of the wind itself has been sufficient to remove large quantities of surface material. As pointed out by Emmons *et al.* (1939), the stony deserts (e.g. large areas of the Sahara) have developed by this process, leaving only the coarser gravels and boulders. More spectacular is the abrasion of softer rocks by windblown sand, providing conspicuous features in arid and semi-arid areas, e.g. buttes, mesas (fig. **17.7**).

DRAINAGE PATTERN

Gullies are probably the most important single class of feature in providing valuable data relating to the parent rock, soil and land-form. Characteristics to be observed on the photographs are the pattern and intensity of the gullies, length, depth,

FIG. 18.2. Diagrams can be prepared which show the basic profiles of common valley shapes and gully shapes in a region. Three examples of gully shape, according to the soil constituents, are illustrated above.

width, cross-sectional area and the profile. Usually photographs at a scale of 1/8,000 to 1/10,000 or larger are required when examining gully shape. Several basic profiles can often be discerned, as shown in fig. 18.2.

Erosion is greatest downstream or at the foot of long slopes. Erosion often seems to be associated with contiguous areas of changing slopes rather than with steepness and length. Aspect and exposure alone do not appear to be usually associated directly with erosion.

The difference in the soil types and vegetation types on 'ravine' slopes are more easily discernible on photographs at 1/5,000 than 1/10,000 (Semenova, 1959). The soil-forming parent material is an important factor in erosion development. For example, loess-like clay loams are able to absorb water readily and therefore do not erode as readily as compact, soft clays, which often record as whitish grey. In colour photographs, at 1/10,000, the land-forms of the soft clay parent material can be more easily traced than on black-and-white photographs between 1/5,000 and 1/25,000 (Semenova, 1959).

P

GEOMORPHIC CONSIDERATIONS

Drainage pattern is provided on the photographs by the arrangement of the surface drainage, involving density, system, orientation and uniformity of the streams and rivers. The surface drainage pattern is largely a reflection of the physical characteristics of the parent and weathered material and the climate. Other factors influencing the drainage pattern include the type and density of the vegetative cover and the natural moisture content of the soil. As mentioned, impervious materials such as clay and shale resist infiltration by surface water and promote run-off. The greater run-off under these conditions provides a well-formed, relatively dense drainage pattern. If the surface clay is replaced by surface silt, then erosion is more rapid and the drainage pattern more spectacular. Over coarse-grained materials, such as sandstones, the drainage pattern is usually poorly developed and on very porous materials, such as chalk, surface run-off is absent or nearly absent and the drainage pattern will be very poorly developed.

The concept of classifying drainage patterns is not new. The basic classification into six broad types and twenty-four modifications was made as early as 1932 (Zernitz). The six basic types are dendritic, trellis, radial, parallel, annular and rectilinear. Their relation to land-forms, soil and bedrock identification from aerial photographs was reviewed by Pervis (1950). Horton (1945) has put forward suggestions for a quantitative approach to the study of drainage patterns. This may help to avoid using only qualitative terms such as youthful or mature, poorly drained or well drained. Quantitative expressions include stream order, drainage density, bifurcation ratio and stream-length ratio. In comparing regions of similar types of parent material it was observed that the drainage ratio obtained for a small area of Tanganyika near Lake Nyasa (Shaw, 1953) was much higher than figures provided by Horton.

Some commonly recorded drainage patterns observed on photographs is given in fig. 18.3 (see Howe, 1960). Those shown include lacunate, which commonly occurs on limestones, kettle type, which commonly occurs in glacial drift, the colinear type of desert areas, radial of granite, dendritic of sedimentaries, braided of alluvial flats, and pinnate of shales. Indiscriminate use of pattern alone for identification can result in gross misinterpretation.

A drainage system may be classified by its sequence of branching. The parent stream is generally considered to be of the first order, and the streams joining this are the second order and their tributaries the third order. Replacing the third order or joining them frequently are short, numerous water courses or rills. Third-order streams are often important in interpretation. Lueder (1960) suggested that primary drainage may be arbitrarily defined as those streams which display a definable expanse of water surface at a scale of 1/63,360.

Zinke (1960) pointed out that the density of the total length of the streams per unit area in the watershed is an expression of rainfall and infiltration capacity, and can be regarded as a quantitative physical description of the watershed. The total stream length can readily be measured on the photograph and, knowing the area, the density can be calculated. Ray & Fischer (1960) concluded that circular sampling areas gave the best determination of the drainage density. They plotted drainage density against various photographic scales and obtained a linear relationship for

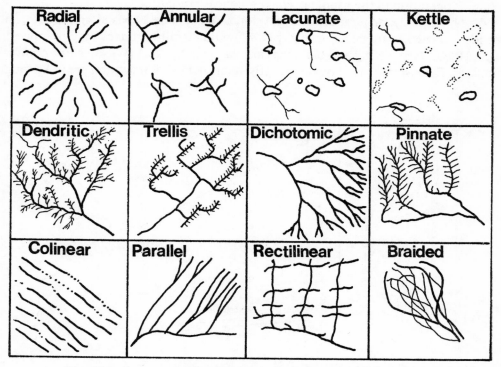

FIG. 18.3. Common drainage patterns observed on aerial photographs.

several rock types as shown in fig. 18.4. The figure also helps to illustrate the considerable variation on drainage density in different parent material (cf. figs. *9.1*, **17.4**, **17.5**).

(b) Land-forms by deposition

AEOLIAN LAND-FORMS

Aeolian land-form units develop from unconsolidated deposits of sand and sometimes loess transported by the wind. In arid climates, the aeolian deposits may be barren of vegetation, but in high rainfall areas, or under conditions providing adequate soil moisture, the deposits may be covered by woodland. Aeolian land-forms are often associated with lacustrine and marinal land-forms.

On photographs, as mentioned in the *Manual of Photo-interpretation*, the deposits have three basic characteristics: (i) a repetition of shape, (ii) an impression of being windswept and (iii) no surface drainage pattern, unless old and stabilized. Under conditions favouring complete vegetative cover on sand-dunes, there is normally a change in vegetative structure between the sloughs (troughs) and crests, which is conspicuous on the photographs. Often it is convenient to classify dunes according to their orientation to the wind, i.e. convex, concave or parallel.

Most commonly, aeolian deposits develop as unstabilized coastal sand-dunes, and are high in calcareous material (e.g. pH 8.0). Old stabilized dunes may be siliceous (e.g. pH 6·0). Both young and old dunes are shown in fig. **22.2** (Wilson's Promontory, Victoria). Extensive areas of dunes recently stabilized by afforestation are to be found

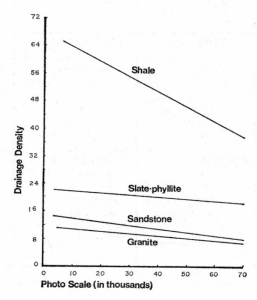

FIG. 18.4. Drainage density determined from photographs at several scales for four different parent materials.

at Formby, Lancashire, Culbin in Scotland, Pembrey and Newborough in Wales, in the Landes in France, along the west coast of North Island, New Zealand (e.g. Palmerston North) and in Jutland.

Aeolian deposits and their shape are a product of climate, topography and source of material. A source of material is provided by an accumulation of relatively fine-grained, discrete particles of sand or silt (loess), at least occasionally in a dry state, and relatively free of encumbrance by vegetation. It occurs continually in an area of coastal deposition, or it may be caused by drought or by land-use, e.g. over-grazing.

Distinctively shaped dunes with well-marked edges can be observed on photographs (e.g. fig. **13.4**). Barchane dunes, typical examples of which are in North Africa and Peru, are large and horse-shoe or crescent shaped and attain heights of

300 ft. They occur where the wind has nearly complete mastery over the sand. In Victoria, similar land units, termed lunettes, have been provided by long-stabilized deposits but these are concave to the wind's direction. They are crescent shaped, are derived from silt, including airborne dust brought down during storms and are on the leeward side of swamps and lakes. Their height is usually under 25 ft. In the Simpson Desert of central Australia, the shape of the dunes has earned them the name 'tuning fork' dunes (fig. **13.4**). Crests are in the direction of the prevailing wind.

Frequently, the crests of inland dunes are conspicuous by being transverse to the prevailing wind. The windward side is gentle with a slope of usually 5° to 15°, and the leeward side is steep, having a slope of up to 30°, and may be scalloped on unstabilized sand. Blow-outs may be observed on the photographs by their whitish tone and shape. In stabilized areas of silt and sand, subject to erosion, the gully pattern is conspicuous with often abruptly ended side gullies.

GLACIAL LAND-FORMS

Associated with the glacial land-forms are features given in standard school geography books, including terminal moraines, morainal deposits (i.e. till) and glacial lakes. Evidence of glaciation in bygone days is extremely important to recognize, as it has the effect of greatly altering the land-form, and of complicating the interpretation of the geomorphology, land units, soils and vegetation.

Three important effects of glaciation to be observed on photographs are the removal of surface material, leaving the subsurface exposed, the carving out by ice blocks of valleys and the deposition of boulders and clay as the glaciers thawed (fig. **18.5**). Glaciated surfaces on the photographs may provide parallel straight ridges resulting from crevasse filling or ice scouring and the deposits often have distinctive forms. Many small depressions may be formed in the path of the moraines providing a scallop-patterned landscape. These shallow undrained depressions are termed kettles in Canada. The term glacial drift embraces all rock material transplanted by glacier ice and includes till plains high in clay and the stony clays of Scotland.

Consolidated drift deposited below the ice as it receded is termed a ground moraine (fig. **18.5**). These may cover extensive areas and are characterized by numerous large depressions occupied by bogs (fig. **18.5**) and lakes having often no obvious relation to major streams and higher land. Terminal moraines are distinguishable from ground moraines by their height and 'knobbly ridges'. Ground moraines provide small smooth oval hills. These are termed drumlins. Their height varies between about 30 ft and 150 ft but they may be 1,000 ft to 3,000 ft long and over 500 ft wide (Emmons *et al.*, 1939). Their gentlest slope is in the direction in which the ice moved. In contrast eskers (fig. **18.5**) are of glacio-fluvial origin, and were formed by streams of melting ice under the glacier. Eskers are snake-like in shape, comprise sorted material, mainly gravel, and are of little width and height; but may be several miles long. Sometimes the gravel was deposited as small hills at the outflow of the glaciers, as kames (fig. **18.5**); and at other times, the ice melted fairly rapidly at the edge of the ice sheet, providing alluvial fans, or outwash plains.

FLUVIAL, LACUSTRINE AND MARINE LAND-FORMS

Fluvial sediments are deposited by flowing water, and provide land units represented by alluvial flats, upper surfaces of deltas and piedmont sediments which accumulate at the foot of mountains as a result of sheet erosion and soil creep (e.g. talus) (fig. **17.7**). Lake deposits are formed by the combination of aeolian action, fluvial action and wave action and may result in extensive deposits of organic matter, e.g. peat. Marine deposits result from fluvial and sea action, forming deltas, from sea and wind action and possibly from partial emergence of the sea floor. Recognition of fluvial, lacustrine and marinal deposits from aerial photographs will suggest to the interpreter that the material is sorted and may indicate the class of material.

Marine land-forms. An initial stage of shore-line emergence is frequently the emergence of a flat coastal plain covered by unconsolidated marine sediments. This is followed by the development of barrier beaches, of off-shore sand-bars and lagoons, and finally the silting of the lagoons to form salt marsh. Other emergent shore-lines include deltas and alluvial plains. Mangroves are usually associated with tropical emergent shore-lines.

Colour photography is being increasingly employed by the U.S. Coast and Geodetic Survey for detailed maps of the low and high water shore-lines and for penetrating the water to 5 to 10 ft in northern waters and 60 to 70 ft in the Caribbean. Rocks below the water-line and surface structure of the sea bed can be adequately recorded on the photographs (Smith, J. T., 1963).

Lacustrine land-forms. Lakes are formed by both degradational and depositional processes. An example of the former are basins formed by glacial ice in geologic time; an example of a combination of the processes are the crescent-shaped oxbow lakes (fig. **18.6**) formed when a loop of a meandering river is cut off as a river erodes a new course on a flood plain. Crater lakes may occur in areas of past volcanic activity (fig. **17.6**) and many of the larger lakes of the world, e.g. Lake Nyasa, were formed in depressions or faults left by the warping and folding of the earth's surface. The slopes of the shore-lines of each are frequently steep; and usually an examination of aerial photographs will suggest the forming process.

Resulting from gradational processes, many smaller lakes gradually disappear from the earth's surface either due to the deposition of materials and diversion of inflow waters or in dry regions by evaporation and deposition (fig. **18.6**). On aerial photographs, during the intermediate stages, lacustrine land units can be recognized in the vicinity of the existing lakes by shape, slope, tone, texture and existing vegetation. Raised beaches are sometimes present indicating the level of the lake in a previous fluvial period or prior to diastrophic movements. In arctic Canada, beaches in granite or gneiss areas are often conspicuous, being formed of sand- and gravel-size particles. Beaches in shale areas are rare and poorly developed. In sandstone areas beaches are intermediate but poorly preserved (Fletcher, 1964). Often the serial change in vegetation from aquatic through marsh to dry-land communities can be recognized on photographs. When deposition is complete, the lacustrine deposits may still be identified on photographs by some of the features mentioned above, and including a 'tracing' of the old shore-line.

In arid regions, often the old exposed lake beds may be accurately mapped from the photographs. Frequently, the soils are saline and conspicuously whitish. In desert areas, playas may be formed. These are flat-bottomed depressions, which flood in the wet season and in the dry season record as snow-like surfaces on photographs due to the high reflectivity of the alkali.

In humid regions, besides the development of organic matter in certain soils, e.g. A_0 or humus layer of podsols and up to about 15% in chernozems, organic matter accumulates in layer 'pockets' resulting from the combined effect of climate, topography and anaerobic conditions. Such deposits of organic matter are provided by glacial action giving kettle-swamps, sag-and-swell marshes and permafrost swamps; by fluvial action giving deltaic swamps and ox-bow swamps; by fluvio-marine action giving mangrove swamps, extensive 'saltings' and lagoon-filling deltaic swamps; and fluvio-lacustrine action giving estuarine and deltaic swamps, lagoon-filling and plain swamps. Identification from the photographs of the causative agent helps in the recognition of deposits of organic matter.

On aerial photographs, exposed organic matter usually records in dark greys or black and the vegetative pattern is often distinctive. On chernozem soils, distinctive tonal differences on panchromatic photographs have been recorded for soils with 2·4% to 3·4% humus.

Fluvial land-forms. Depositional fluvial land units are not nearly as important as those of degradational origin; and are frequently associated with other land-forms. Fluvial land-forms (fig. **18.6**) are often conspicuous by river terraces, deltas and deltaic plains and by fluvial fans (fig. **17.7**) and ox-bow lakes (fig. **18.6**). Deltas and alluvial deposits are conspicuous by their flatness and gentle slope (e.g. of 1° to 5°) and possibly the presence of braided streams. Braided streams on aerial photographs provide an 'anastomosing' pattern of streamlets, (fig. 18.3) which are separated by bars of gravel, sand or silt. They result from the overloading of streams with sediment at times of flood.

In arid and semi-arid areas, the erosion products sometimes provide filled valleys due to the products being moved downward by the torrents at times of heavy rain. Colluvial material, mostly rock detritus, dislodged from cliffs and precipitous bare mountain-sides, is often deposited on lower slopes as talus/screes, resulting from the action of gravity, wind and water (fig. **17.7**). Slopes are usually steep (e.g. 25° to 35°) and the vegetation stunted, sparse or absent.

GEOMORPHIC CONSIDERATIONS

(c) Soils

For convenience, soils have been placed in Chapter 18. Soils are defined by their type and profile; and as it is impossible to recognize profiles on the aerial photograph, they have to be identified by examining the geology, geomorphology, vegetation and tone or colour of the surface soil. Photographs help in providing soil boundaries, prior to soil classification in the field.

Soils comprise both the weathered material of the parent rock and associated organic matter derived from plants. Sometimes in photo-interpretation it is helpful to consider each separately. In glacial areas the weathered materials will contain both transported and local material. As soils are often hidden by trees and herbs in humid regions, interpretation is frequently achieved only by an indirect study using vegetation to indicate soil type. For example, in Ceylon, on 1/15,840 photographs the high ridge forest between 2,500 ft and 4,000 ft was observed to be associated with shallow soils over slab-rock (de Rosayro, 1958).

In black-and-white photographs, the tone of two differently coloured soils may record similarly and differences can only be shown up by the taking of colour photographs. Colour photographs help not only in correct identification, but also in more accurate and quicker boundary delineations (fig. **18.7**). This in turn is helpful to the photo-ecologist.

When comparing soil maps with photographs, it is often helpful to the photo-interpreter to ascertain what was the purpose of the soil map. It may have been prepared for engineering, agricultural, geologic or pedologic studies; and for each of these purposes a soil map with different boundaries would result. The engineer is primarily concerned with soil for cut and fill. He is interested in the particle size of the sand and gravel and the percentage of clay and silt. The geologist tends to classify soils in relation to their geologic origin, the weathering process and the method of transportation of the material. The agriculturist and forester are interested primarily in site quality or soil fertility.

If the soil can be observed on the photographs, it will usually be found that tone and hue on the photographs depends on the colour of the soil, humus content, moisture content, salinity and possibly structure. Sands and gravels usually have a light tone (see e.g. Fletcher, 1964). When exposed, the B-horizon usually records photographically in distinctive tones. In practice, this can be helpful in detecting erosion.

In tropical areas, where for a very long time the vegetation has not been disturbed, the ground vegetation may be a good indicator of soil conditions, as was recorded in Malawi (Muir, 1955). In Canadian forests (Losee, 1942), in tropical woodlands (e.g. Bourne, 1931a; Robbins, 1931, 1934; Howard, 1965) and tropical rain forest (e.g. Swellengrebel, 1959), the vegetative types have been observed to be closely associated with the soil types (fig. 18.8). In south-eastern Australia, krasnozem soils often occur under wet sclerophyll forests, whilst podsols are commonly associated with dry sclerophyll woodland. Bourne (1928) suggested there were possibly 1,000,000 square miles of woodland in Africa with a close association between vegetation and soils. However, in the Shire Valley, Malawi, differences of tone and texture, on photographs (1/35,000) of the cultivated areas, were found to reflect not the type of soil, but differences in land-use history (Muir, 1955).

Platonenko (1963), in an attempt to quantify photographic soil-studies, carried out a mathematical analysis of the correlation between soil, relief and photographic tone in the vicinity of the Omsk. The maximum correlation was obtained between soil and photographic tone for cultivated land ($r_{sp} = 0.858$). This was due to the maximum contrast being provided by the humus content and degree of 'solodization'.

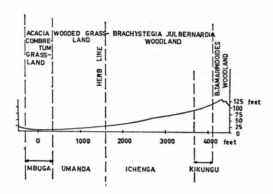

FIG. 18.8. Catena diagram showing the sequence of the principal soil types and the associated ecological types from valley bottom (mbuga) to hilltop (kikungu), near Tabora, Tanzania.

Correlation coefficients of less than 0·5 were obtained for soil and relief. It was also reported that in Siberia the darkness of the tone increased regularly with the increasing columnar structure of the 'solenetzic' soils.

Differences in relief and in land-use portrayed on the photographs of areas in England have only occasionally been helpful in recognizing soil conditions. For example, in Lancashire peat or alluvium can sometimes be identified by the level terrain, open ditches and fields without hedges, and adjoining areas of till are conspicuous by the undulating landscape and marl pits. After ploughing in Cambridgeshire, 'Fen clay' can be clearly distinguished from the calcareous loam (Curtis, 1966). Tonal differences provided by the vegetation, particularly with spring photography, have been found useful at higher elevations in the United Kingdom. For example, in Galloway and Exmoor (Curtis, 1966) and west Wales, *Juncus* is associated with low humic or peaty gleys, *Pteridium* with acid brown earths, *Eriophorum* with peat and *Molinia* with peaty gley podsols.

19

REGIONAL BIOLOGICAL SURVEYS

INTRODUCTION

IN the ensuing chapters, interest will be directed towards the qualitative and quantitative description of units of vegetation, whether these are forest stands or plant communities. In recent years, some plant ecologists have tended to use the term 'stand'. This may be unfortunate as 'stand' is a long-established term used by foresters. Greig-Smith (1964) defined stand ecologically as denoting 'any area, the vegetation of which has been treated as a unit for purposes of description'. Whilst Husch (1963) in *Forest Mensuration and Statistics* defined a stand as 'an aggregation of trees having some unifying characteristics, which occupies a specific area of land'. In giving examples, Husch quotes stands as being defined by age or size of a group of trees, drainage lines or legal boundaries and mentions that a forest stand defined by its tree species composition is referred to as a *forest type*. Husch's forest type seems to correspond approximately to Greig-Smith's 'stand'.

In the present chapter, description and discussion will be directed not towards local studies but regional surveys. The term 'survey' is used as being interchangeable with cruise, enumeration and inventory, although this is not always etymologically correct nor conforming to nomenclature as developed by local custom. For example, inventory may suggest a detailed stocktaking, whilst enumeration may simply imply a record of species by numbers in each of a few size classes. Attention will be paid to aspects of identification and classification not covered elsewhere. These include agricultural surveys, entomological surveys, game surveys and pathological surveys. It should be appreciated that much of what will be presented can be applied to local studies, and that the basic planning of regional surveys is similar, whether the survey is primarily concerned with agriculture, forestry, land-use or phytogeography. Usually a survey fulfils two functions, namely, to provide a map of the region with important strata delineated on it, and quantitative data for present policy decisions and future planning. Under certain circumstances the preparation of a map can be avoided.

A regional survey may cover all the earth's surface within a geographical region, as in the land-use survey of the Ryukyus (Cunningham, 1952). It may encompass certain land types only within a region, as in forest surveys. Such surveys include a teak forest survey in Thailand (Loetsch, 1957), a tropical rain forest survey in

216

the Amazon basin (Heinsdijk, 1961), a tropical woodland survey in northern Iran (Rogers, 1960, 1961), the recent survey of Maine (Young, 1963), a red gum survey of the Murray river (Davies, 1954), an Araucaria assessment on Norfolk Island (Cromer and Aitken, 1948) and the regional inventories in Canada (Seely, 1960). Of ecological interest is Curtis's *The Vegetation of Wisconsin* (1959), Floret's paper on ecological mapping in Tunisia (1966) and the state-wide ecological survey of the national parks in Victoria which was initiated in 1966. Finally, surveys may be undertaken for specific purposes and encompassing entire national territories (i.e. national inventories). These include national forest inventories in France (personal communication, Huguet, 1961), Spain (1965 onwards), Portugal (personal communication, Stridsberg, 1966), Australia (Carron & Hall, 1954; Sims & Hall, 1955; Cromer, 1960), Cyprus (Polycarpou, 1957), the exotic plantation survey in New Zealand (personal communication, T. Wardrop, 1964) and the indigenous forest survey of New Zealand (Masters *et al.*, 1957). A paper on surveys applicable to extensive forest areas was presented at the Fifth World Forestry Congress (Wilson, 1962).

(a) Identification and classification

Probably in the initial stages of any survey, an attempt will be made to identify directly from the photographs the objects of interest, whether these are tree species, agricultural crops, specific insect damage or disease. Failure to attain acceptable results by direct identification will lead to classification of the photographic data into recognizable groups within which sampling will be carried out in the field. The overall operation is often referred to as typing and coding or simply typing. From a statistical viewpoint, typing helps to reduce variations within the classes. When classifying the data contained in the photographs, relatively few criteria should be chosen, as the combinations provided by even a few important criteria can easily result in a hundred or more different strata. Usually a minimum size is fixed for type delineation, below which areas are not broken down. For example, in forest inventories in Canada, the minimum area of stand delineation varies between 5 and 100 acres (Seely, 1960).

Normally, in regional survey, the photographs are at a small or very small scale. Probably the largest scale available would be 1/15,840, and frequently scales are between 1/20,000 and 1/63,000. In certain categories of survey (e.g. pathological studies), large to very large scale photographs are essential if success is to be achieved; and may be achieved by 'pin-point' photography (see Chapter 1) of randomly or systematically selected areas. Preferably the region should be covered already by normal smaller scale photographs.

LAND-USE SURVEYS

Land-use surveys call for the objective application of special skills acquired in other fields of 'natural resource' studies. Factors affecting land-use are directly determinable or can be inferred from the aerial photographs. Often virgin land achieves an ecological equilibrium, which is only disturbed by man's intervention. Man then has to establish a new equilibrium. As the original or native vegetation of a virgin area is therefore possibly the best representation of the combined effects of all the factors (e.g. climate, parent material, topography, soil), it is very important to identify and delineate the original vegetation recorded on the photographs, and a unique opportunity exists where the negatives of very old photographs are available on which the former vegetation is recorded. Frequently in past decades, land-use surveys have been based on field studies, using a single environmental feature or factor, and in recent years, soil surveys have often provided such a basis for land-use determination with little reference to other features (Downes, Gibbons *et al.*, 1957). The use of aerial photographs, however, helps considerably to correct the single-factor approach, as it is possible to evaluate the area using the maximum number of characteristics recorded on the photograph.

On photographs at scales as small as 1/90,000 (fig. **19.1**), induced vegetation can readily be separated from the natural vegetation. Land currently used for agriculture can normally be separated from forests and 'scrub' covered land, by tone, texture and presence or absence of man-made boundaries and indications of man's activities (e.g. hedges, barns, dams). The general pattern of land-use can be observed and recommendations made on future land-use. For example, poor drainage, rock outcrops or extensive erosion can all be observed on the photographs and so help to explain the reason for abandoned farms, or the need to change the pattern of land-use from agriculture to forestry. At larger scales the existing forest types may indicate that an area is economically better suited to agriculture than forestry, e.g. wet two-storey montane forest near Amani, Tanzania, more suitable for tea. In areas where agriculture is being discontinued, it is often helpful to examine the areas carefully under the stereoscope to detect photographic characteristics of the plant succession.

AGRICULTURAL SURVEYS

Normally, scales larger than 1/15,840 are preferred for crop identification, although interpreters have worked with enlargements of scales of 1/20,000 (e.g. Bomberger, Dill *et al.*, 1960), and occasionally scales as small as 1/31,000 (Howard, 1965) or 1/40,000 (Beard, 1941; Miller, 1960). In Iran 1/30,000 photographs were successfully enlarged to 1/10,000 for the identification of common crops, e.g. barley, wheat, alfalfa. On small scale photographs, specific agricultural crops frequently cannot be recognized; but grazing lands can often be classified as seasonal and according to incidence of erosion, and similar crops can be grouped (e.g. small grains, row crops, hay, permanent pasture). Near Nzega (Tanzania), grazing lands on photographs at 1/31,000 were classified according to stock-carrying potential as all-year grazing (erosion minimal), all-year grazing (erosion conspicuous), dry-season grazing, wooded grassland (at least 50% grassland) and regeneration/thicket (under 10% grassland) (Howard, 1965).

Woody-stemmed perennial agricultural crops may be considered for identification as being intermediate between annual farm crops and forest crops. They include bush crops (e.g. coffee, tea), small trees (e.g. citrus fruits, stone fruits) and vines (e.g. hops, grapes). As shown in fig. **19.2**, woody-stemmed crops are characteristically arranged in rows and are usually dark against the background of soil or drying grass in summer. It is helpful to their identification to measure the distance between rows. For example, the spacing of vines does not usually exceed 10 ft, whilst apples and pears and citrus may be at 20 ft centre to centre. Coffee may be planted at 8 ft by 8 ft; and in Tanzania may be protected by shade trees (e.g. *Grevillea robusta*) at a spacing of 30 ft or 40 ft.

The seasonal and yearly change of tone, texture and pattern of annual agricultural crops and leys may vary markedly within a few weeks, and much more so than temperate deciduous forests in spring and autumn. Therefore, in the interpretation of annual crops, it is important not only to relate the data of the photography to spring and autumn, but also to the customary times of sowing, cultivation and harvesting and local techniques of crop cultivation including degree of mechanization and irrigation, practice of animal husbandry and crop rotation. Ploughing, harrowing, planting, cultivation and harvesting will each provide a marked difference in tone, texture and pattern for the same area. It often helps in the recognition of a crop to appreciate the local influence of geology, land-form and slope on soil cultivation (see Brunnschweiler, 1957).

Cereal crops (e.g. wheat, rye, oats, barley) are either broadcast or drill sown, and develop a close and continuous cover at an early age. Unless it is local practice to drill sow one crop and to broadcast the other, or to autumn sow one crop and spring sow the other, it is normally impossible to identify the specific cereal. Exceptions occur and as Brunnschweiler (1957) observed, oats is usually the darkest cereal crop recorded on the photographs; but it is normally lighter in tone than ungrazed leys. Clovers and alfalfa usually record darker than cereals of the same age; and permanent sward may be darker than short-term leys. Using colour transparencies Colwell (1956) recorded that rye can frequently be identified up to maturity on account of its unique bluish green hue.

Texture, on photographs at 1/20,000 or larger, is normally visible until the crop is a few inches tall. When grain is sown across the ploughing or harrowing, a 'plaid' pattern may be produced (Bomberger *et al.*, 1960). As the cereal grows, the rows can be seen only if the distance between the rows is usually 2 ft to 3 ft; but maize rows on the other hand are 3 to 6 ft apart and rice is usually surrounded by bunds. Potatoes can be separated from the cereals, excluding maize, by their ribbon-like pattern and the tone of the soil not yet covered by the foliage. Other crops resembling potatoes are beans, peas, sweet potatoes and cassava. Frequently a root crop can be identified by a system of elimination.

REGIONAL BIOLOGICAL SURVEYS

ENTOMOLOGICAL SURVEYS

The successful identification of insect damage depends largely on specialist training, local knowledge and the association of 'photo-identifiable' plant species with specific insects. Strip photography or pin-point photography of high population centres are often adequate, as the boundaries of the infected areas are usually continually changing. The information obtained from the photographs is not used so much for mapping but to assess the seriousness of the outbreak and to plan control measures.

Scales used in the United States have ranged from 1/800 to 1/15,840 using black-and white film and colour film, e.g. Anscochrome, Super Anscochrome and colour infrared. Probably 1/2,500 to about 1/8000 have been preferred as a compromise between cost and successful identification of insect damage. Spurr (1960) commented that, in general, visual reconnaissance from low-flying aircraft is often preferred to aerial photography. Insect infestations photographed in the United States have included spruce bud-worms in Maine, gipsy moth in Massachusetts, bark beetles in California and the south-eastern states, weevils in New York State and balsam woolly aphid on silver fir in Oregon. For the latter, 1/10,000 colour photographs have been used to locate the boundaries of outbreaks, and 1/2,500 colour photographs for density counts of dead and dying trees. Little improvement was achieved by increasing the scale from 1/5,000 to 1/2,500 for damage to *Pinus coulteri* in California. The recording of damage on photographs may be possible within a few days of attack by defoliators and somewhat longer for sucking insects; but may not be discernible for one or two years in the case of bark beetles. In general, defoliated groups of trees tend to photograph in contrasty tones on photographs. In south-eastern Australia, phasmid damage (*Didymuria* spp.) could not be detected on black-and-white photographs; but this was probably due to the film–filter combination. On small scale Kodachrome 35 mm oblique transparencies, both light and heavily defoliated stands of eucalypts were detected by the reddish hues in the transparencies which contrasted with the greenish hues of healthy trees (fig. **19.3**). In Tasmania, the browning crowns of *Pinus radiata* attacked by *Sirex noctilio* (pine wood wasp) are quite distinctive from the air. In the United States, following change in the colour hue of the foliage of pines and Douglas fir attacked by bark beetles, it has been found that the damage can be recorded on black-and-white aerial photographs at a scale as small as 1/10,000 (Wear & Dilworth, 1955).

FOREST SURVEYS

Photographs are useful in providing rough estimates in answer to general questions, for more efficient control of ground work, for delineating sampling strata and, in conjunction with field studies, as a stock-map.

The first question arising in the application of aerial photographs to the quantitative estimation of the forest growing stock in regional surveys is what are the principal characteristics of the individual trees and groups of trees which can be evaluated from the aerial photographs. From Part I, it will be recalled that in species identification, the film–filter combination, quality of the photographs, season of the year in which the photographs were taken and photographic scale can all be important; but usually the photographs in regional forest surveys are at small or very small scale and taken with a panchromatic film and minus blue filter. In Canada, however, under a Federal–Provincial agreement, photographs have recently been at 1/15,840. The development of local forest inventory using photographs taken specially at the normal scales for a forest project will be pursued again later in Chapter 21.

Having unsuccessfully attempted to identify individual trees by species, it is common practice to classify the forest into stand classes or statistical sampling strata, using as the principal criteria forest types (i.e. tree species grouping), stand height, crown closure and sometimes tree counts and site classification. For example, in Iran (Rogers, 1961) vertical photographs at 1/50,000 were used to delineate four forest types by species (beech, oak, mixed hardwoods and cypress), three density classes and four stand height classes. When the photographs are at a very small scale, difficulties will be experienced in measuring tree height. Ten foot height classes should be practicable below 1/15,840, 20 ft height classes below 1/23,760 and 30 ft height classes below 1/40,000. Much depends on the characteristics of the forest, experience of the interpreter and quality of the photographs. In the case of eucalypt sclerophyll forest, even at 1/10,000, it may be necessary to use 20 ft height classes.

A few methods of stratification commonly used in forest surveys will now be briefly commented on. Stratification methods used by the U.S Forest Service for volume inventory surveys and management surveys were concisely summarized by Moessner (1963b). Seven methods are recorded as being based on field measurement of sample plots, eight methods rely on stratification of aerial photographs prior to field sampling, and three methods require the provision of a forest type map from which to sample. The latter is prepared from aerial photographs, previously stratified; and sampling involves division of the type map into 9, 4 or 3 strata. In the second group of eight methods, the forest areas on the photographs are divided into 28, 13, 7, 6, 4 or 3 strata. Some of the methods require photographs at scales between 1/10,000 and 1/15,840. The 28 strata method calls for the division of the forest area by topographic site into 1,000 cu. ft timber volume classes, estimated stereoscopically in conjunction with aerial volume tables. If stratification is by 1,000 cu. ft classes only, 7 strata are used. The 13 strata method divides these stands into 500 cu. ft volume classes using total height, crown diamter and crown closure, in conjunction with an aerial volume table. Stratification by utilization classes (saw logs, poles, saplings) and stocking (poor, medium and well stocked) have provided 9 strata, and by tree species 6 strata. In the methods using only 3 or 4 strata, the forest is divided into classes based only on crown closure, volume or utilization types.

When preparing the classification maps from the photographs for a national inventory of Australia, forest lands were delineated and classed as productive, semi-productive (i.e. capable of producing few logs) and low-grade woodlands. The productive forests were then classified into broad forest types: eucalypt forest, cypress pine, rain forest, mixed coniferous plantations, mallee or dwarf tree woodland and miscellaneous (e.g. mangroves). The productive forests were also stratified into stand size classes (regeneration, saplings, poles, mature and over-mature); height classes (170 ft and over, 130 to 169 ft, 90 to 129 ft, 50 to 89 ft, 10 to 49 ft and under 10 ft) and density classes (below 10% crown cover, 20% to 40%, 50% to 70% and over 80%). Photographic scales were frequently at 1/31,000 or smaller.

GAME SURVEYS

With the advent of game management providing a valuable source of revenue in countries with limited natural resources and providing important recreation in the highly industrialized countries, the use of aerial surveys is increasing. Surveys include counts of animals, birds and fish from light aircraft and helicopters, oblique photography from light aircraft using a hand-held 35 mm camera and pin-point and random strip photography with a conventional aerial camera. As animals are easily disturbed and continually moving, conventional strip photography with normal side-lap is usually not as satisfactory as oblique or randomized pin-point or strip photography. In a recent survey by helicopter of elephant in the Tsavo National Park (Kenya), it was found that previous ground census had under-estimated the number of elephant by 50%. Frequently, visual estimates are made at altitudes of 300 ft to 1,000 ft. Many animal species are disturbed by low-flying aircraft below about 500 ft.

Animal counts and population estimates, since the first aerial photographic surveys in the United States in the 1930s, have included sea-lions (Anon, 1946), caribou (Banfield, 1956), beaver (Knudson, 1951), deer (Leedy, 1948, 1953), duck (Chattin, 1952), seals and rabbits in Holland, greater snow goose (Spinner, 1949), pheasant (Miller, 1946), salmon (Kelez, 1947), herrings and sardines (Leedy, 1948, 1953) and trout. When counting the animals or birds, stratification into density classes, use of stereograms, use of transparent grids and pinpricking each image will all help in improving the speed and accuracy of the survey. Chattin (1952) recorded that more than 1,000,000 ducks may be located on 1,000 acres of water. On suitable photographs of large game animals young and old, male and female can sometimes be separately identified and counted (fig. **19.4**).

Besides being useful in animal counts, aerial photographs can be used to locate and delineate breeding areas, important grazing areas and migratory routes, and to provide sketch-maps both for protection and management. For example, in the vicinity of the Ugala river (Tanzania), kudu were observed to prefer the white ant-hill areas and reed-buck the *Acacia–Combretum* grassland. Both land units were readily identified and delineated on photographs at 1/31,000 (Howard, 1965). Also within the same land-system, it was noted that many species of the larger game animals congregated in the dry season on the wooded grassland, which was also recognizable on the photographs.

PATHOLOGICAL SURVEYS

In pathological surveys, photographs have been used since the early days of aerial photography. Photographic scales in the successful identification of disease have varied between about 1/800 and 1/20,000. Sometimes insect damage is mistaken for disease. Neblette (1927) published a paper on the application of aerial photographs. Tabenhaus *et al.* (1929) used aerial photographs in the detection of cotton plants killed by *Phymatotrichum omnivorum*. The photographs were taken at 250 ft to 400 ft using panchromatic film and a light yellow filter. In 1933, Bowden observed that terrestrial infrared photography was preferable to panchromatic film to detect virus disease of the potato leaf. In the infrared, the virus showed as dark streaks. In 1936, van Atta suggested that for panchromatic films a deep red filter (Wratten 70) was best for showing diseased tissue, and about the same time Eggert observed that damaged tissue of leaves of fruit trees caused by spraying or drought showed clearly on infrared film but not on panchromatic. Recently (1962) Brenchley & Dadd used aerial photographs successfully in the United Kingdom for the detection of potato blight (fig. **19.5a**); and colour photographs have been used for detecting diseased sugar beet. In the South Dandalup catchment area, Western Australia, die-back of *E. marginata* has been identified on panchromatic photographs (1/15,840) due to a reduction in the canopy and the death of the understorey (fig. **19.5b**). In the U.K., areas of Corsican pine damaged by *Fomes annosus* are easily identified in colour. In East Germany, 'spectrozonal' photographs provided five recognizable classes of needle cast disease of conifers. Also, conifers treated with nitrogen fertilizer recorded as orange and the non-fertilized plots as green (Wolff, 1966) (fig. **19.6**).

Using aerial infrared film and a Wratten 89 filter, Colwell (1964) was able to identify unhealthy orange trees in advance of detection on the ground. In recent years, Colwell (1956, 1964) has contributed greatly to an increasing appreciation of the use of aerial photographs in disease detection. He has shown that with the appropriate film–filter combination and at the appropriate seasonal state of development of the host plant and pathogen it is possible for the photo-interpreter to identify healthy cereals from a group of plants infected by virus and fungus. Black stem-rust of oats (*Puccinia graminis avenae*) can be detected on infrared and colour photographs in its early stage by tonal and colour differences. Rust (*Puccinia graminis trictici*) on wheat in heavy infestation, is readily detectable on large scale aerial Ektachrome film and on infrared film. On very large scale colour photographs (e.g. 1/2,000) the severity of the rust can be readily assessed as accurately or more accurately than on the ground by an experienced agronomist. Colwell also found that yellow dwarf virus on both oats and barley causes a reduction in the reflectivity of infrared 'light' from leaves similar to that caused by rust. Diseased areas show up on smaller scale infrared photographs and where these have been detected, very large scale photographs will possibly enable the specific disease to be identified.

(b) Survey design

The first decision will relate to the question of whether or not to group the vegetation as shown on the photographs. The division of the area under investigation into several parts or classes or types is termed stratification; and within each stratum a convenient number of samples will be dispersed using either a random or a systematic design. Only the effective area of each photograph should be used and delineation of strata is carried out stereoscopically.

The purpose of the stratification is to reduce the variation within each stratum and to increase the variation between strata by separating the area into more homogeneous classes. Alternatively, stratification may be viewed as giving a reliability similar to unstratified sampling but with fewer samples. The value of stratification cannot be over-emphasized. For example, in parts of Thailand, Loetsch (1957) found that the results obtained using stratification would have needed four times as many plots without stratification to have given comparable results. If the area is not covered by photographs, it is usually not possible to take advantage of the increased efficiency to be obtained by stratification prior to sampling; but statistical analysis after random sampling may suggest that the data should be grouped into strata prior to calculating standard errors and confidence limits.

Assuming the same number of samples are used to test the relative efficiency of stratifying and not stratifying an area, a useful indicator is provided by the variance ratio or F-test (see Husch, 1963; Snedecor, 1956), which is obtained by dividing the variance of the between strata by the variance of the within strata. The variance (S^2) is obtained by squaring the standard deviation (S). If the several strata represent a single population, the two variances will be similar. Other techniques using variance have been suggested. For example, Moessner (1963b) has suggested an approximate method of rating a number of stratification schemes and providing an estimate of the number of samples needed to give similar reliability using different methods of stratification. The pooled within stratum variance of each scheme ($\Sigma P_i S_i^2$), obtained by weighting the variance of each stratum within the scheme by its area (P_i expressed as a percentage of total area), is divided by the variance of the unstratified scheme (S^2). This is then subtracted from one, i.e.

$$1 \cdot 00 - \frac{\Sigma P_i S_i^2}{S^2}.$$

The remainder is then used as a rating of the efficiency of the method of stratification and as an estimate of the reduction in the number of plots needed to give the same reliability as an unstratified scheme.

Avery (1964) gave three methods of determining the number of samples required in a survey by calculation of the variance of each stratum or type. The methods assume photographs are used and that strata can be recognized. The most obvious method involves laying out a series of random samples from photo-strata in the field and using these to calculate the variance of each stratum prior to a full survey. This, however, is costly and often impracticable due to the terrain conditions. A modification of the method involves the allocation of field plots directly to photo-strata according to the sampling precision of similar surveys elsewhere; but there are statistical objections to this approach. The second method involves the full computation of the variances from photo-plots; but can only be used provided large scale photographs and other relevant data are available. For example, in forest surveys of timber volume, not only are photographs needed at a scale larger than about 1/16,000, but aerial volume tables are also necessary.

A third method calls for the delineation of strata on the photographs, and then by stereoscopic examination an estimation is made of the range of values in each stratum. The range in each stratum is divided by four to indicate what the standard deviation is likely to be; and this is squared to provide a crude estimate of the variance. The method assumes the samples are normally distributed; and as pointed out by Avery, the reason for dividing by four is readily appreciated if one recalls that a dispersion of 95% of the samples encompasses ± 2 standard deviations.

The crude estimations of the standard deviation and variance so obtained are then used to calculate how many plots should be allocated to each stratum. For sampling in the field to give approximately the same standard errors of the mean, the calculated variances are summed (ΣS_i^2) and the variance of each stratum is expressed as a percentage of the total

$$\left(\frac{S_i^2 \times 100}{\Sigma S_i^2}\right).$$

For a predetermined number of plots (n), the number (n_i) to be allocated to the ith stratum to provide a similar standard error of the mean will be

$$\frac{S_i^2 \times 100 \times n}{\Sigma S_i^2}.$$

The standard error of the mean will be

$$\sqrt{\frac{variance}{n_i}}.$$

In practice this method may result in too few plots being allocated to a stratum estimated as having the greatest range. Also it may be desired to place varying confidence limits on the mean of each stratum according to other criteria. For example, in forest surveys, each stratum based on tree species and stand volume in cubic feet may have a different market value; and, as a result, it is desirable to vary the standard error of the mean with value of the forest stand as ascertained by stereo-scopic examination of the photographs. Under these circumstances, the crude esti-mate of the standard deviation (S_i) is used to calculate the number of plots to be allocated to each stratum for a specified sampling error. *Sampling error* is the standard error $(S_{\bar{x}})$ expressed as a percentage of the mean $(\overline{X})$, i.e.

$$\frac{S_{\bar{x}}}{\bar{x}} \times 100.$$

In forest volume surveys, sampling errors of 5%, 10%, 15% and 20% are commonly used. The estimated mean of the stratum multiplied by the sampling error will give a crude estimate of the standard error of the mean (i.e. $S_{\bar{x}} = \overline{X}$. (sampling error)), and hence an estimate can be made of the number of plots (n) to be allocated to each stratum.

For the ith stratum

$$n_i = \frac{(S_i)^2}{(S_{\bar{x}_i})^2}$$

i.e. at the 68% probability level
or

$$n = \frac{(S)^2}{(S_{\bar{x}})^2}$$

for an unstratified population
or

$$n = \frac{(S)^2}{(S_{\bar{x}})^2} \cdot \frac{(N-n)}{N}$$

for an unstratified small finite population where N = maximum number of samples in the population.

If the probability level is increased to 95%, 3·84 times the number of plots will be required, as t^2 for a large number of samples is $1·96^2 (= 3·84)$.

The method of weighting the allocation of samples by area or sometimes by the expected volume of the product contained in the stratum has been commonly used in forest inventories; but has the disadvantage of providing standard errors of the mean which may be out of keeping with the objects of the survey. Obviously strata weighted by area and covering large areas will be allocated more sample plots than strata covering smaller areas; and as a cursory examination of a t-table will show, the 't' values used in fixing the confidence limits of the standard error of the mean vary according to the number of samples.

A factor often overlooked is that efficiency, using standard error as the criterion, will not be greatly increased by using more than forty to sixty plots unless the number of plots is greatly increased. Possibly the error will be halved by quadrupling the number of sample plots. The hypothesis may be formulated that within any one stratum the number of plots should normally not exceed forty to sixty. The word 'normally' is used since if the population is suspected of being heterogeneous the number of plots may be increased beyond these limits in expectation that upon being statistically analysed the data will be broken down into two or more strata. For example, if a stratum is suspected of being heterogeneous and likely to provide three homogeneous strata then up to 120 plots (i.e. 3×40) might be allocated initially.

SAMPLING METHODS

In the above discussion it has been assumed that the samples were usually dispersed randomly. This, however, is not always practicable; and frequently a systematic design is resorted to or the photo-plots are randomly allocated and field plots systematically allocated. It is desirable, therefore, to define what is implied by random and systematic sampling and the limitation of systematic sampling. Husch (1963) concisely defined *systematic sampling* as a type of sampling in which the first sampling unit is chosen randomly, but all following units are at a constant sampling interval. This has also been termed systematic sampling with a random start, so as to distinguish it from systematic sampling with a subjective start. In the latter case as the starting-point is chosen in the field on arrival in the area, personal bias is likely to be introduced; and on this account the subjective method should not be used. In *random sampling*, each sample has an equal chance of selection; and to achieve this, an objective method of selection is adopted, e.g. selection of sample plots from tables of random numbers. If the interpreter selects what he considers to be representative, then the sample is no longer at random and the method would again be termed subjective sampling.

Present-day consensus of opinion suggests that systematic sampling will provide as good or sometimes even a better estimate of the mean for a specified number of samples than random sampling. However, as probability theory is based on random selection, the theory and techniques of random sampling no longer apply; and the calculation of standard error of the mean, variance ratio, fiducial limits and other statistical parameters are no longer valid and therefore remain unknown. In practice, as it is often too expensive or time consuming to locate random plots in the field, systematic sampling is to be preferred. According to Seely (1964) an estimate of standard error of the mean can be made from systematic samples which are only moderately biased. The final decision, however, on whether to use random or systematic methods must depend on local conditions, the way in which the collected data is to be analysed and used and on the objectives of the survey. De Rosayro (1959) compared random sampling and samples taken along lines drawn between prominent map features (i.e. *selected line sampling*) and found there were no important differences. Commonly, in systematic photographic forest surveys $\frac{1}{4}\%$ to $2\frac{1}{2}\%$ of the area is covered by plots in the field as compared with $2\frac{1}{2}\%$ to 20% when photographs are not available. In stratified random photographic surveys, it is often possible to reduce the sampling percentage in the forest to well under 1%. De Rosayro using various methods found the sampling intensity varied between ¼% and 3%. In the Murray river survey, New South Wales, about ¼% of the area was sampled on the ground.

As sampling on photographs is far quicker than ground sampling, it is advantageous to sample in the field only a fraction of the plots sampled on the photographs and to collect field data which cannot be otherwise obtained. This is termed *sub-sampling*.

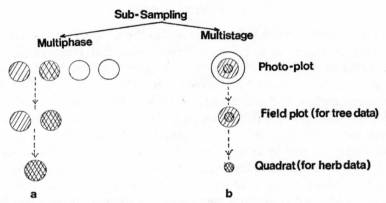

FIG. 19.7. Diagram illustrating multiphase and multistage sampling (see text).

Two methods of sub-sampling are used. If the size of the individual photo-plots is retained, but only a percentage of the plots is sampled in the field (fig. 19.7a), the method has been termed *multi-phase sampling* (Spurr, 1952; Husch, 1963), or *double sampling* (Bickford, 1952; I.U.F.R.O., 1963) or *triple sampling* (Wilson, 1962). The latter can occur when the plots are initially used to determine land-use classes or the area of forest as compared with the total area. If the sub-samples in the field form but a part of each photo-plot, then the term *multi-stage sampling* has been used (Mahalanobis, 1944; Spurr, 1952; Husch, 1963). In the field the process may be repeated by sampling the herb layer or part of the plot used for tree measurements (fig. 19.7b). This again provides triple sampling. Sometimes multiphase and multi-stage methods are combined. For example, when double sampling in Maine, 39,817 one acre photo-plots were examined and of these 2,146 ¼ acre plots were visited in the field (Young *et al.*, 1963). Only 476 acres were actually measured on the ground for 17,100,000 acres of forest. A forest type map was not prepared.

In Iran (Rogers, 1961), Chapman's triple sampling design was used. The triple sample comprised 2,381 photo-points for determining the area of forest land and non-forest land. Then 2,308 photo-plots were used for taking forest stand measurements and finally fifty-four ground plots were used for obtaining normal forest inventory data. Point sampling aerial photographs directly to determine areas of forest classes has been termed *photo-mapping* and has the advantage of not requiring the preparation of a stock map. Other workers who have made use of point sampling photographs in order to determine forest area and forest classes include Spurr (1952), Masters *et al.* (1957), Huguet *et al.* (1958) and Wilson (1962).

Two useful tools of double sampling are the separate calculations of the *correlation coefficient* and of the *regression* between data from the photographic sample-plots and from the same plots in the field. Wear *et al.* (1966) recommended for forest entomological surveys that the photo-plots and field plots should be large (e.g. 8·20 acres) and identical in size, shape and location. Regression is further considered on page 232 and in chapters 20 and 21.

The determination of the correlation coefficient between photo-plots and field plots is helpful in indicating whether a double-sample survey is likely to be as acceptable as a field survey without photographs. The correlation coefficient provides an estimate of the (linear) association between the photo-plot data and the corresponding field plots. If the association is poor, then the coefficient tends towards zero; and conversely if the correlation approaches 1, the correlation is high. Wear *et al.*, for example, obtained coefficients of 0·8 and 0·9 for counts of dead trees (colour photographs).

In conjunction with multi-phase and multi-stage sampling in difficult terrain, or where there are a large number of tree species, it is advantageous to carefully consider the size and design of sampling units. Greig-Smith (1964) should be referred to concerning interrelations in small areas of plot size and pattern provided by the under- and over-dispersion of plant species. In the Brazilian tropical rain forest, according to Heinsdijk (1960), tree species seem generally to be randomly distributed, but group sizes of 2 to 5 hectares were detected. In forest surveys, using aerial photographs, the sizes of sample plots are normally $\frac{1}{10}$, $\frac{1}{5}$, $\frac{2}{5}$, $\frac{1}{4}$, $\frac{1}{3}$ and $\frac{1}{2}$ acre, although the smaller plots have been most frequently used for young stands and 1 acre plots for old/overmature forest, e.g. *E. regnans* in Tasmania. In ecological studies, quadrat size is commonly 1 sq. m or subdivisions or multiples of this unit.

It is sometimes convenient to vary the size of the sample plots according to species and/or utilization classes in an attempt to provide normal distribution curves. This may be termed *variable plot* sampling. For example, during the collection of field data relating to the diameter classes of Japanese larch, a fixed plot size resulted in a large number of measurements for small and medium-sized trees but few for the sawlog-sized trees in the same compartments. By increasing the plot diameters for the larger trees, zero readings at many of the plots could be avoided and so an adequate number of samples obtained for statistical analysis. In Tanzania (1957), the writer used in conjunction with photographs a plot design of four concentric circles (0·1 to 0·25 acres). Sampling intensity varied between 0·1% and 1·2%; and the size of the plots used for each tree species depended on its estimated frequency and market value/market potential. Plots were located in clusters on line transects and these were randomized in pairs within blocks of 15 to 30 square miles. The length of each transect was 3 to 6 miles so as to coincide with a day's work for one survey team. A survey group comprised two to four teams operating from a temporary camp.

In Mexico (Huguet *et al.*, 1958), rectangular plots (0·12 hectares) were initially located at random along randomly selected lines of a grid and later were transferred to a planimetric map prepared from the photographs. Field crews then located the lines on the ground and used the line bearings to traverse from plot to plot. For statistical analysis, it should be appreciated that the plots in relation to each other are only randomized in one direction (i.e. along the transects or *y*-axis), although the individual transects are also randomized in relation to each other (i.e. along the *x*-axis). This in effect is a cluster design; and results in a loss of the number of degrees of freedom as compared with the maximum dispersion of plots. The latter is, however, impracticable in many forest surveys.

If a group of plots provides a fixed pattern in relation to each other, and the group as a unit provides part of another pattern formed by other units, then each group of plots is referred to as a cluster. Frequently plots in a cluster are systematically arranged and the clusters as units are randomly arranged. Plots in a cluster are often arranged in a line or on the sides of a square, rectangle or triangle or in the form of a cross. In Sweden for the third national forest survey, the clusters of plots were arranged on a square (Hagberg, 1956). There were sixty-four evenly spaced plots for recording thinning/felling data (termed stump enumeration), and also at the four corner plots tree heights and terminal shoots were measured. Basal area and diameter increment were recorded at the corner plots and in three plots equally spaced on each of the four sides of the square. In Thailand Loetsch (1957), using photographs at 1/48,000 employed clusters of seven systematically arranged squares (camp units), as a compromise between maximum dispersion of plots and forest conditions. One random photo-point only was needed, being that of the centre-square. Forty-eight ½ hectare sampling areas were located around the periphery of each square.

Provided each cluster of plots is randomized in relation to the other clusters, statistical parameters can be calculated even though the plots within a cluster are systematically arranged. The number of degrees of freedom, however, will be one less than the number of clusters (i.e. $m-1$) and not one less than the number of plots. Assuming there are m clusters and n plots in each cluster, then the variance (S^2) is:

$$S^2 = \sum \frac{(\overline{X}_j - \overline{X})^2}{m-1}$$

where X_j = mean of the *j*th cluster and $\overline{X}$ = mean value for all plots i.e.

$$\overline{X} = \frac{\Sigma X}{mn}$$

The standard error of the mean for all plots in all clusters will be:

$$S_{\bar{x}} = \sqrt{\frac{S^2}{m} \cdot \frac{M-m}{M}}$$

where M equals the maximum number of clusters in the population and corresponds to N samples in a finite population. If this is compared with the standard error, assuming all plots are randomized, i.e.

$$S_{\bar{x}} = \sqrt{\frac{S^2}{n} \cdot \frac{N-n}{N}}$$

it will be observed that degrees of freedom are lost and therefore, statistically, the method is not as efficient.

Circular plots arranged systematically along randomized line transects provide a further example of cluster design, and so do the 'photo-triangles' which are being used in the national forest inventory of Portugal (Stridsberg, 1966, personal communication). A 'photo-triangle' is formed by the flight line and radial lines from each of

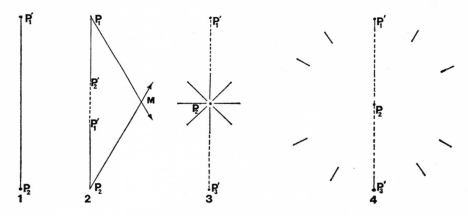

FIG. 19.8. Radial line sampling (see text).

the two contiguous principal points on a pair of photographs. These three lines when plotted on one of the photographs provide a triangle the sides of which will also be straight when located on the ground by chain and compass. It will be observed that the technique avoids the problem of radial displacement of a straight ground line on a photograph. The actual sample plots are arranged at equidistance along the sides of the triangle. As in Loetsch's camp plot design, a traverse of the plots on the sides of a single cluster brings the field worker back to his starting-point.

Radial line sampling has considerable merit for agricultural, ecological and forest surveys, and should be given greater attention by field workers in the future. Figure 19.8 shows four ways by which line transects drawn on the photographs will also always be straight lines on the ground (or vice versa). Sample plots/quadrats can be arranged along the lines to form cluster samples. (1) shows the flight line between the principal point (P_2) and the transferred principal point (P'_1) being used as the line transect. (2) illustrates the method described in the previous paragraph. The radial

231

line P_1m is drawn on the photograph with centre P_1 and the radial line P_2m on the photograph, centre P_2. (3) A '*cart-wheel*' design tried by the author in western Tanzania in 1957. The design has the disadvantage of greatly increased sampling intensity towards the principal point. The name was given to the design as the plotted transects vaguely resemble the spokes of a cart wheel. (4) '*Exploded cart-wheel*' design in which relatively short transects are plotted away from the centre of the photograph to overcome the disadvantage of method (3). This design is readily elaborated and random sampling can be introduced.

Regression analysis. A number of statistical techniques are useful when comparing photo-data with field data. These include the t-test, chi-square test, variance analysis and regression analysis. For the procedures used in the calculations of each, reference should be made to a standard statistical text (e.g. *Statistical Methods*, Snedecor, 1956). In a land-use study related to the survival of forest in settled areas of south-east Brazil, Haggett (1964) used multiple variance analysis; but otherwise regression analysis has received most attention in the several published statistical studies using photographs (e.g. Rogers, 1960, 1961; Stellingwerf, 1963; Wear *et al.*, 1966). Each of the techniques has been tried experimentally in sclerophyll forest near Mount Disappointment (Victoria), e.g. regression data in graphical form for determining stand height from very small-scale photographs.

Regression analysis applied to photo-interpretation depends on (a) the recording of measurements from a relatively large number of photo-plots (e.g. tree height, stand volume, crown diameter, number of dead and dying trees) and (b) for a portion of the photo-plots, the corresponding field measurements of the same plots. The measurements of the smaller number of plots in (b) are used in calculating a regression constant and regression coefficient. The commonest form of the regression equation, assuming a linear relationship, is as follows:

$$y = a + b(\bar{x}_1 - \bar{x}_2)$$

where y: adjusted mean photo-measurement of all photo-plots.

a: regression constant; mean of all field plot measurements.

b: regression coefficient; calculated for field on photo measurements.

$\bar{x}_1$: unadjusted mean of measurements of all photo-plots.

$\bar{x}_2$: mean of the measurements of the photo-plots also sampled in the field.

In an example related to tree mortality, Wear *et al.* (1966) used 100 photo-plots, each of 20 acres, and twenty of these were carefully located in the field. The mean number of dead trees for all photo-plots was 4·52, the mean of the photo-x plots, also counted in the field, was 4·15, the mean of the field plots was 2·80 and the estimated number (y) for all photo-plots was 3·03 (regression coefficient: 0·63).

20

FORESTS OF THE TROPICS AND SOUTHERN HEMISPHERE

ON making a perusal of literature relating to forest photo-interpretation, it will be found that, with few exceptions, North American methods have been applied directly to forest surveys in other parts of the world with varying degrees of success. In general, results as satisfactory as in North America have been obtained throughout the northern hemisphere for both softwoods (i.e. conifers) and temperate hardwoods (i.e. angiospermous tree species) and for forest plantations throughout the world. Probably the best results have been obtained in the boreal forests. Examples of U.S. techniques in forest studies include the taking of photographs at scales between 1/10,000 and 1/15,840, the long-standing use of vertical photography and the use of infrared photography to separate conifers from hardwoods. Examples also occur which indicate the development of distinctive regional techniques, as for example the simultaneous taking of infrared and panchromatic photographs at 1/20,000 in France and the use of two-layer colour film in the U.S.S.R.

When North American techniques have been applied to tropical forests and indigenous southern hemisphere forests, the benefits derived from using aerial photographs have varied greatly. As an example, it may be noted that up to the present time infrared film has not proved useful in separating angiosperms from conifers, e.g. *Callitris glauca* in Australia and *Podocarpus* spp. in New Zealand. Even in the hardwood forests of the most south-eastern parts of the United States, the techniques so successful farther north are not fully satisfactory; and the interpretation of these forests is more akin to the problem of interpretation in some tropical forests.

(a) The natural vegetation

Probably half of the world's forests are contained in the tropics (Haig, 1958); and as suggested by Richards (1952), it is reasonable to 'regard the broadleaved evergreen forests of the tropics as the generalized type, and the deciduous forests of the temperate regions as fitted to adverse winter conditions, just as the thorny vegetation of deserts is suited to a dry climate'.

FORESTS OF THE TROPICS AND SOUTHERN HEMISPHERE

When examining photographs of forests in the tropics, the photo-interpreter is immediately confronted with a great range of forest types, which vary from single-storey pure coniferous woodlands at high elevations, e.g. *Juniperus procera* in East Africa, to typical multi-storeyed tropical rain forest. Over extensive areas of Africa and South America, the stable vegetation is tropical woodland rather than tropical rain forest. The only major characters common to tropical rain forests and tropical woodlands is that they are generally angiospermous and are rich in tree species of many families. Some writers imply tropical rain forest when they refer to tropical forests.

As pointed out by Troll (1947) there is asymmetry in the vegetative zones of the northern and southern hemispheres; and for the purpose of the ensuing examination of photo-interpretation in this and the next chapter, it has been found convenient to use this asymmetry in an initial breakdown of the subject into two. Ideally, a further breakdown is desirable, but at the present stage in the development of the subject this was judged to be unwarranted.

In the southern hemisphere there is a far greater representation of the temperate floras in the higher elevation floras of the tropics than is the case in the northern hemisphere. This may be observed in eastern Africa, eastern Australia, New Guinea and Chile. At the same time there are extensions of the tropical forest species into sub-tropical eastern South America, eastern Australia and south-east Asia. As suggested by Richards (1952) these extensions are preferably termed sub-tropical rain forest as they differ to a limited degree from the true rain forest of the tropics and from the less luxuriant evergreen forests of the mountains in the tropics. They also require to be distinguished from the temperate (sub-antarctic) rain forests of south-eastern Australia, Tasmania and New Zealand. Temperate rain forest, although several-layered at maturity, as is tropical forest, is commonly dominated within a land unit by one or a few truly temperate species (e.g. *Nothofagus* spp., *Podocarpus* spp., *Dacrydium*) (fig. **20.3**)

The (tropical) montane forests differ structurally and floristically from the tropical rain forests of the lowlands by a reduction in the number of tree species, decreasing tree-and-stand heights with increasing elevations, the disappearance of emergent tree species above the top or A-storey, less interlacing of the crowns and fewer or an absence of lianes and a reduction in tree storeys from three to two. Tropical rain forest can be characterized as having three tree storeys, a well-developed shrub layer, a sparse herb layer and interlaced crowns. Richards (1952) quoting Schimper (1903) defined rain forest as evergreen, hygrophilous in character, at least 30 m high, but usually much taller, rich in thick-stemmed lianes and in woody as well as herbaceous epiphytes. At low elevations where environmental factors are limiting to the development of rain forests, the forest may have a single dominant tree species. For example, in the Congo Republic and Uganda, areas of rain forest have a top storey dominated by a single species (*Cynometra alexandri* or *Macrolobium dewevrei*). In Trinidad *Mora excelsa* forms a consociation over extensive areas and provides up to 85% of

234

the trees in the closed A-storey. Other species of smaller size are codominant or common in the B- and C-storeys. Tropical montane forest is shown in fig. **20.2**.

Over extensive areas of the Australian continent favourable to tree growth, the flora is dominated by upwards of 500 tree species of a single evergreen genus, i.e. *Eucalyptus*. Both wet and dry sclerophyll forest formations develop (figs. **6.3b, 17.4, 20.3, 22.2**); and in north-eastern New South Wales formations of three flora adjoin. The wet sclerophyll forests with a single storey of co-dominant *Syncarpia laurifolia* and *Eucalyptus saligna*, and an understorey of mesophyllous shrubs, occur on the drier siliceous soils; and are replaced by subtropical rain forest of Indo-Malayan origin on the basaltic soils of eastern aspect. At the highest elevations, these two formations are replaced by a temperate (sub-antarctic) rain forest consociation dominated by microphyllous *Nothofagus morei*.

In New Zealand, the natural vegetation provides an unusual and challenging complexity of patterns for temperate regions. In parts of North Island the mosaic of vegetative patterns provided by an abundance of tree species (fig. **20.1**) has been accentuated over the centuries by the influence of volcanic cataclysms. In South Island the last major advance of Pleistocene ice and possibly the relatively recent development of a drier climate have also provided a vegetative pattern which often changes markedly within a very short distance. Often several stages of tree succession may occur in what would normally be identified as a single land unit. 'Site' as an important interpreter's tool cannot be reliably used in helping to recognize the species. Of possibly two dozen important tree species in South Island, five are of the genus *Podocarpus* and five of the genus *Nothofagus*. Holloway (1954) pointed out in relation to the *Nothofagus–Podocarpus* forests southwards of the Haast river that there is 'a reversal of pattern on a grand scale, the more mesophytic species, red beech (*N. fusca*), being the dominant species in the forests close to the main ice-fields and the cold climate species, silver beech (*N. menziesii*), the dominant species in the warmer belt sites close to the coast . . . One interfluve may carry pure stands of silver beech; the next . . . devoid of any silver beech.'

(b) Photographic identification and forest typing

When black-and-white conventional-scale photographs of forests in the tropics or in the southern hemisphere are examined, the interpreter will frequently be impressed by the repetition of tone and shape of the majority of the recorded tree species, and may be inclined to conclude that it is virtually impossible to type on a floristic basis. He would be correct in many cases concerning the identification of individual tree species on the scale of photographs commonly used; but would be incorrect to assume it to be impossible to type communities at larger scale. Nominal photographic scales used in forest studies include 1/10,000 (Western Australia, Guyana), 1/15,840 (Australia, Ceylon, New Zealand, Portugal, Tasmania, Thailand), 1/20,000 (Indonesia, Ceylon), 1/23,760 (Burma, Tasmania), 1/25,000 (Brazil), 1/30,000 to 1/40,000

(Australia, many parts of the tropical Commonwealth, Spain), 1/40,000 (Brazil, Cambodia, Ceylon, Guatemala, New Guinea—Papua, Surinam, Vietnam), 1/48,000 (Thailand), 1/50,000 (Sabah, Indonesia).

The common use of scales between 1/30,000 and 1/40,000 in the tropics in no way implies that these are the best; but as Francis (1957b) has pointed out, in the initial studies of tropical rain forest photographs at a scale considerably smaller than 1/20,000 are of value as they provide an overall view of large areas. Several writers have tended to favour the use of larger scale photographs for photo-interpretation particularly in intensive surveys (Hannibal, 1962; Loetsch, 1957; Swellengrebel, 1959; Nyyssonen, 1961). De Rosayro (1958, 1960) concluded from work at Kirindi-oya, Ceylon, that a scale of 1/20,000 was to be preferred for larger scale mapping and interpretation; but is possibly not justified economically due to the paucity of readily marketable species. Francis (1957) commented that for tree species identification and the examination of the tree crowns a minimum scale of 1/20,000 is needed and 1/10,000 may be preferable. In Victoria, photographs at scales between 1/4,000 and 1/15,840 have been used for ecological studies and at 1/8,000 to 1/15,840 for volume studies (Chapter 21). In Tanzania photographs at 1/10,000 and 1/15,000 were tried in experimental 'inventory' studies of deciduous woodland species; but were of little help in the identification of tree species (Howard, 1959).

Research carried out in eucalypt forest suggests that species identification using crown texture, shape and tone is considerably improved by the larger scale photographs and that colour photography will further facilitate forest typing. The hues of the crowns are distinctive and often characteristic, particularly in the seasons of new leaf development in the spring and autumn. In tropical rain forest, colour photography has not proved so far as helpful in identifying the dominant species in the top storey or consociations, as it was at one time forecast. Little use has been made of infrared black-and-white photography. In Victoria and New South Wales the use of modified infrared photography at 1/8,000 to 1/20,000 has not aided species identification. In the Congo Republic, infrared photography was used in 1957 for mapping, and it was claimed that better tonal differences of the vegetation were achieved (Simonet & van Roost). With all film–filter combinations, it is important to check in the field, at the time of the aerial photography, the condition of the vegetation both in respect to new growth and flowering and fruiting. Failure to do so may result in the interpreter being unable to explain differences in photographic tones or in his not taking advantage of phenological differences between species.

Up to the present in studies of tropical forests and tropical woodlands, emphasis has been placed normally on the use of small scale aerial photographs as maps in conjunction with a ground survey of one or a few currently economic species out of the many tree species present. However, as pointed out by De Rosayro (1963), stratification of economic forest types on 1/15,840 photographs in Ceylon bear a very close relation to previously recognized ecological types (e.g. *Mesua–Doona–Shorea* association).

PHOTOGRAPHIC IDENTIFICATION AND FOREST TYPING

Following Schimper's classification as modified by Tansley and as later developed by Greenway (1943), the writer successfully applied Greenway's physiognomic classification to photographs of tropical woodland in western Tanzania using principally site and the photo-pattern of the crown cover (1959). The terms savannah woodland or miombo (woodland) are often used as synonyms for tropical woodland. 'Miombo' is a most unfortunate term as it is the Kinyamwezi name for a single species (*Brachystegia boehmii*), which is seldom locally as common as *B. spiciformis* and *Julbernardia globiflora*.

TABLE 20.A

Photo-key—tropical woodland (Tanzania)

HILL-TOP

I. *Woodland*—50% or more of the ground covered by the crowns of medium and large trees. Stand height is usually 35 ft to 55 ft.

    A. Hill-top woodland (H/W)—on hill-tops. In Tabora District: *Brachystegia tamarindoides woodland.*

    B. Brachystegia–Julbernardia (Isoberlinia) woodland (BJ/W)—In Tabora District: *Brachystegia spiciformis–Julbernardia globiflora* woodland. Productive of sawlog-size *Pterocarpus angolensis.*

HERB-LINE

(H/L)—delineated on photographs as a thick line.

    C. Valley woodland (V/W)—always below the herb-line. Often as islands in the wooded grassland. In central Tabora District: *Brachystegia boehmii–Julbernardia globiflora* woodland. Sometimes *B. longifolia* is locally common.

II. *Wooded grassland*—under 50% of the ground covered by the crowns of large and medium-sized trees.

    D. Wooded grassland without frequent termitaria—trees scattered (W/G)—tree height is usually 30 ft to 45 ft.

    E. Wooded grassland with frequent termitaria—trees grouped (W/GT)—tree height is usually 30 ft to 45 ft.

III. *Grassland*

    F. Grassland with bushes and dwarf trees. Usually gall *Acacia* and *Combretum* spp. (AC/G)—under 10% of the ground is covered by large and medium-sized trees. Small trees sometimes attain 2 ft girth and a height of 30 ft.

    G. Palmstand grassland (P/G)—usually single/groups of *Borassus flabellifer*. It is not an important sub-type.

IV. *H. Riverine* (Riv.)—not always present in the catena and especially absent in hill areas. It is usually best described as riverine thicket; but riverine forest may occur along perennial streams.

VALLEY BOTTOM

V. Other main types/sub-types.

    I. Regeneration/thicket (R/T)—stand height is under about 30 ft and usually large trees are absent.

    J. Induced vegetation—normally absent from forest reserves.

    K. Permanent swamp—considered absent from forest reserves.

As the vegetation recorded on the photographs emphasizes certain characteristics not conspicuous on the ground, some modifications were made to Greenway's classifications. For example, in the prepared photo-key (table 20.A) emphasis was placed on the herb-line (fig. **22.1b**). This is conspicuous on many photographs of the woodland formation, aids the overlapping of contiguous photographs and defines the lower limit of the *Brachystegia–Julbernardia* woodland. On the ground, the herb-line is often difficult to locate without the aid of aerial photographs and even then sometimes remains inconspicuous. Ideally, the herb-line of the photographs is a narrow ecological zone (e.g. 5 to 30 m wide) of herbs/grasses, lying between the woodland and wooded grassland. The soil is usually a clayey alluvial loam covered with a layer of hill-wash sand. Field observations indicated that the herb-line is associated with the subsurface hard-pan of the valleys and the wet-season water table. Frequently, the ground-layer vegetation of the woodland burns earlier in the year than the ground vegetation of the wooded grassland. This produces a conspicuous line along the edge of the woodland and is seen on aerial photographs as the herb-line. Again, if the woodland stops abruptly and continues as *Acacia–Combretum* grassland, the herb-line on the photographs may be formed by the edge of the woodland.

In general, when typing tropical forests, the normal procedure should be first to see if the forests of interest can be separated as a group from other land patterns within a land-system and then to examine the forests separately to see if any species of interest can be identified either in mixture or as a consociation, and what attributes of the forests as recorded in the photographs can be or should be used for stratification. For example, in tropical South American, Heinsdijk (1952) could identify six non-forest types and four forest types (dryland forest, marshland forest, swamp forests and mangroves) in conjunction with a key prepared during field studies. In Guatemala, using 1/40,000 photographs, a general classification of forests into seven types was made in a survey for mahogany (Harrison & Spurr, 1955). In West Irian, on photographs at 1/40,000, the vegetation types were identified as dry land forests, fresh-water swamp, seasonal swamp, peat swamp, grass and reed swamp, swamp forest with sago palm (*Meteroxylon* sp.). Some of these were subdivided according to height and crown class. A similar type of swamp classification has been employed in other parts of Indonesia.

It soon becomes obvious from studying tropical rain forest that it is not usually possible to identify an individual species from small scale photographs and the identification of photo-communities is difficult. For every niche in the extensive regions of tropical rain forest on the deep lateritic soils there seem to be many species competing with each other for occupation of the site. As stated by Aubreville, the composition of the tropical forest is constant in neither space nor time. In addition, similar groups of species may develop different crown sizes and shapes and have a different stand height with a change in aspect. Hannibal (1952) using 1/40,000 photographs, found that even when a species formed small pure groups, identification may be impossible. This, however, it not always the case, and is judged to be due partly to the scale of the photographs. Sometimes a commercially valuable species can be associated with other species which can be identified. In Ghana, areas containing silk cotton (*Ceiba pentandra*) were recognizable on the photographs by their association with the umbrella tree, *Musanga smithii*. Associations containing *Raphaea* spp. can usually be readily delineated and excluded from tropical forest, as they occur in a swamp land-system. That plant formations and associations often occupy distinctive sites can be important. In North Borneo, Francis & Wood (1954) used site to help identify mangroves, and forest liable and not liable to flooding.

Sometimes typing is improved by a topographic stratification of the photographs (Francis, 1957). Particularly in the tropics, in conjunction with aspect, altitudinal zonation may be distinct. For example, Loetsch (1957a, 1957b) recognized two groups by elevation: (a) mixed deciduous teak-bearing forests, dry *Dipterocarpus alata* forests, and semi-evergreen forests below 3,250 ft and (b) inaccessible hill evergreen forest above 3,250 ft. Teak was not recorded above 3,250 ft.

Occasionally a genus or a species is photographically distinctive, although growing in association with many others. *Dipterocarpus* as a genus is often recognizable by its large crown and characteristic light photographic tone produced by shiny leaves (e.g. Loetsch, 1957; Merritt & Ranatunga, 1958; Wheeler, 1959). In New Guinea, at high elevations Boon (1955) was able to identify the conifers *Agathis* and *Araucaria* by their distinctive crown shapes; and in New Zealand, *Agathis* has similarly been identified (figs. **20.1, 20.2**). Swellengrebel (1959) in Guyana using photographs at a scale of 1/10,000 was able to identify successfully a number of species occurring in groups in mixed forest and *Mora* forest. These included *Mora excelsa*, *Mora gonggrijpii* and *Ocotea rodiaei* (greenheart). Successful recognition depended on 'group' recognition and not on individual trees, as the growth of foliage on different trees of the same species and different species was always at different stages in development.

Consociations are normally readily recognizable on aerial photographs. They are frequently associated with limiting edaphic factors at low elevations in the tropics, or with altitude. Several examples will be quoted. In Dutch Guyana, baboon trees have been successfully identified and counted as they provide a light-toned image and form the dominant crown canopy. In Brazil, consociations of *Mora excelsa* and pure upper storeys of *Goupia glabra* were identified on 1/40,000 photographs (Heinsdijk, 1957). In Sumatra, on photographs at 1/40,000, pure stands of camphor have been identified (Hannibal, 1952). In Guyana, *Eperua* was readily recognizable as a transition zone between 'mixed' forest and *Mora* forest. The gregarious *Doona congestiflora* consociation was recognized in Ceylon on 1/15,000 photographs, helped by the prior knowledge that it forms a dominant layer and by its characteristic cauliflower-like densely packed crowns having a convex appearance and a light grey tone (De Rosayro, 1959). On alluvial flats, the forest was chiefly associated with sands and gravels and a high water table. In Victoria, *Eucalyptus pauciflora* var. *alpina* (fig. **17.4**) is associated on the photographs with high elevations and often

R

provides a distinctive pattern in contrast with *E. delegatensis*. In Western Australia, pure stands and groups of jarrah (*E. marginata*) (fig. **19.5**), Karri (*E. diversicolor*) and sometimes marri (*E. calophylla*), blackbutt (*E. patens*), red tingle (*E. jacksoni*), and wandoo (*E. redunca*), are successfully identified and delineated on photographs at 1/15,840 and larger.

An interesting example of typing an extensive formation is provided in the forest assessment of 200,000 acres of *Eucalyptus camaldulensis* (red gum) along the Murray river. Much smaller areas of cypress pine (*Callitris glauca*), box and mixed red gum-box within the formation could be identified and delineated on the photographs so as to leave virtually pure *E. camaldulensis* for further studies. The red gum was then divided into classes according to characteristics recognizable on the photographs which normally influence the criteria used in a ground volume assessment (i.e. height, stocking, crown diameter).

(c) Measurement of the forest

In general the measurements which can be made directly from the aerial photographs are stand height and crown closure, and sometimes tree number, tree height and crown diameter. Under favourable conditions, diameter at breast height may be estimated from the crown diameter. These parameters are most often used for stratifying the forest in advance of a ground inventory and not used directly for calculating the volumes of merchantable timber. The calculation of volumes from aerial photographs will be briefly discussed in the next chapter.

It will often be found that despite high-quality photography accurate measurements cannot be made; and therefore, as compared with the same measurements discussed in the next chapter, the results are expressed in broad height classes, broad crown closure classes, etc. Sometimes variation in tone and texture on the photographs represents considerable differences in site conditions, and these in turn reflect considerable differences in the volumes of one or more species per unit area (see De Rosayro, 1959). For example, along the river Murray, two site quality classes of *Eucalyptus camaldulensis* could be identified by differences in tone. This was probably due to differences in the reflection of the ground vegetation. In western Tanzania, *Pterocarpus angolensis* did not attain sawlog size in two land units, ('mbuga', 'umanda'), covering up to 50% of the total area; and these were readily delineated on the aerial photographs (Howard, 1959). Heinsdijk (1952) similarly recorded in the Amazon that *Ceiba pentandra* is a small tree in marsh forest but attains large size in dry land forest.

In any region in which it is known that serious fire damage occurs or white ants are active in the forest it is likely that there will be defects in the timber of the mature standing trees. This limits considerably any prospect of estimating accurately the merchantable volume of timber from aerial photographs, and even causes inaccurate results in ground survey. Sometimes crown characteristics and tree height, as recorded on the photographs, can be used to stratify the forest into two or three utilization classes with significant differences in defect. Heinsdijk (1958) points out that the range of volume in South American tropical forest is relatively small and therefore easier to estimate than in (European) temperate forests. Three volume classes were readily assessed by an experienced interpreter: (a) less than 150 cu. m per hectare, (b) more than 250 cu. m per hectare, (c) 150 cu. m to 250 cu. m per hectare. He was also able to develop a linear relationship between the timber volume of the upper storey trees and total timber volume of the stand.

Height. In rain forest, it is normally impossible to measure by parallax bar tree height and stand height as the ground is not seen (e.g. Heinsdijk, 1958; Haig, 1958). Frequently the measurement of height from the photographs is found to be not as important as in temperate forests, as the canopy is much more even heighted and acceptable results may be obtained without stratifying by height (e.g. Haig, 1958). In Thailand, Loetsch (1957) found that the correlation between top height and diameter at breast height is considerably lower than in temperate coniferous forest.

In the *Eucalyptus camaldulensis* forest on the Murray river, stand height was at first determined by parallax bar measurements of randomly selected measureable trees on the photographs; and site quality was expressed as the ratio of diameter at breast height to stand height to provide three quality classes. However, later on gaining experience in determining stand height, the interpreter assessed it subjectively from the stereo-pairs of photographs. This was found to be much faster and to be as accurate.

The separation of forest from woodland on photographs is probably best achieved using stand height as the principal criterion (e.g. Haig, 1958; Howard, 1959); or dominant tree height in conjunction with crown diameter. Woodland is normally single storeyed and the trees have non-interlacing distinct crowns, are short-boled and have a low $\dfrac{\text{height}}{\text{crown diameter}}$ ratio and a well-developed herb layer. As a general rule, rain forest attains a maximum tree height of 140 to 180 ft, whilst other evergreen and semi-evergreen forest, excluding eucalypt, has a tree height of about 80 to 120 ft, and woodland trees seldom exceed a height of 70 ft and commonly occur between 45 ft and 60 ft. Tropical woodland, excluding eucalypt and some *Acacia* spp., is normally deciduous for a major part of the dry season. In the vicinity of the river Styx, Tasmania, *Eucalyptus regnans* forest has attained a stand height of 250 ft to 300 ft and has understoreys of 100 ft to 150 ft and 15 ft to 30 ft (fig. **20.3**).

Number. Sometimes the number of stems per acre is used as a measure of density, but for it to be an effective description of stand density some other measurement of tree size must also be included, e.g. crown diameter. Tree counts can be made fairly accurately on photographs of open-grown stands, e.g. wooded grassland and the emergents of tropical rain forest; but are of limited value when examining closed forest. Paijmans (1951) working in Celebes pointed out that separate crowns belong mostly to large or marketable trees, and trees where crowns are not visible do not

241

play a role in the total timber production. A similar comment can be made concerning annual volume increment in relation to visible crowns in temperate forests.

Successful counting of trees from photographs depends on skill of the interpreter, scale of the photographs, physiognomy of the forest, species and age of the stand. In Guyana, Swellengrebel (1959) found at first when studying the photographs that several crowns were mistaken for one, but with experience they could be separated. In eucalypt forests, even the most skilful interpreter frequently fails to make accurate counts due to recording two trees as one. However, in the mature/overmature eucalypt forests in Tasmania, crown counts have been found to be effective in separating the lowest volume classes from the higher volume classes. Lawrence (1957) pointed out that as in overmature forests the largest trees have relatively small crowns, crown closure may prove to be no better than tree counts.

Crown closure. Whereas tree counts have a limited application as mentioned above, crown closure estimates have been widely used in temperate regions for stratifying forests into volume classes. Using crown closure (density) cards (fig. 16.3) an interpreter can quickly place a stand, as observed on a stereopair of photographs, in its appropriate crown closure class. Crown closure refers to the percentage of the ground area covered by the tree crowns; and is one of several methods of estimating stand density. In Western Australia (McNamara, 1959), canopy closure classes are estimated for the upper storey and the total stand. Three basic stand structures are recognized from the photographs in classifying regrowth forest.

As pointed out by Husch (1963), the weakness of crown closure as a descriptive characteristic of forest stand volume is that crown closure is not closely correlated with the number of trees or tree sizes. It is important to appreciate that the crown closure percentage of a single-storey stand often does not cover 100% of the ground area even when fully stocked. It is unusual, for example, for a stand of *Eucalyptus obliqua* at maximum density to exceed 80% to 85% of the ground area; and if this is adjusted for actual ground area shadowed by leaves, it is considerably less. When studying *E. regnans* on the ground, the (net) crown cover has been recorded as differing considerably from the gross crown cover, i.e. crown closure (see table 20.B, D. H. Ashton, 1965). Tall trees, polar-facing slopes, short focal length, infrared photographs, early morning or late afternoon photography and areas towards the periphery of the photographs all favour the over-estimation of crown closure by an interpreter when he is examining the photographs under the stereoscope.

In the tropics, the stratification of forests into crown closure classes has been used for woodland (Paelinck, 1958; Howard, 1959), and for montane forest (De Rosayro, 1959; Merritt & Ranatunga, 1959), but not as far as is known for the emergents of rain forest (Nyyssonen, 1961). In Ceylon, the forests were divided into three productivity classes using 1/40,000 photographs (De Rosayro, 1959), and virgin forests were divided into eight classes using 1/15,840 photographs. In Brazil, crown density has been used successfully to stratify parana pine (Heinsdijk, 1960). In *Eucalyptus camaldulensis* forests, five crown classes (0 to 20%, 20% to 40%, 40% to 60%, 60% to 80%,

80% to 100%) were recognized and were related to basal area per acre by crown ratio and the establishment of plots in the field.

Crown diameter. Crowns may be measured using a crown-wedge; or the interpreter may visually separate the crowns in the stereo-model into crown classes. The latter technique was used in the Murray red gum assessment and verified by field checks. In tropical rain forest, the crowns other than emergents are often so interlaced that it is impossible to measure crown diameter; but the crowns of dominant trees can be measured at a scale of 1/10,000 (Paelinck, 1958) and possibly at scales down to 1/20,000. Working with 1/40,000 photographs of Amazon forest, Heinsdijk (1958)

TABLE 20.B

The development of the canopy of E. regnans *stands with age of stand (viewed from the ground)*

Age of stand	Crown closure %	Crown cover %	Crown density (% of crown area)	Cover breakdown		
				Leaves	Branches	Trunk
8	95	57	60	44	11	2
16	93	58	63	47	11	1
26	90	59	63	41	16	—
40	87	64	74	43	19	2
80	84	57	68	39	21	—
150	70	43	61	31	11	—
220	66	38	58	27	11	—

was able to recognize four crown diameter classes (10, 15, 20, 25 m). For three vegetation types in Guyana, Swellengrebel (1959) recorded crown widths on the photographs (1/10,000) and on the ground. For photographic measurement, best results were obtained with a crown diameter gauge, which provided class intervals of 5 ft. This was better than could be obtained by measurement on the ground due to difficulty of locating the visible edge of the crowns.

Diameter at breast height. A number of workers have established a correlation between crown diameter and diameter at breast height. These include Lane-Pool in Australia (1936), Farrer (1953) in Tanzania, Paijmans (1951) in the Celebes, Howroyd (1954) in Sabah, Heinsdijk in Surinam and the Amazon, Davies (1954) in New South Wales, Weir (1959) in Victoria, Dawkins (1963) in Uganda and Curtin (1964) in Victoria. Davies found when working from photographs that six diameter classes could be visually estimated as shown in table 20.C.

That crown diameter should be related to diameter at breast height by tree height classes has also been examined by Ilvessalo (1950), Howard (1957), Dilworth (1959). Normally, the statistics collected in the course of study of crown width to stem diameter ratios have been pooled before deriving the regressions, but Weir (1959) and Dawkins (1963) derived regressions for individual stands. Weir observed that the ratio of crown width to stem diameter (for a given stem diameter) decreased with an increase in the stand density and site quality based on dominant height.

In fig. 20.4 the regressions of diameter (b.h.) on crown diameter for Swellengrebel (two vegetation types) is shown and compared with similar regression lines of

TABLE 20.C

Diameter classes (E. camaldulensis)

Tree class	Diameter (in.)
Seedlings	Under 4·0
Saplings	4·0 to 11·9
Poles	12·0 to 19·0
Piles	20·0 to 27·9
Mill	28·0 to 43·9
Veteran	44·0 and over

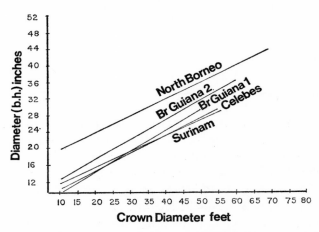

FIG. 20.4. A comparison of the tree diameter at breast height on crown diameter in several tropical forests.

Heinsdijk (1952) in Surinam, Paijmans (1951) in Indonesia and Howroyd (1954) in Sabah. The fact that Swellengrebel's has two regression lines agrees with Dilworth's (1959) observation that for the same diameter at breast height the crown diameter is higher on poor soils. Swellengrebel pointed out that although a straight-line relationship existed between crown diameter and diameter at breast height, this is not always so. Two exceptions are overmature trees, when the crown may either form a compound curve or at a later stage diminish in size due to die-back or loss of large

244

branches, and small trees, of which the crowns are sometimes larger than expected when not overshadowed. Using ground survey in conjunction with photo-interpretation, Paijmans found that on poorly drained sites in Celebes (1951) and northern Papua (1966) that the forest is lacking in trees of the lower girth classes; and this results in a relatively high average girth.

21

TEMPERATE FORESTS OF THE NORTHERN HEMISPHERE

INTRODUCTION

ALTHOUGH the earliest important studies relating to the application of aerial photographs to temperate forestry were made in Germany (e.g. Krutzsch, 1925; Zeiger, 1928; Jacobs, 1932) subsequent development was confined mainly to the United States and Canada. In the last 20 years, however, there has been a reawakening of interest at forest management level in western Europe. For example, in Finland a detailed study has been made of the role of aerial photographs in the estimation of the growing stock with special reference to height and crown characteristics of a single species (Scots pine) (Nyyssonen, 1955, 1962). In general in Scandinavia, however, the primary value of aerial photographs to forestry has been in providing maps as an aid to ground inventory. The scant information available from the U.S.S.R. suggests that, excluding the war years and the aftermath, forest studies using photographs have been developed continuously since the mid-1930s. The application of photo-interpretation to temperate forestry has also been extended to Iran (Rogers, 1960, 1961), Turkey, Cyprus (Polycarpou, 1957), Japan (Nakayama, 1958; Nakajima, 1962) and Mexico (Huguet *et al.*, 1958).

With the exception of the higher latitudes of Canada and the U.S.S.R., photo-interpretation in North America and Europe is concerned with either extensively or intensively managed forests as distinct from many of the forest regions considered in the last chapter, which are either extensively managed or as yet not under sustained forest management. Emphasis in forest studies from photographs in the temperate northern hemisphere is at the intensive management level; and the majority of the forests are even-aged either by stands or by compartments. Obviously, man as a biotic factor has greatly influenced and continues to influence the phytosociology of the forest. Most of the tree species are either deciduous angiosperms (deciduous hardwoods) or conifers (softwoods) such as pine and spruce.

The pines with three exceptions do not occur naturally south of the equator nor in Africa. *Pinus merkusii* occurs in Indonesia, *P. sylvestris* in North Africa and *P. canariensis* in the Canary Islands. All other species, excluding *P. caribaea* var. *hondurensis* are present in the temperate northern hemisphere although some (e.g. *P. patula* in

Mexico) occur also in the tropics. For convenience the exotic pine plantations of the southern hemisphere have been included in this chapter.

Excluding virgin conifer stands in western North America, stand height seldom exceeds 100 ft and the forest is normally single storeyed. Sometimes two storeys in a forest occur naturally as stages in succession, e.g. hemlock under Douglas fir in Canada; or by deliberate management, e.g. chestnut coppice and oak standards in France. Most commonly the shrub-layer is poorly developed or absent and the development of the herb-layer depends on present and past management.

At higher latitudes, higher elevations and in the Mediterranean area, stand height is considerably less than mentioned above and the ground or herb-layer is frequently conspicuous on the photographs. This facilitates the measurement of tree height by shadow length. In fact, around the Mediterranean many of the forests comprise widely spaced short-boled gnarled trees, interspersed with grazing lands, which have been managed over many centuries primarily for grazing and minor forest products (e.g. cork, tannin, resin, fruits). These forests are quite distinctive on aerial photographs, resembling tropical wooded grassland or English parklands; and due to rapidly declining markets for many minor forests products are presenting a difficult economic problem to the governments of countries concerned.

In the temperate native forests, especially in the more northern latitudes of tree growth, it is possible to identify the tree species on the photographs, provided the film–filter combination and scale have been carefully chosen. Hardwoods and conifers can usually be separated using pure infrared photography (fig. **21.1**). Normally, a geographic region contains only a few tree species, and it is likely that both conifers and hardwoods will be present. Possibly not more than two or three common tree species will occur in mixture and frequently individual species are associated with recognizable land units. In Sweden, for example, with the exception of oak and beech in the south, there are only four common forest species (*Populus tremula, Betula alba, Pinus sylvestris* and *Picea excelsa*). Braun (1941) pointed out that the number of tree species in a region seldom exceeds twenty to twenty-five. In an area of 14,000 acres near Brechfa (west Wales), only four indigenous hardwood forest species were recorded, although twenty-four exotic coniferous species had been planted.

It is thus easier to separate the natural forest into uniform stands by species, density and height or age classes; and as the photographic scale is relatively large and the crowns are not interlaced, individual trees can be measured in the stereo-model. In forest inventories of natural forest, typing is most commonly based on utilization classes (e.g. saplings, poles, sawlogs). For example in Japan, using 1/20,000 photographs, this procedure was followed after separating the forest into hardwoods (e.g. beech), mixed and conifers (e.g. *Tsuga Sieboldii, T. diversifolia, Thujopsis dolabrata*) (Nakajima, 1962).

The increasing application of aerial photographs to intensively managed forests is indicated by Hildebrandt (1960), who reported the use of aerial photographs in Germany for stand description, site quality assessment and division of compartments into sub-compartments for management. He also pointed out that the use of photographs had resulted in a 40% reduction of time and expenses as compared with conventional field practice. Even in an area such as Brechfa forest, containing a considerable number of species, it was practicable to identify individual species and to delineate species boundaries and poorly stocked areas by species, since old planting records by compartments were available.

(a) Direct photographic measurement of the forest

The attributes of the forest which can be observed and measured from the photographs depend considerably on the scale of the photographs; and as the photographs are mainly to be used at management level, scales are normally larger than in the last chapter. This, in conjunction with the physiognomy of the forest, enables measurements to be made by species of both the individual trees and of the stand. Height and crown diameter can frequently be measured with reasonable accuracy; and crown closure can be an important parameter when stratifying the forest into merchantable volume classes. There seems little doubt the development of accurate volume measurements from the photographs lies in improving the accuracy of measuring the tree height, crown width and crown closure; and using these in regression programmes as single or combined variables or as ratios (e.g. height/crown width). Under favourable conditions, tree counts may be combined with one or more of these parameters.

Scales range between 1/5,000 and 1/20,000, although scales between 1/8,000 and 1/16,000 are most commonly used in forest inventories. The U.S. Forest Service for many years have preferred a scale range between 1/10,000 and 1/15,840; and in Canada, photographs have commonly been at 1/15,840 when studying site, tree height and stand height, and recently at 1/3,960 or larger for special purposes (Seely, 1964). Scales have been used at 1/5,000 in Finland for *Pinus sylvestris* (Nyyssonen, 1955) and in strips across 1/50,000 photographs in Iran (Rogers, 1961); at 1/8,000 and 1/10,000 for mapping and volume studies of *Pinus radiata* (Cromer & Brown, 1954; J. Dargavel, personal communication, 1965); at 1/10,000 in Finland (Nyyssonen, 1955), in Iran (Rogers) and in Germany for amending stock maps (by Stereotope); at 1/15,840 for *Pinus radiata* (Cromer & Brown, 1954)), for *Pinus brutia* and *Quercus* species in Cypress (Polycarpou, 1957); for *Pinus pinaster* and *Eucalyptus globulus* in Portugal (Stridsberg, personal communication, 1966); at 1/20,000 for conifers and hardwoods in Japan (Nakajima); in the U.S.A. (e.g. Avery & Myhre, 1959) and in France for species identification using panchromatic and infrared film (Huguet, personal communication, 1961); at 1/25,000 in Italy and at 1/31,000 in Spain (1965). In Austria, Lackner (1964) obtained similar results in the identification of tree species at photographic scales of 1/6,000, 1/10,000 and 1/15,000; and Stellingwerf in Holland (1966) concluded from identifying trees at 1/10,000 and 1/20,000 that correct interpretation depends on species and not scale. However, in the United States, Avery & Myhre (1959) were unable to adequately separate southern pines and hardwoods on panchromatic photographs at 1/20,000 when preparing an aerial volume table.

In southern Sweden Welander (1952) observed that the identification of tree species was better by infrared film than panchromatic. He found, however, that in mixtures of pines (20%), spruce (65%) and birch (15%), in every plot for every species there was incorrect identification of species, being 1% to 3% for pine, 0·0% to +1% for birch and 1% to 4% for spruce.

Tree height. Of the two parameters of the individual tree which can be measured directly from aerial photographs, tree height has proved to be the more important. However, tree height measured on the ground has been observed to be more closely related to volume when preparing aerial volume tables (Avery & Myhre, 1959; Pope, 1962). Avery & Myhre (1959) in Arkansas and Nyyssonen (1955) in Finland, concluded that tree height is the photographic parameter most closely related to location or stand volume. Under Finnish conditions the standard error of the volume estimate was ±13%, provided the mean stand height and crown closure were accurately determined; but otherwise, resulting from an interaction of errors, the precision could be as low as ±28%. Nyyssonen (1955) obtained conditions most favourable to height measurement from photographs in the north of Finland, as the ground is clearly visible. Stellingswerf (1963) in Holland found that the influence of a standard error of ±5·5 ft in height resulted in an error of about 14 cu. ft in volume; and was relatively large compared with the influence of stems not counted.

In *Pinus radiata* plantations, the canopy closes at an early age (e.g. 10 to 12 years in warm temperate climates); and as a result stand height classes and not individual tree height should be determined. In radiata pine plantations, ideally two sets of photographs are required; one at time of planting to provide a topographic map and/ or an ecological site map and the other at time of crown closure to provide a site quality or site index map. It may be possible with the two sets of photographs to measure tree heights even in rugged terrain.

Crown diameter. As pointed out earlier, individual crown diameter measurements obtained from the photographs will frequently not agree exactly with measurements made on the ground, as some branches may not resolve on the photographs. Also the perspective view of the crown is different as seen from below on the ground and as seen from above on a stereo-pair of photographs. Generally, aerial crown measurements of the same tree will be less than measurements taken on the ground (Spurr, 1960). Obviously some crowns may not be measured due to being hidden, other crowns being partly hidden cannot be measured and some crowns may coalesce and provide the impression of a single large crown.

According to various workers the following crown classes have been determined, and vary with photographic scale: 2 ft classes (1/7,000, Nash, 1949; 1/10,000 and 1/15,000, Nyyssonen, 1955; 1/12,000, Spurr, 1960), 3 ft to 4 ft classes (1/12,000 Worley & Meyer, 1955; 1/15,840, Spurr, 1960); 5 ft classes (1/20,000, Spurr, 1960).

Crown closure. This may be defined as the proportion per unit area of the ground covered by the vertical projection on to it of the overall tree crowns (i.e. gross crown cover). The numerical result is expressed either as a percentage (e.g. 80%), sometimes termed crown coverage, or as a decimal coefficient (e.g. 0·8), sometimes termed crown density. As ecologists (see Greig-Smith, 1964) use the term (crown) cover to express as a percentage the vertical projection of the various individual aerial parts of the plant on to the ground (i.e. net crown cover), it has been found necessary to distinguish between these two approaches to cover; and with this purpose in mind the term 'crown closure' is used when referring to gross crown cover. It is gross crown cover which is measured from aerial photographs; and which provides the forest photo-interpreter with an important parameter when determining the volume of timber per unit area of forest.

As pointed out by Spurr (1960), crown closure can generally be estimated better from the bird's eye view provided by the photographs than on the ground; and possibly crown diameters are also more readily measured from large and very large scale photographs than on the ground. Most frequently crown closure is measured with a crown (density) scale (see fig. *16.3*). Preferably this type of scale should be prepared within 1/1,000 to 1/2,500 of the scale of photographs; and with a texture equivalent to the crowns of the species on the photographs.

Crown closure is possibly a satisfactory expression of stand density until maximum density is attained. For some species, a crown closure of 100% is achieved; but as mentioned in the last chapter a forest stand may be at maximum stocking when the ground is not fully covered by the crowns. In the case of eastern white pine (*Pinus ponderosa*) on an average site, crown closure has been reported as 100% at 40 years, 90% at 70 years and 65% at 100 years (Spurr, 1960).

Stellingswerf (1963) determined crown closure for *Pinus sylvestris*, using 500 points or about one point per 160 sq. m, and concluded that although crown closure varied between 45% and 75%, it was not generally correlated with volume. Using a grid of points, an error of ±10 points in 100 was obtained by Nyyssonen (1955) for full crown closure; and this was observed to be equivalent to 17% error in timber volume for stands with dominant trees 65 ft high.

Number of stems. In ecological terminology, density implies a measure of the number of individuals or plant units per unit area (see Greig-Smith, 1964); but to many foresters timber volume per acre is the ultimate expression of density (i.e. stand density) and is measured not only in terms of individual trees per unit area but also in terms of other stand parameters including volume and basal area (see Husch, 1963, and Forest Terminology of the Society of American Foresters). Irrespective of whatever parameters are used to measure stand density, it is important to recognize the effect of density or the crowding of individuals within an area on timber volume, crown width, crown length, the $\dfrac{\text{crown width}}{\text{diameter b.h.}}$ ratio and the $\dfrac{\text{tree height}}{\text{crown width}}$ ratio.

INDIRECT PHOTOGRAPHIC ESTIMATES OF THE FOREST

The use of 'number of trees' *per se* as an estimate of stand density has two disadvantages. Firstly, crown counting from photographs is normally difficult or inaccurate. For example, in making crown counts of 'plantation *P. radiata*', using $\frac{1}{10}$ in. squares on 1/8,000 photographs, it was found that a spacing of about 12 ft by 12 ft was needed for successful counting, and stand volume per acre was not closely correlated with number of trees per acre (Cromer & Brown, 1954). In general, crown counts under-estimate the actual number of trees per unit area as some trees may be hidden, others appear as one, and small crowns do not resolve.

Secondly, as at a specified age natural stands at full crown closure frequently have a wide range in the number of trees per unit area, it is necessary to use this measure of stand density in association with some other parameter (e.g. height, basal area, diameter). Smith (1965) commented that even with 100% crown closure, the number of stems in stands which were open-grown may be only 25% of the 'normal' number of stems and the timber volume may be only one-sixth of that in an average stand. A height/crown width ratio of 5·0 was considered 'normal' and ranged from 2·5 for stands which had been open-grown to 8·0 for very dense stands. When number of stems is combined with another parameter, the two-dimensional measurement may reveal correlation with age and site quality.

An easy method of estimating the number of stems (N) per acre from the photographs is to use average crown width (C_w), crown closure (C_c) and the following formula as an alternative to attempting to count the number of stems:

$$N = \frac{43,560 . C_c\%}{0·785 . (C_w)^2}$$

(43,560 = sq. ft in 1 acre, and 0·785 is the decimal coefficient of a square and circle of the same width/diameter).

(b) Indirect photographic estimates of the forest

As discussed in the last chapter, crown diameter has been shown to be correlated with diameter at breast height, e.g. correlation coefficient for *P. elliotii* is 0·82 (Willingham, 1957); but unfortunately the precision of the estimates has not been high enough to satisfy the normal objectives of intensive ground inventory in plantations. Zieger as long ago as 1928 demonstrated the correlation between crown diameter and diameter at breast height for Scots pine. Other workers, who have shown similar relationships for northern hemisphere conifers and hardwoods, include Ilvessalo (1950) for Scots pine in Finland, Eule (1959) for beech in Germany, Dilworth (1959) for Douglas fir in the western United States, Feree (1953) for hardwood species in the north-eastern United States and Minor (1951) and Willingham (1957) for *Pinus taeda* and *P. elliotii* in the south-eastern U.S.A.

In general, the regression of stem diameter on crown diameter tends to be linear for middle-diameter classes, but somewhat curvilinear for young and old trees.

Frequently, but depending on species, density and diameter classes, the stem diameter in inches may be found to approximate to the crown diameter in feet. In British Columbia, the ratio expressed as a decimal coefficient varies between 0·7 in dense stands to 2·0 in open stands (Smith, 1965). According to Eule (1959), thinning does not greatly influence the regression. As pointed out by Spurr (1960), the inclusion of height as an independent variable considerably improves the precision of the estimate of stem diameter as measured by crown diameter on the photographs. Ilvessalo (1950), using 2 m height classes for Scots pine, Norway spruce (*Picea excelsa*) and birch (*Betula alba*) and 30 cm stem diameter classes, obtained sampling errors of $\pm 5\%$ to 10%.

On the basis of a mathematical relationship existing between tree diameters on the ground and crown diameters on the photographs, it should be possible to calculate from stereo-pairs of photographs the area of tree foliage, branch dry weight, leaf dry weight, leaf green weight and total dry weight in a forest stand. The estimates may be expected to be no more accurate than the estimating of bole volumes by aerial volume tables.

In establishing correlations there is a choice of two techniques. Material may be collected and weighed as is customary in crop plant and fruit tree studies and as used by Ovington (1957) in the study of forest stands in the United Kingdom. Alternatively a mathematical relationship between the size of different branch members and foliage, etc., may be determined and expressed in the general form:

$$\log_{10} y = a \log_{10} x + \log b$$

where a and b are constants and x and y are weight or dimensions (e.g. upper diameter of stem).

A useful review of literature of studies carried out on individual trees on the ground was provided by Kittredge (1944), who also studied ten species and twenty-eight stands in California. He concluded that the relationship between leaf weight and diameter/(b.h.) is applicable to trees of different sizes, densities, crown classes and ages at least up to the culmination of growth and beyond that age for tolerant species in all-aged stands. Satoo *et al.* (1956) and Cable (1958) extended Kittredge's studies to leaf area per tree (fig. 21.2) and oven-dry weight using aspen poplar and ponderosa pine respectively by a 'least-squares' solution; and showed that surface area in square centimetres can be estimated from dry weight of needles. Later, Attiwill (1962) calculated both the branch dry weight and leaf area of *Eucalyptus obliqua* from branch girth. He demonstrated that the girth of a branch depends on the weight of material supported by the branch.

The stem. Three further attributes of the forest stand of importance to the forester are defect, stem quality and stem form. The impracticability of estimating stem defect and quality was commented on in the last chapter. It should be borne in mind, however, that defect is not so much a problem in well-managed plantations. Stem form, like stem diameter, cannot be estimated directly from the aerial photographs, but the interpreter who is familiar with local forest conditions can with varying success roughly estimate stem form or stem taper from the photographs.

Taper and tree form are influenced by age, site and stand density. Successful recognition of differences in age and site is a skill to be acquired as an art by local experience. It seems probable that the tree height/crown width ratio is a better indicator of stand

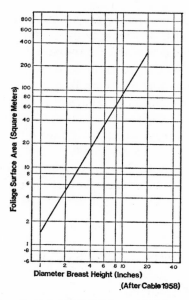

FIG. 21.2. Graph showing the relationship between leaf area and the diameter of the tree (see text).

density than it is of site quality. As outlined by Smith (1965) the tree height/crown width ratio can be used to indicate the crown width/diameter (b.h.) ratio, which in itself is an indicator of tree form.

VOLUME ESTIMATES

Normally in forest inventory, the most accurate method of estimating sawlog or bole volumes is provided by a combination of aerial photo-interpretation and field measurement. The photographs are used as a map or to prepare a map and for stratifying the forest into utilization classes so as to reduce error variance. Measurements are then taken in the field for the purpose of calculating stand volumes. Whenever possible basal area at breast height should be used as one of the variables in volume determination, since it is readily measured using the Bitterlich principle of angle counts, and the derived precision of the volume estimates is usually satisfactory. Spurr (1952), using basal area and height in field studies, obtained a standard error of only 5% in volume per acre for Douglas fir. Several workers have commented on the low accuracy of aerial volume tables (e.g. Nyyssonen, 1955, 1962; Nakayama, 1958).

The principal methods of estimating volumes in conjunction with aerial photographs may be summarized as using (1) photo-plots and ground survey, (2) stereograms with photo-plots, (3) mean tree, (4) aerial tree volume tables, (5) aerial stand volume tables.

(1) As a comprehensive example of photo-plots combined with ground measurement, a forest inventory carried out in the vicinity of the Caspian Sea may be referred to (Rogers, 1961). Basically this combined Chapman's triple sampling design, Meyer and Worley's method of estimation of volumes from aerial photographs and Chapman and Schumacher's regression sampling technique. As photographic variables, total tree height and crown closure were used.

Regression constants and coefficients were computed from the data collected from fifty-four ground plots and the corresponding photo-plots, using the 'method of least squares'. The ground plots were mechanically selected at every forty-third photo-plot. The 2,308 photo-plots provided the average stand height ($\bar{X}_1 = 18 \cdot 7$) and the average crown closure ($\bar{X}_2 = 63 \cdot 0$). These values were then used in the following regression formula to provide the average volume ($\bar{V}_t$) per hectare for the gross land area.

$$\begin{aligned}
\bar{V}_t &= P_f \,(a + b_1 \bar{X}_1 + b_2 \bar{X}_2) \\
&= 0 \cdot 8151 \,(21 \cdot 8297 + 3 \cdot 6506 \,\bar{X}_1 + 1 \cdot 3565 \,\bar{X}_2) \\
&= 143 \cdot 09
\end{aligned}$$

where P_f = proportion of forest land to gross land from aerial photographs

a = regression constant

b_1 = regression coefficient for height

b_2 = regression coefficient for crown closure.

The average volume per hectare was multiplied by the forest area to give the total volume. The total volume was then distributed by forest types according to the volume classes of the ground plots.

(2) As the timber buyer estimates the standing timber in the field subjectively, so can the photo-interpreter use his art in estimating timber volumes and total volumes (including branchwood) from the photographs, in conjunction with an adequate number of stereograms classified by volumes according to measured field plots. Given practice and suitable stereograms the photo-ecologist could probably estimate, as accurately as estimates of timber volume, the leaf dry weight of trees per unit area, total cellulose production, total organic matter and possibly the leaf area.

The precision of timber volumes from photographs using stereograms is frequently as good or better than volumes obtained by using aerial volume tables; but the precision cannot normally be expected to be as high as the precision obtained by ground measurements in conjunction with aerial photographs. Stereograms have been used in volume estimation in the U.S.A., Canada, Sweden and Holland.

Frequently each stereogram comprises a stereopair of 1 sq. in. cut-outs at a distance of about 2·2 in. centre to centre.

In Holland (Stellingwerf, 1963) attention was given to *Pinus sylvestris*, which covers 60% of the forest areas and 90% of the coniferous forest. Stereograms had volume differences of 50 m³ per hectare and the area of each photo-plot was 500 m² at an intensity of one per hectare. About 500 hectares per day were sampled on the photographs when estimating volumes in 10 m³ units per hectare. Some of the plots were double-sampled on the ground so that a regression equation between photographic volume and ground volume could be prepared. The mean photographic volume was 153 ± 3 m³/ha. compared with a terrestrial volume of 149 m³. About fifteen ground plots representing the entire range of stand volumes were found to be sufficient for determining the equation. The regression equation was calculated as:

$$y = 86 + 0·61x.$$

where y = adjusted photo-volume

x = unadjusted photo-volume

(3) An alternative to the determination and use of regression equations and aerial tree or stand volume tables lies in the direct measurement of stand variables from the photographs and using these and their combinations and ratios to determine stand volumes. Stand height, crown closure and mean crown width can under favourable conditions be measured from the photographs. As mentioned elsewhere, crown width and crown closure can be used to estimate the number of trees per acre, and crown width and height can be combined to provide an estimate of stand density. This in turn indicates the crown width/diameter (b.h.) ratio, which is required for estimating the mean tree diameter of the stand from the mean crown width. With these estimates of mean tree diameter, mean height and number of trees, stand volume can be calculated.

(4) An *aerial tree volume table* is prepared by correlating the volume of selected trees measured on the ground with 'top height' and bole diameter via apparent crown diameter of the same trees on the photographs. Details of the construction of a tree volume table were given by Feree (1953) and as early as 1928 by Zieger (Scots pine in Germany).

The completed aerial volume table is used to estimate the volumes of forest stands by the measurement of trees on selected plots located on the photographs. From the volume of these plots and the areas of the stands, the total volume of timber in the forest is calculated. Several writers (e.g. Spurr, 1960) have pointed out that the standard errors obtained when using tree volume tables are high; and in consequence aerial volume tables involving photographic measurement of forest stands and not trees are to be preferred.

(5) *Aerial stand volume tables.* Since World War II, over twenty publications are

available on aerial stand volume tables in the United States and a few are also available from other countries (e.g. Taniguchi in Japan; Tomasegovic, 1956, Nyyssonen, 1955, in Europe). In western Europe, the existence of detailed and intensive management data contributed to the reduced application of photographs to forest problems.

Tables based on two variables (i.e. mean total height and crown closure) have been prepared by Gingrich & Meyer (1955). As a third variable crown diameter was used by Moessner, Brunsen & Jensen (1951) and Moessner (1957). Spurr (1960) pointed out that the addition of the third variable does not decrease the standard error of the estimate. Willingham (1957) found that radial growth at breast

TABLE 21.A

Composite aerial volume table for southern Arkansas (Avery & Myhre, 1959)

Average total height (ft)	Crown closure									
	5%	15%	25%	35%	45%	55%	65%	75%	85%	95%
					(Gross cu. ft per acre)					
40	175	215	250	290	325	365	400	440	475	515
45	240	295	345	400	450	505	555	610	660	715
50	330	395	460	525	590	655	720	790	855	920
55	395	490	580	675	765	860	950	1,045	1,140	1,230
60	480	600	715	830	950	1,065	1,180	1,300	1,415	1,530
65	585	725	860	1,000	1,135	1,275	1,410	1,545	1,685	1,820
70	715	865	1,020	1,175	1,330	1,480	1,635	1,790	1,945	2,100
75	860	1,025	1,195	1,360	1,530	1,695	1,865	2,030	2,200	2,365
80	1,020	1,200	1,380	1,555	1,735	1,910	2,090	2,270	2,445	2,625
85	1,205	1,390	1,575	1,760	1,945	2,130	2,315	2,500	2,685	2,870
90	1,410	1,600	1,785	1,975	2,165	2,350	2,540	2,730	2,915	3,105
95	1,635	1,820	2,010	2,200	2,385	2,575	2,765	2,950	3,140	3,330
100	1,875	2,060	2,245	2,430	2,615	2,800	2,985	3,170	3,355	3,540
105	2,140	2,315	2,495	2,675	2,850	3,030	3,210	3,385	3,565	3,745
110	2,420	2,590	2,755	2,925	3,095	3,260	3,430	3,600	3,765	3,935
115	2,725	2,880	3,030	3,185	3,340	3,495	3,650	3,805	3,960	4,115
120	3,045	3,180	3,320	3,455	3,595	3,730	3,870	4,005	4,145	4,280

height and age for long-leaf pine in Florida was significantly related to height and crown closure. Meyer & Worley (1957) recommended using three or four times as many photo-plots as ground plots, as both sampling and photo-interpretation errors occur in sampling the photographs. Pope (1962) suggested that aerial volume tables should be prepared using the heights of dominant trees measured in the field and crown closure obtained from aerial photographs.

Most aerial stand volume tables have been prepared for a single species or a group of similar species for the purpose of statistical accuracy; but Avery & Myhre (1959) provided a 'composite' volume table for conifers and hardwoods in southern

Arkansas, as these could not be separated on the local photographs (scale 1/20,000). Part of the table giving volumes by stand height (5 ft intervals) and crown cover (10% intervals) is shown in table 21.A. In preparation of the table, nine stand variables were used, being combinations of the three basic variables: height, crown cover and crown diameter. From 511 regressions computed, using an IBM 704 regression programme, a five-variable equation was selected as being best for compiling the table:

$$V = 939 \cdot 065 + 1 \cdot 258 \ HC + 0 \cdot 426 \ H^2 - 39 \cdot 35 \ C - 34 \cdot 155H - 0 \cdot 007 \ H^2 \ C$$

where V = gross volume per acre
H = average total height of three tallest trees
C = crown cover per cent.

As average total height was shown to be the most valuable single variable for predicting gross stand volume, the table was compiled with 5 ft height classes. Crown cover was found to be the next most important variable; and crown diameter was omitted from the table, since none of the variables contributing to the volume estimate in the five-term regression included it. However, other workers (e.g. Moessner, 1957, for conifers in the Rocky Mountains; Smith, 1957, for western hemlock and Douglas fir; Losee, 1953, for spruce, jack pine and mixed species) have found that crown diameter as a third independent variable improves the volume equation. As pointed out by Meyer and Worley (1957), three types of error contribute to errors of aerial stand volume tables: (a) systematic and random errors of the photographic measurements, (b) error of the estimate in the aerial volume table, (c) sampling error of the selected plots.

22

THE PHOTO-COMMUNITY

INTRODUCTION

As early as 1935, Tansley emphasized the need for a holistic approach when studying the ways in which nature is organized (McIntosh, 1963); and previously Gleason (1926) had drawn attention to the need for plant ecologists to study the vegetation relationships of entire regions. Today, the regional approach is encouraged by applying aerial photographs to the study of both vegetation and terrain.

Rowe (1961) has pointed out that it is the neglect of the geographic aspect which has resulted in the drawing of the erroneous conclusions that forest communities can be studied as entities, and that concepts derived in one region can be transplanted *in toto* to another. Several other workers, including Bray (1956), Curtis (1959) and Loucks (1962), have drawn attention to the existence and nature of both physiographic and phytosociological gradients under varying regional conditions.

Even so, some ecologists tend to view plant ecology as being confined to the study of small scale distributions of plants in relation to local environmental factors (e.g. climate), and plant geography as the study of large scale distributions of plant species in relation to major climatic factors and important historical factors. Based on this concept of ecology, then, many foresters are concerned with 'plant geography', and not with plant ecology. This, however, would be incorrect and unacceptable, as ecology continuously plays an important role in both management and silviculture.

No doubt confusion of thought can arise through confining ecological studies to small areas, plant geography to regional studies, and ignoring the interplay of factors and plant distributions over an extensive range of intermediate-sized areas. If, therefore, the concept of ecology as given above is termed *micro-ecology*, and studies relating to the intermediate areas *macro-ecology*, possible confusion can be eliminated. The writer, however, prefers to consider both as being facets of plant ecology.

It is at the level of macro-ecology that aerial photographs are most valuable and should help to correct the imbalance between plant geography and plant ecology. Commonly, photographs at conventional photogrammetrical scales can be used to identify and delineate plant assemblages at the physiognomic or formation levels. Photographs at conventional forest scales can be used similarly, and under favourable conditions may be used at the floristic or association level. However, special very large scale photographs, using carefully chosen film–filter combinations, are far more satisfactory, and are essential for studies of small assemblages of species. Occasionally, photographs can be used to examine interspecific problems, as distinct from geographic or communal patterns. For example, Viktrov (1947), working on the vegetation of dried lake beds in the Kara-kum desert, used photographs when determining the distance between species (e.g. *Tamarix* spp., *Haloxylon aphyllum*). It is interesting to note that McIntosh (1963) has drawn attention to using in the field an attribute of the physical environment (e.g. topography), instead of vegetative pattern, to identify ecosystems. Frequently this attribute is relatively easy to recognize from photographs. Features contained in the photographs, which suggest changes in climate or soil conditions, are often helpful to preliminary studies.

It will be appreciated that site expresses a combination of a number of important factors represented particularly by climate, geology, topography, aspect, soil and the soil moisture regime. Bourne (1928) and Robbins (1934) were the earliest workers to draw attention to the study of site quality from photographs, and later studies included Losee (1942), Moessner (1948) and Johnson (1962).

TERMINOLOGY

The term '*photo-community*' is introduced as an abbreviation for 'aerial photographic community' and will be used when referring to the smallest distinct assemblages of plant species discernible on stereo-pairs of aerial photographs at a specified scale. In continental European nomenclature, it is a phytosociological grouping. English-speaking botanists normally use the term plant ecology to include plant sociology.

The term is open to criticism, however, as the photo-community will vary, not only with the scale but also with other factors influencing the photograph (see Part I). At scales of 1/20,000 to 1/80,000, the unit conforms to the plant formation or sub-formation; at scales between 1/10,000 and 1/15,840, it normally conforms to the sub-formation or the association; and at scales below about 1/5,000, it may be practicable to recognize smaller assemblages of vegetation. These smaller units conform to the terms plant community, stand or society, according to how the term plant community is defined.

The conventional use of the terms *formation*, sub-formation and association has been followed both for description and discussion; but this does not infer that the writer fully agrees with the definitions now to be given. The most extensive and readily observed vegetative unit recorded on aerial photographs is the formation. This may be concisely defined as 'a synthetic structural unit to which are referred all communities exhibiting the same structural form' (Beadle & Costin, 1952). A formation has its own characteristic physiognomy and internal structure and usually comprises two or more associations. The term *sub-formation* is used when it is necessary to make a major physiognomic subdivision of the formation. The *association* is a floristic unit containing (two or more) plant communities which have qualitatively in common a dominant stratum that is floristically uniform (Pryor, 1939; Pidgeon, 1942; Beadle & Costin, 1952). It will be appreciated that this infers uniform structure as in the formation, but that it does not imply quantitative uniformity and does not extend to habitat. Some continental workers, for example Af Rantzien (1951, in Goodall, 1953), appear to be more limiting in defining an association: the unit of lowest rank which can be distinguished by the presence of indicator species. Often workers group closely associated associations into *alliances*. In the study of vegetation from photographs, the alliance can be used, but this has not been found necessary in this chapter.

In defining '*plant community*', ecologists have differed greatly. On the one hand, as a general term the plant community has been viewed simply as a group (or list) of species growing together. The group may be large or small or of indeterminate size. It is common practice to use the term as a collective noun, in place of a pronoun or as a substitute for 'formation' and 'association' (e.g. Patton, 1955). Under these circumstances it is conventional to introduce the term 'society'; and each association will be synthesized from a group of societies. Alternatively the plant community may be viewed as the smallest recognizable unit of vegetation. This unit may have recognizable common tolerances but with the individual species having little or no effect on each other (e.g. Gleason, 1926); or the individuals within the plant community can be considered to function for their mutual benefit as an organism or quasi-organism (e.g. Clements, 1936; Tansley, 1935; Phillips, 1934, 1935; Curtis 1959). Concerning differences in the concept of the plant community, it is pertinent that the photo-interpreter should recognize these differences and should relate photo-interpretation to the most acceptable concept for the region in which he is working.

In recent years, the probability of communities having discrete boundaries has been increasingly questioned (e.g. Whittaker, 1953; Bodenheimer, 1958). Both the Zurich–Montpellier and Uppsala schools appear to have accepted discrete boundaries in their grouping of species into communities. According to Tuomikoski (1942), the Uppsala school's approach to quadrat survey in the field was initially to divide subjectively by eye the area into 'associations' and then record qualitatively the species in each association. The Zurich–Montpellier school did not stratify prior to the systematic collection of quasi-quantitative data, but used the field data from numerous stands to recognize 'associations' and then assumed floristic homogeneity in each. Poore (1955), having adapted the Zurich–Montpellier approach to his study, was able to demonstrate that vegetative units selected subjectively may be placed in groups of similar composition (termed noda) and that the proportion of intermediate stands is relatively small. Recently at Montpellier, aerial photographs have been used to delineate homogeneous zones in which photo-characteristics were examined and coded, and the zones are then grouped according to their coding into 'isolines' (Poissonet, 1966).

Maycock & Curtis (1960), after an exhaustive study of boreal deciduous forest, clearly favour a 'vegetational continuum'. They consider that the continuum, which exists in time and space, must defy all attempts to segregate distinct associations. Quoting W. T. Williams (Lambert & Dale, 1964), continuous variation may be defined as follows: 'If we have a set of sites such that for every site in the group there is at least one other site with which it has one or more species in common and if this group cannot be divided into two or more groups such that all the sites of any one group have no species in common with those of another, then the vegetation may be said to be continuous.'

(a) The physiognomic classification of vegetation

The vegetative patterns recorded on aerial photographs are due to both hetero-geneity between plant communities and homogeneity within plant communities. In examining the photo-physiognomy of vegetation, those characteristics are evaluated which emphasize the differences between formations and similarities within the formations. As Poore has observed for community 'noda', namely that there are relatively small intermediate stands, so also relatively large photo-physiognomic units can be recognized and delineated on the photographs as formations or sub-formations and the intermediate plant structures will usually be relatively few.

When suitable aerial photographs are examined in conjunction with a field study, it becomes apparent that what is referred to commonly as structure or stand struc-ture or community structure should be termed 'internal structure' to distinguish it from the 'external structure' of the vegetation as observed on the aerial photographs. Structure observable on aerial photographs is usually confined to the top vegetative stratum but may extend to the second stratum (e.g. fig. **20.3**) and occasionally to a third stratum. The term *photo-physiognomy* will be used when examining the external structure of vegetation as recorded on aerial photographs. It will be appreciated that when vertical photographs taken with a short focal length are examined, two distinct perspective views of the same tree may be obtained: one within the vicinity of the principal point and the other near the edge of the photographs. The photo-physiognomy is possibly as valuable an aid in plant studies as stand profiles and crown projection sketches prepared in the field.

Vegetation of two distinct regions of the world have been examined by the writer using small scale photographs. These covered the vegetation of western Tanzania and areas of south-eastern Australia. Also a selected area of tropical Queensland has been studied from photographs only. Appendix III should be referred to.

In each of the studies, the writer was impressed by the ease of associating visual observation of the photo-physiognomy with the classification as provided by ground observation. At least some of the photographic characteristics are rapidly measured and expressed quantitatively. These may then emphasize an important characteristic not readily observed on the ground. This is due particularly to the ease of measure-ment and the repetition of the characteristic pattern over a large area in the model provided by a stereo-pair of photographs.

Of the several measurable parameters, stand height at or near maturity (or combined with crown closure) may be initially the most useful criterion. Often it is convenient to apply woodland and forest as two distinct terms for stands with marked differences in height at maturity. This arbitrary distinction, in which height need not be precisely measured by differences of parallax, is not only helpful to the ecologist in the classification of forest formations, but is also helpful to the forester in utilization surveys and to the geographer and geomorphologist in identifying land-systems and land units. No hard and fast rule can be provided for the maximum height of mature woodland, but over vast areas of southern Africa *Brachystegia–Julbernardia* woodland (Greenway, 1943; Howard 1959, 1965) is a distinctive formation with a top height of 45 to 60 ft (fig. **22.1**). Only occasionally under extremely favourable conditions does it attain 70 ft. Crowns do not form a thickly interlaced canopy and may only occasionally touch. Forest, on the other hand, can be described as 'continuous stands of trees which may attain a height of 150 ft or more, with crowns touching and intermingling' (Greenway, 1943). Similarly Haig (1958) has used height as one criterion of the physical features of the tropical formations. He suggested a top height of 140 to 180 ft for rain forest, 80 to 120 ft for moist forest (i.e. seasonal or monsoon) and 60 to 75 ft for dry forest (i.e. woodland). (Cf. Appendix III).

In Australia, a similar distinction can frequently be made using stand height as observed on aerial photographs. A top stand height of about 60 to 80 ft may be often used as one important criterion for separating woodland from forest. In some geographic regions two subdivisions may be required: one on height (tall and low woodland) and one on crown cover. In Victoria, *Eucalyptus hemiphloia–E. sideroxylon* associations conform structurally to tall sclerophyll woodland. With increased rainfall, change in aspect, improved soil conditions and geological differences, woodland on the Hume plateau (e.g. *E. macrorrhyncha*, *E. elaeophora*, *E. dives*) is replaced by dry sclerophyll forest with *E. obliqua* as the dominant species or as co-dominant with *E. radiata*. Stand height at maturity varies between 100 and 150 ft; and with increasing rainfall, this formation is replaced by wet sclerophyll forest with stand heights at maturity of 200 ft or more (fig. **20.3**). There is also a change in understorey species, their height and leaf form. This contributes to texture and tonal differences on the photograph. Similarly in Tasmania, with the change in species and leaf morphology, there is a change in photo-physiognomy which enables the rain forest formation to be separated from the adjoining wet sclerophyll formation, by height, crown closure, tone and occasionally tree shape (e.g. sassafras: narrow pointed crown).

An examination of photographs of Tidal River, Victoria (fig. **22.2a**), following a detailed ground study by R. F. Parsons (1965), indicated that in association with terrain classification, several sub-formations should be photographically identifiable in advance of future detailed field studies. On the granite hill, conspicuous by its light tone, massiveness and weathering pattern, are two woodland sub-formations (*Eucalyptus baxteri* and *Casuarina stricta*), thicket (e.g. *Leptospermum laevigatum*) and heath (e.g. *Kunzea ambigua*) and a stand of sclerophyll forest (*E. obliqua*). Northwards of the granite, but on parallel sand dunes of recent origin, small areas of woodland (*Banksia integrifolia*) can also be distinguished stereoscopically.

Using Raunkiaer's terminology, with slight modification to his height classes, the dominant woody strata recorded on photographs may be referred to as megaphane-rophytic (i.e. forest), mesophanerophytic (i.e. woodland), microphanerophytic (e.g. thicket 2 to 8 m), and nanophanerophytic (e.g. heath, under 2 m high). It is judged as preferable to extend the maximum height of the microphanerophytes to at least 30 ft. Stands conforming to this terminology can often by subjectively identified stereoscopically without precise parallax measurements. In western Tanzania, the microphanerophytes are represented by scattered *Combretum* species and *Acacia pseudofistula* (gall acacia) on grassland (fig. **22.1**) and by dense Itigi thicket. Thicket species include *Albizia petersiana* (deciduous) and *Craibia burtii* (evergreen). All attain heights of 25 to 30 ft and their formation is recognizable in Central Province on photographs at 1/31,000 by region, site, texture, shadow length and apparent height. Raunkiaer's size classes have been incorporated in Appendix III.

In eastern Australia, microphanerophytes are well represented by the mulga, the mallee and coastal thicket; and nanophanerophytes by alpine and coastal heath. As mentioned above, formations of thicket and heath were present on the granite, at Tidal River. Similar photo-physiognomic types are recognizable on calcareous parallel dunes, parabolic (siliceous) dunes and swamp deposits (fig. **22.2**). According to Jennings (1959) the siliceous dunes originated in the late Pleistocene (Parsons, 1965). Thicket was observed to attain a height of about 25 ft on the parallel dunes and on the estuarine (peaty) soils. The dominant species are *Leptospermum laevigatum* on the parallel dunes and *Melaleuca ericifolia* on the peaty soils. *Melaleuca squarrosa* and *M. ericifolia*, which occurred as thicket on a poorly drained swale between the dunes could be photographically separated from *Leptospermum* thicket. The peaty soils were known to flood at times of spring tides. Tall heath (*M. squarrosa*) was identified photographically farther upstream and heath of negligible height (under 2 ft) was identified on the parabolic dunes by site, tone, texture and apparent height. Also present on the parabolic dunes was (dwarf) tree heath with scattered emergent *Eucalyptus baxteri* up to 6 ft tall. Structurally tree heath is intermediate between heath and sclerophyll woodland.

Subsequent to height being used as the initial aid to the recognition of formations, crown closure, crown diameter and the tree height/crown diameter ratio should be examined and possibly also tone, texture and pattern. As mentioned elsewhere, rain forests provide extensive areas with 100% crown closure. Sclerophyll forest frequently has a crown closure in excess of 85%. Woodland is conspicuous by its much lower crown cover. For example, *Eucalyptus baxteri* woodland was observed as having a crown closure of 40% to 60%. In woodland the crowns are no longer interlaced and may be only slightly in contact. In the stereo-model, the ground layer of vegetation is clearly seen. In *E. hemiphloia* woodland, with an annual rainfall under 20 in., the crowns are narrow, seldom touch and trees are relatively widely spaced. Crown closure seldom exceeds 50%. Near Mount Disappointment State Forest (Victoria) 40% to 60% crown cover was recorded for woodland as against 85% to 95% for wet sclerophyll forest. The tree height/crown width ratios were 2·3, 2·8, 3·0, for woodland, dry sclerophyll and wet sclerophyll forest.

When crown cover falls below 30% to 50%, it is desirable to introduce a further descriptive term. Probably the most readily understood is 'grassland'. Thus an area with 25% crown closure may be referred to as grassy woodland or grassy forest. With less than 5% to 10% crown cover, it may be convenient to classify a formation as grassland (Greenway, 1943; Howard, 1959). Where grassland is replaced by forbs or shrubs two further terms are required. Savanna is frequently used to describe grassy woodland or grassy forest or grassland with scattered trees, dwarf trees or shrubs. The only objection to the use of savanna is that it may bring to mind any one of these definitive physiognomic terms. To some, the combining of grassland and woodland (or similarly tree and heath) into a single term may seem paradoxical, but a careful examination of suitable photographs will provide evidence of a uniformity of patterns over extensive areas with a photo-physiognomy between the two. Examples from western Tanzania include grassland with scattered trees (*Afrormosia angolensis, Albizia harveyi*) (fig. **22.1b**), grassland with grouped trees on termitaria (*Mimusops densiflora, Tamarindus indica*) (fig. **22.1b**) and grassland with palms (*Borassus flabellifer*).

Natural grassland, using summer photography and panchromatic film, is frequently recognizable on photographs by its lighter tones, lack of texture and uniformity or lack of minute detail. Sometimes communities dominated by forbs record in darker tones, have a coarser texture and have a repetition of minute detail without distinctive shadows. Local expertise, applied to site, slope and tone, is important in helping to identify different grassland photo-communities. With practice fire patterns can be separated from species patterns. A knowledge of the flowering and fruiting habits of dominant grass species is useful. In the vicinity of Lake Victoria, *Papyrus* swamp was identifiable by site, tone and absence of slope in photographs at 1/31,000. On the photographs of Tidal River, salt marsh, liable to frequent flooding (*Juncus maritimus* and *Samolus repens*) and grassland with a height of 4 to 6 ft (*Cyperus lucidus, Phragmites communis*) were separated from each other by differences in tone, texture, slope and site. The salt marsh was obviously on land liable to seasonal inundation and was conspicuously level.

In tropical areas of high rainfall, the classification of vegetation from photographs has usually been based on criteria of forested and non-forested, with a subdivision of the forests according to site into mangrove swamps, brackish-water swamps, freshwater swamps, peat swamps, riverine forest and dry-land forest. If deciduous woodlands or deciduous forest are also present, then these have been separately identified. For example, Sabah, Howroyd (1954) separated dryland forest from mangroves, Nipah palm and second-growth forest. Francis & Wood (1954) recognized sixteen types on photographs at scales of 1/25,000 and 1/30,000 including twelve predominantly woody. They separated forest from 'non-commercial' vegetation and cultivated land and then divided the forest into salt-water forest, swamp forest, transitional forest, forest liable to flooding and inland forest. Francis (1955) was able to identify ten mangrove species. Again formation typing was used in Sarawak (Browne, 1958). He recognized four swamp types, heath forest (Kerangas), riparian forest, heath, Dipterocarp and mossy forest. In New Guinea on 1/40,000 photographs Taylor and Stewart were able to recognize mature rain forest, secondary forest, grassland, regrowth, swamps, mangroves (four types): *Casuarina* forest, *Octomeles sumatranus* forest. In Cambodia Rollet (1959) used the following: dense evergreen, mixed moist deciduous; dry Dipterocarp secondary growth, savanna, scrub, swamp, marsh and cultivated land. Within dry-land forest, formations on the poor 'podzolic' soils (e.g. *Eperua falcata* in Surinam, Guyana) can be separated from the forests on the deeply weathered lateritic soils (Boon, 1964). Other examples are given in Chapter 20.

(b) The floristic classification of vegetation

In striking contrast to the study of the photo-community at the formation or sub-formation levels from small scale or very small scale photographs, floristic studies of the photo-community at the association or society levels usually require normal forest scale to very large scale photographs, and probably will benefit by carefully selected film–filter combinations according to season of flying. The only exception occurs when the principal species of the dominant stratum are found to be faithful to the plant formation. Then the photo-physiognomic unit, as observed on smaller scale photographs, can be used to define the plant association (see Chapter 20). Provided the species of the dominant stratum can be identified, plant associations may be recognized and delineated on the photographs. With very few exceptions, the identification of species has been confined to trees. Occasionally, attempts have been made to identify a shrub or tree layer of the understorey or the dominant grass or sedge species in tree-less areas.

Whether the classification of vegetation should be based on the dominant species of the highest stratum depends on the purpose of the study. Obviously this is favoured, using photographs; but it may provide plant classes too broad for micro-ecological studies. The classification, however, is usually adequate for terrain classification, land-use and forest surveys and the planning of detailed ecological studies. Alternatively, if the classification of vegetation is provided by a combination of dominants in each stratum, this may result in an inconveniently large number of small units for ecological studies (Goodall, 1953a). A compromise followed by some ecologists is to use groups of characteristic species to classify the quadrats. The doctrine of characteristic species was introduced by Gradman (1909), expanded by Braun-Blanquet (1921, 1928) and developed quantitatively and statistically, using floristic criteria by Goodall (1952, 1953).

Serious pitfalls await the unwary in attempting to use aerial photographs without ground studies of plant societies. The canopy-forming species may give little indication of the plant communities within the association provided by the floristically uniform dominant cover. There will exist within the association one or more groups of species with their own relative homogeneity. As suggested by Goodall (1954), the homogeneity within the group may not be complete, but at least it is greater than between the groups. Homogeneity is provided by floristic composition, abundance of each species, growth and life form, physiognomy and pattern (see Greig-Smith, 1964).

On large scale photographs, there may be sufficient differences in the patterns provided by the communities to enable their identification or at least partial identification and where occasionally discrete boundaries exist, these may be located. This is helpful in the planning of quadrat surveys in the field and should improve the efficiency of the sampling. It seems likely that photographic studies will favour 'noda' as outlined by Poore (1955) and the presence of communities based on indicator species as described by Horn Af Rantzien (1951): 'units of lowest rank which can be distinguished by the presence or absence of indicator species'. This also favours classification techniques; but as has been pointed out by Lambert & Dale (1964) 'it cannot be too strongly emphasized that there is no *a priori* reason why the use of either method (classification or ordination) should be restricted. Continuous systems can be efficiently classified if classification is desired, while discontinuous systems can be ordinated if ordination is thought to be more more useful.' A study using rank correlation coefficients for eighty-four species at Mount Disappointment (Victoria) provided nine groupings with significant positive rank correlations within the groups and negative correlation between the groups. All quadrats were randomly distributed. Of the nine groups, three have different tree species and five different shrubs as members of the group. Some of these can be identified on very large scale photographs (1/4,000). Work continues, as it may be found possible to identify the remaining communities by other characteristics of the photograph. In Mauritania, Rossetti (1962) used a captive balloon for obtaining photographs at scales from 1/300 to 1/3,000 of open-grown herbs and woody plants; and concluded from his brief study that photographs at these scales have certain advantages, particularly in studying the pattern and spatial structure of the vegetation. Other studies have been carried out in arid areas of the U.S.S.R. (Vinogradov, 1962, 1964).

As mentioned in an earlier chapter, several photo-communities could be recognized on photographs of the tree-less upland areas in western Wales. Site, slope, tone and sometimes texture are the most important criteria; and a knowledge of plant succession in the region is helpful or essential. Similar sites may have different communities according to their grazing and fire histories. Photographs taken in the spring, prior to the development of the new season's growth, are far the best as the tonal differences of the photo-communities are maximized. When identifying the associations from the photographs, forbs and grass-type communities were first separated. The *Erica–Calluna* heath was rough textured and dark and the *Vaccinium myrtillus*

266

photo-community was intermediate in tone and texture and located on steep slopes or in association with stunted oak coppice. Bracken (*Pteridium aquilinum*) also occurred on steep slopes but was very dark. The *Molinia caerulea* photo-community is conspicuous by its very light tone, smooth texture and association with site characteristics suggesting sub-soil water flow. In contrast, *Scirpus caespitosa–Eriophorum* associations are darker toned and located on level ground with no evidence of water flow. *Nardus* and *Sphagnum* photo-communities were also identified. On the island of Rhum, west cost of Scotland, the Nature Conservancy has completed a vegetation survey in which photographs were used for identifying the plant associations. From panchromatic photographs at 1/10,000 up to ten types could be directly identified and in the field the photographs were useful for delineating the boundaries of several other types.

IDENTIFICATION OF TREE SPECIES

In section (a) no attempt was made to identify the dominant species. When species were quoted, their identification had been obtained in conjunction with ground reconnaissance or on the basis that a recognizable photo-community was known to contain one or more of the dominant species. A considerable number of papers have been published, particularly by foresters, which relate to the identification of economic tree species on conventional and small scale photographs. Usually the forest stands are divided into 'homogeneous' units or forest types. The procedure is referred to as forest typing; and relies to varying degrees on differences in the photo-physiognomy. Although in the typing of natural forest there is no attempt deliberately to delineate plant associations, it will sometimes be found that this has been achieved or could be achieved with only a slight modification. When the forest typing includes a number of age classes, it may be worth while to examine these as a first step in the study of vegetation succession.

Examples of forest typing in which the species identification suggests that plant associations could have been recognized from the photographs include studies in Canada, western Europe, Russia, United States, several tropical regions and Australasia. Several examples will be quoted. In Scandinavia, spruce stands (*Picea excelsa*) can usually be separated from pine stands (*Pinus sylvestris*), provided one or other of the species comprises 75% of the crop and are older than 5 years (Francis, 1957) In the U.S.S.R., using spectrozonal (colour) film at a scale of 1/25,000, spruce has been reported as being recorded in dark green, birch in yellowish green or greenish brown, oak in yellowish brown, aspen poplar in brownish red and willow in dull brown (Mikhailov, 1961). In Cyprus, *Pinus brutia* was identified on 1/10,000 and 1/15,000 photographs (Polycarpou, 1957) and in British Honduras *Pinus hondurensis* (Miller, 1957). In the tropics, *Dipterocarpus* spp. can often be recognized on small scale photographs by its large crown (e.g. Sabah, Francis & Wood, 1955; Thailand, Loetsch, 1957; Ceylon, De Rosayro, 1959). Occasionally in tropical rain forest a species can be identified due to its being gregarious (e.g. *Morus excelsa* in Brazil; *Goupia glabra* and *Hymenolobium petraeum* in Surinam, Heinsdijk, 1957; *Ocotea rodiaei* in British Guiana, Swellengrebel, 1959).

Tree species are, however, best identified by using large scale photographs. Characteristics are then tone, shape, texture and size of the tree and a knowledge of its geographic distribution and site requirements. A grey-scale (fig. *16.3*) may be used to correlate the tone of the species recorded on black-and-white photographs (fig. 22.3).

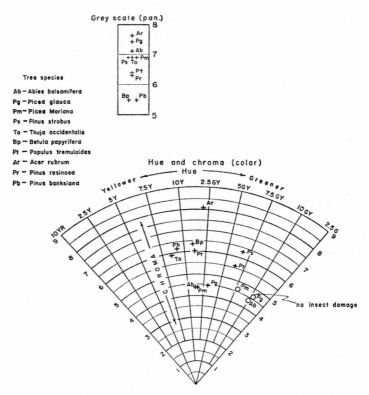

FIG. 22.3. Tree species depicted by a grey scale (part) and the Munsell colour notation.

Considerable variations in tone may be expected for the same species across the photograph and particularly on the line about the axis joining the specular and reflex reflection zones. What is more important than an exact tonal definition for a species is its relative tone in relation to other species.

If colour transparencies are being examined, then the tree image represented by its hue, chroma and value may be related to standard Munsell colour chips as shown in fig. 22.3 for several hardwood and conifer species in northern Minnesota. Heller *et al.* (1963) described a technique for comparing transparencies with the Munsell chart, whilst being able at the same time to view the tree image in the stereo-model provided by the colour transparencies. At scales between 1/1,188 and 1/3,960 they obtained highly significant differences based on colour and tree shapes for eight tree species. A ninth species, *Betula papyrifera* was at times incorrectly confused with *Populus tremuloides*. Their results for Pan-Aerographic (black-and-white) film and Super Anscochrome (colour) film are summarized in table 22.A. It was observed that each species had a definite central tendency to a recognizable colour. In Finland, using 1/10,000 enlargements from 1/28,000 panchromatic photographs, about 80% of the pine and spruce stands were correctly identified, but only 50% to 60% of the hardwoods. The identification of species in pure stands was noticeably better than mixed stands; and in the lower volume stands pine was over-estimated to the detriment of spruce (Nyyssonen, 1966). As pointed out by Stellingwerf in Holland (1966), a higher percentage (70% to 90%), correctly identified, can be expected for pure stands than for mixtures by groups of species (60% to 80%) and single tree mixtures (40% to 60%).

TABLE 22.A

Tree species—identification by colour and tone
(Percentages—successful identifications)

Film	Pinus strobus	Thuja occidentalis	Pinus banksiana	Pinus resinosa	Picea spp.	Abies balsamifera	Acer rubrum	Pupulus tremuloides	Betula papyrifera
Pan (black and white)	95	81	80	80	77	73	70	59	44
Colour (Anscochrome)	97	96	88	88	84	84	73	64	56

In the identification of tree species using shape it will be appreciated that photographic scale is important; and, as a general rule, the scale requires to be much larger than 1/10,000. Recent studies in North America indicate the need for a very large scale (e.g. 1/1,584) (Sayn-Wittgenstein, 1960, 1961; Heller *et al.*, 1963). The trees should be examined individually and as members of a group, both vertically at or near the principal point and if possible in side profile towards the periphery of the photograph.

Botanical terminology as used in the description of leaf morphology will be found convenient to describe the crown apex, the crown profile, the crown edge in side view and the crown in plan view. Thus as shown in fig. 22.4 the crown apex can be described as acuminate, acute, obtuse, etc.; the crown in side view as hastate, lanceolate, ovate, obovate, etc.; the crown edge in side view as simple or compound, serrate, sinnuate, lobed, etc.; and in plan view as entire or compound, sinnuate, serrate, lobed, etc.

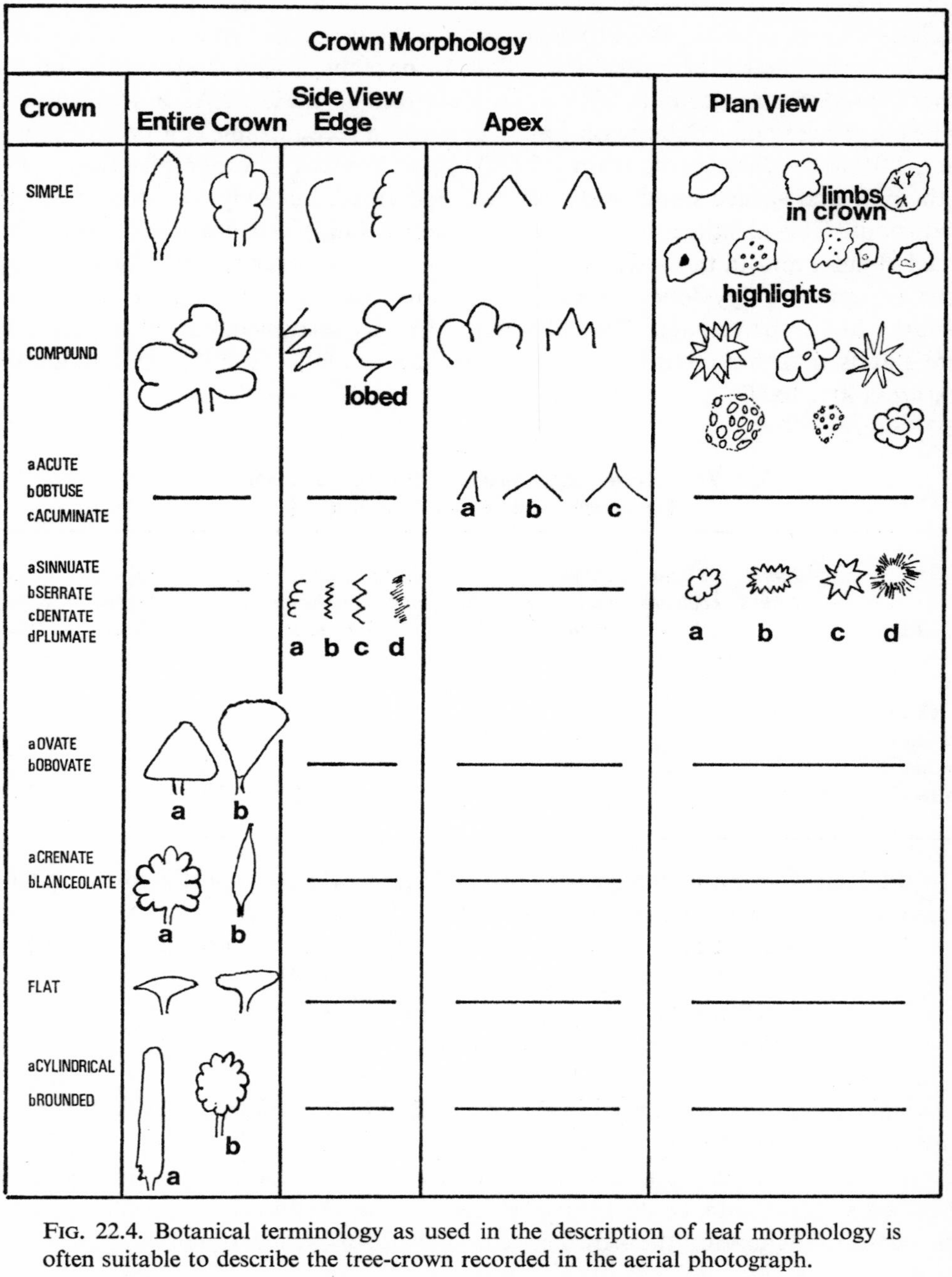

FIG. 22.4. Botanical terminology as used in the description of leaf morphology is often suitable to describe the tree-crown recorded in the aerial photograph.

THE FLORISTIC CLASSIFICATION OF VEGETATION

As outlined by Sayn-Wittgenstein (1961), identification begins by eliminating those species whose presence in the area of the photograph is impossible or unlikely due to climate, physiography or locality. Next, by a knowledge of the associations formed by the common species and of their requirements, the group or groups of species which are likely to occur in the area are established. The identification of similar species in a group depends finally on the tree-form, particularly crown shape and branching

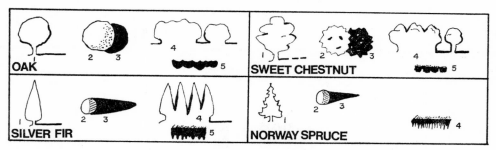

FIG. 22.5. Isolated trees, trees at the edge of plantations, trees in stands recently heavily thinned and boreal forests with snow on the ground, all favour recording which may be distinctive for the species of the region. This is illustrated above for four common European species. Shadow further requires to be combined with crown shape, crown tone and crown texture.

habit. The use of tree-form to identify a species becomes less and less valuable as the scale of the photographs is reduced, until eventually these characteristics are replaced by photographic tone, texture and shadow-pattern. Shadow within the crown and shadow on the ground are also helpful, but there are often not sufficiently wide openings in the canopy to perceive the latter. Dark shadows cast on the ground indicate a compact crown and dense foliage, and a 'light' shadow the opposite. As early as 1952, the Société d'Histoire Naturelle de Toulouse drew attention to the value of shadow in identifying the tree species (i.e. identification spécifique (fig. 22.5)).

Appendix Ia
KEY TO FLOATING-CIRCLES
STEREOGRAM TEST SHEET

(Showing parallax differences of the floating circles)

BLOCK A	inches	BLOCK B	inches
A5	0·045	A1	0·016
B7	0·040	B8	0·012
C2	0·035	C3	0·010
D4	0·030	D1	0·008
E1	0·025	D5	0·006
E6	0·020	E6	0·005

BLOCK C	inches	BLOCK D	inches
A2	0·005	A1	0·002
A5	0·004	B4	0·002
B7	0·004	B7	0·002
C3	0·003	O2	0·001
D1	0·003	C5	0·001
E8	0·003	D3 less than	0·001
		E6 less than	0·001

Appendix Ib
DEPTH ORDER OF DETAILS IN THE EIGHT RINGS: ZEISS TEST

1 Marginal ring
Square
Triangle
Point

2 Carl Zeiss
No. 2

3 Circle lower left
Marginal ring
Cross; marginal ring
Square; No. 3
Circle upper centre

4 No. 4
Centre top circle
Marginal ring; lower circles
Top right circle
Top right penultimate circle

Top left penult-
imate circle
Top left circle

5 Marginal ring;
No. 5
Square
Diagonals
Triangle

6 Circle upper right
Marginal ring
Circle upper left
Circle lower left
Circle lower right

7 Marginal ring
Double cross with arrowhead
Black triangle
Black circle
Flag with ball (black)
Black rectangle
Tower with cross and ring
White triangle
White rectangle

8 Marginal ring
Steeple and the
two triangles
(The right triangle
demonstrates
a height parallax)

Note: Within each ring, the images (as seen under the stereoscope) have been placed in order of increasing distances from the observer.

Appendix II
LIST OF SYMBOLS USED

α: alpha, any angle.

Å: Ångström units.

$A, B \ldots G$, etc.: ground stations or objects or the base and top of objects.

$a, b \ldots g$, etc.: image points in the photograph corresponding to $A \ldots G$.

b.h.: breast height. b: photo air base; B: ground air base

d: displacement of an image on a photograph.

Δp: Delta, parallax difference.

D: ground distance.

d: the ground distance D on the photograph; the image displacement of an object on a photograph.

f: focal length (of a camera); more precisely the equivalent focal length.

F: f-stop (exposure number).

γ: gamma, slope of the characteristic curve.

h: elevation of a ground station above sea-level or height of an object.

H: height of the camera or flying height above sea-level datum; frequently $H-h$, height above ground datum, is approximated to 'H'.

I, i: ground isocentre, photograph isocentre.

λ: lambda, wave-length.

M: magnification.

μ: mu, microns ($1\,\mu = 1{,}000\,\mathrm{m}\mu = 10{,}000\,\text{Å} = 10^{-8}\,\mathrm{cm}$).

N_1, N_2: nodal points in lens.

N, n: ground nadir, photograph nadir.

O: interior perspective centre.

$P_1 p$: ground principal point, principal point on photograph.

ϕ: phi, polar scatter angle: pitch, longitudinal tilt.

Σ: sigma, sum of.

s: standard deviation.

s^2: variance.

$S_{\bar{x}}$: standard error of the mean.

t_x, t_y: x-tilt, y-tilt.

274

θ: theta, an angle commonly the angle between the camera axis or camera angular field and the ground datum.

$\overline{X}$: mean.

ω: omega, lateral tilt (x-tilt).

$\cong$: approximately equal to.

Appendix III

An example of two-way photo-physiognomic classification of vegetation using primarily dominant height and gross (ground) cover—based on photographic observations of formations/sub-formations in eastern Australia* and/or East Africa (Tanzania†) (pages 259-65).

Life-form:	Megaphanero-phytes	Meso-phanerophytes	Micro-phanerophytes	1. Nano-phanerophytes 2. Chamaeophytes
Size class (metres)— Raunkaier	>30	8–30	2–8	1. 0·25–2·0 2. <0·25
Recorded height range (ft.)	>85	25–80	7–30	<7
Photo-physiognomy	Forest	Woodland	Shrubland	'Herbland'
Gross cover (Dominant woody strata)				
c. 100%	Rain F. (e.g.Trop., Temperate*)			Heath
Very dense e.g. >50%	Closed F. (e.g. Trop., Wet Montane†) Wet Sclero-phyll F.* Tall W.	Layered W.*	Thicket (e.g. coastal)	
e.g. 10–50%	Open F.	Shrub W. Low W.	Thicket (e.g. Itigi† Mallee*)	Steppe (e.g. salt bush*)
		Wooded heath* W. grassland	Shrub grass-land	
Open e.g. <10%				Grassland (e.g. Hummock Tussock)
Topographic types	Riverine F. Swamp F.	Riv. W., Swamp W., Mangroves†	Riv. thicket Mangroves	Swamp (seasonal or permanent) Salt marsh

Appendix IV. SUMMARY OF CLASSIFICATORY LAND UNITS USED BY VARIOUS WORKERS

Brink et al. (General)	Christian & Stewart	Downes & Gibbons (Australia)	Coaldrake	Beckett Webster (U.K.)	Ollier et al. (Uganda)	Hills (Canada)
Landzone						
Land division						Site region
Land province	(Geomorphic unit)	Land zone		(Physical region)		
Land region		Land zone				
Compound land system. Land system (i) Complex (ii) Simple	Land system (i) Complex (ii) Simple	Land system	Land-scape	Recurrent landscape pattern	Land system	(Site district)
	Land system	Land unit	Land system			Landscape unit
Land facet (i) Cognate (ii) Non-cognate	Land unit		Land unit		Land facet	Land type
		Land component		Facet		Land type component
Land element	Land unit	Land component	Land unit	Facet		Physiographic Site type

BIBLIOGRAPHY

ABELL, C. A. (1939). A method of estimating irregular shaped and broken figures. *J. For.* **37**, 344–5.

ADELSTEIN, P. Z. & LEISTER, D. A. (1963). Non-uniform dimensions of changes in topographic aerial films. *Photogramm. Engng.* **29** (1), 149–61.

ALDRED, A. H. (1964). Wind-sway error in parallax measurements of tree height. *Photogramm. Engng.* **30** (5), 732–4.

ALDRICH, R. C. (1955). A method of plotting a dot grid on aerial photographs of mountainous terrain. *J. For.* **53**, 810–13.

ALDRICH, R. C. (1966). Some potential forestry applications of 70 mm colour photography. ACSM-ASP Convention, Washington, D.C.

ALDRICH, R. C., BAILEY, W. F. & HELLER, R. C. (1959). Large scale 70 mm colour photography techniques and equipment and their application to forest sampling. *Photogramm. Engng.* **35** (5), 747–54.

ALISON, G. W. & BREADON, R. E. (1960). Timber volume estimates from aerial photographs. *For. Surv. Note* No. 5. *B.C. For. Serv.*

ALLUM, J. A. (1961–62). Photogeological interpretation of areas of regional metamorphisms. *Photogramm. Engng.* **28**, 418–38.

AMBROSE, W. R. (1965). Recent developments in high-performance stereoscopes. *Photogramm. Engng.* **31** (3), 505.

ANDERSEN, H. E. (1956). Use of twin low-oblique aerial photographs for forest inventory in south-east Alaska. *Photogramm. Engng.* **22**, 930–4.

ANDREUCCI, E. (1964). The behaviour of vegetation in infrared photography. *Ferrania*, **12** (1). Milan, 13–19.

ANDREWES, G. S. & TROREY, L. G. (1933). The use of aerial photographs in forest surveying. *For. Chron.* **9**, 33–5.

ANON. (1946). California sea-lion census. *Calif. Fish Game*, **33**, 19–12.

ANON. (1955). The avoidance of shadow point on aerial photographs. *Leafl.* 74 *For. Timb. Bur. Aust., Canberra.*

ANON. (1955). *Introducing Victoria* (ed. G. W. Leeper), ch. 5, 52–68. Vegetation in Victoria (R. T. Patton). A. N. Z. A. A. S., Melbourne Univ. Press.

ANON. (1962). Forstliche Photogrammetrie und Luftbildauswertung. Special Edition. *Forstzeitschr*, No. 17 (1/2).

ANON. (1965). Modulation transfer data for Kodak films. *Sales Service Pamphlet* **49**. Kodak Research Laboratories, Rochester, N.Y.S.

ANON. (1966). The camera. *Sunday Times Magazine*, 18th September, 66 pp.

ANSON, A. (1959). Significant findings of a stereoscopic activity study. *Photogramm. Engng.* **25**, 607–11.

AQUINO, R. R. (1954). Glossy photographs versus positive transparencies for tree height measurements. *Philipp. J. For.* **10**, 1–13.

ASCH, VAN P. (1961). The growth of aerial survey and its effect on the work of the New Zealand private surveyor. N.Z. Institute of Surveyors.

BIBLIOGRAPHY

ASCHENBRENNER, C. M. (1952). A review of facts and terms concerning the stereoscopic effect. *Photogramm. Engng.* **18**, 818–23.

ASCHENBRENNER, C. M. (1954). Problems in getting information into and out of aerial photographs. *Photogramm. Engng.* **20**, 398–401.

ASHBURN, E. V. & WELDON, R. G. (1956). Spectral diffuse reflectance of desert surfaces. *J. Opt. Soc. Am.* **46**, 583–6.

ASHTON, D. H. (1965). School of Botany, Univ. of Melbourne, personal communication.

ASHTON, P. S. (1964). A quantitative phytosociological technique applied to tropical mixed rain forest vegetation. *Malay Forester*, **27** (4), 298–317.

ATTA, G. R. VAN (1936). Filters for separation of living and dead leaves in monochromatic photographs. *J. Biol. Photogr. Ass.* **4**, 177–91.

ATTIWILL, P. M. (1962). Estimating branch dry weight and leaf area from measurements of branch girth of Eucalyptus. *Forest Sci.* **8**, 132–41. Also Ph.D. thesis, School of Forestry, Univ. of Melbourne.

ATTWELL, B. J. (1955). Some factors affecting the physical quality of the image in the aerial photographs. *Photogramm. Rec.* **1** (6), 13–31.

ATTWELL, B. J. (1959). Image movement in aerial photography. *Photogram. Rec.*, **3**, 4–15.

AUNG MYINT, U. (1958). The uses of aerial survey in managing Burma forests. *Burm. Forester*, 181–8.

AVERY, T. E. (1957). Forester's guide to aerial photo-interpretation. *U.S. For. Serv. Occ. Pap. Sth. For. Exp. Sta.* 156.

AVERY, T. E. (1958). Helicopter stereo-photography of forest plots. *Photogramm. Engng.* **24**, 617–24.

AVERY, T. E. (1960a). A checklist for airphoto inspections. *Photogramm. Engng.* **26**, 81–4.

AVERY, T. E. (1960b). Identifying southern forest types on aerial photographs. *U.S. For. Serv. Pap. Stheast For. Exp. Sta.* **112.**

AVERY, T. E. (1964). To stratify or not to stratify. *Journal of Forestry* **62**, 106–8.

AVERY, T. E. (1966). Identifying forest vegetation on aerial photographs. World Forestry Congress, Madrid.

AVERY, T. E. & MYER, M. P. (1959). Volume tables for aerial timber estimating in northern Minnesota. *U.S. For. Exp. Sta., Pap.* No. 96, 37 pp.

AVERY, T. E. & MYHRE, D. (1959). Composite aerial volume table. *U.S. For. Serv. Sth. For. Exp. Sta.*, paper 175.

AXELSON, H. (1956). Effect of photo scale on the use of aerial photographs in Swedish forestry. *Committee of Photogrammetry Bull.* **2**, Stockholm.

BABEL, A. (1935). Infrarot-Photographie im Pflanzen Schutz. *Angew. Bot.* **17**, 45–53.

BACHMANN, W. K. (1946). *Études sur la Photogrammetrie Revue.* Wild, Heerbrugg (Switzerland).

BACKSTROM, H. E. (1956). Some investigations about film and filter in aerial photography. Int. Congress Photogrammetry, Stockholm.

BACKSTROM, H. E. & WELANDER, E. (1953a). An investigation of the possibilities to distinguish different tree species on aerial photographs. Royal Institute of Technology, Stockholm.

BACKSTROM, H. E. & WELANDER, E. (1953b). An investigation in diffuse reflection capacity of leaves and needles of different species. *Norrlands SkogsvFörb. Tidskr., Stockh.* **1**, 141–69.

BANFIELD, A. W. F. (1956). The caribou crisis. *Beaver*, 1–7.

BARBER, H. (1955). Adaptive gene substitution in Tasmanian eucalypts. *Evolution*, **9**, 1–14.

BIBLIOGRAPHY

BARRINGER, A. R. (1963). The use of audio and radio frequency pulses for terrain sensing. *Proc. 2nd Symposium on Remote Sensing*, 201–14, Report 4864–3–x, Univ. Michigan, Ann Arbor.

BARROW, C. H. (1960). Very accurate correction of aerial photographs for the effects of atmospheric refraction and earth's curvature. *Photogramm. Engng.* **26**, 798–804.

BAUMANN, H. (1959). *Forstliche Luftbildinterpretation.* Forstdirektion Südwürtemberg. Hohenzollern, Tübingen, 109 pp.

BAUR, G. N. (1957). The nature and distribution of rain forests in New South Wales. *Aust. J. Bot.* **5**, 190–233.

BAWDEN, F. C. (1933). Infrared photography and plant diseases. *Nature*, **132** (July), 168.

BEADLE, N. C. & COSTIN, A. B. (1952). Ecological classification and nomenclature. *Proc. Linn. Soc., N.S.W.* **77**, 61–82.

BEARD, J. S. (1941). Land utilization survey of Trinidad. *Caribb. Forester*, **2** (4).

BEARD, J. S. (1953). The savanna vegetation of northern tropical America. *Ecol. Monogr.* **23**, 149–215.

BEARMAN, W. E. (1957). An improved technique of forest type transfer. *Photogramm. Engng.* **23**, 599–600.

BECKET, P. H. & WEBSTER, R. (1962). *The Storage and Collation of Information on Terrain.* Military Engineering Experimental Establishment, Christchurch, U.K.

BECKING, R. W. (1957). The Zurich–Montpellier school of phytosociology. *Bot. Rev.* **23**, 411–88.

BECKING, R. W. (1959). Forestry applications of aerial colour photography. *Photogramm. Engng.* **25**, 559–65.

BEDARD, P. W. (1956). Reconnaissance classification and mapping of Philippine forests. *Symposium on Tropical Vegetation. UNESCO, Ceylon,* 1956, **8**.

BEGEROO-CAMPAGNE, B. (1955). Inventaire des forêts tropicales par photographies aériennes. *Bois Forêts Trop.* **43**, 3–7.

BELCHER, D. J. *et al.* (1951). *A Photo-analysis Key for the Determination of Ground Conditions*, vol. **1**. U.S. Office of Naval Research, Cornell University, Ithaca, N.Y.S.

BELOV, S. V. (1959). *Aerial Photography of Forests.* Moscow–Leningrad, Izdan. Akad. Nauk SSSR, 214 pp.

BELOV, S. V. & ARCYBASEV, E. S. (1957). An investigation of the reflectivity of tree species (Russian only). *Bot. Z.* **42** (4), 517–34.

BELTMAN, B. J. (1952). Shadows on aerial photographs. *Photogramm. Engng.* **18**.

BENNET, W. E. (1956). *Summary of Characteristics of Camouflage Detection Film.* Bibliography, *Aerial Photographic Interpretation* (D. R. Lueder).

BERNSTEIN, D. A. (1958). Does magnification improve measurements of stand heights on aerial photographs? *For. Chron.* **34** (4) 435–7.

BERNSTEIN, D. A. (1962). Two-storey, forest type mapping. *U.S. For. Serv., Region 6.*

BICKFORD, C. A. (1952). The sampling design used in the forest survey of the northeast. *J. For.* **50**, 290–2.

BICKFORD, C. A. (1959). A test of continuous inventory for national forest management based upon aerial photos, double sampling and remeasured plots. *Proc. Soc. Am. Foresters,* 1959, 1960, 143–8.

BILLINGS, W. D. & MORRIS, J. (1951). Reflection of visible and infrared radiation from leaves of different ecological groups. *Am. J. Bot.* **38**, 327–31.

BILLINGS, W. D. & MORRIS, J. (1952). The environmental complex in relation to plant growth and distribution. *Q. Rev. Biol.* **27**, 251–65.

BITTERLICH, W. (1952). Das Spiegel Realaskop. *Öst. Forst- u. Holzw.*

BLACHUT, T. J. (1957). Photogrammetric long distance bridging. *Can. Sci.* **13**, 644–9.

BIBLIOGRAPHY

BLACK, G. A., DOBZHANSKY, T. & PAYAN, C. (1950). Some attempts to estimate species diversity and population density of trees in Amazonian forests. *Bot. Gaz.*, **III**, 413–25.

BLACKMAN, G. E. (1935). A study by statistical methods of the distribution of species in grassland associations. *Ann. Bot. Cond.* **49**, 749–77.

BLACKWELL, H. R. (1946). Contrast threshold of the human eye. *J. Opt. Soc. Am.* **36**, 624–43.

BLAKE, S. T. (1938). The plant communities of western Queensland. *Proc. Roy. Soc. Queensland*, **49**, 156.

BLANDFORD, H. R. (1924). The aero-photo survey and mapping of the Irrawaddy delta. *Indian Forester*, **50**, 605–16.

BODENHEIMER, F. S. (1958). *Animal Ecology Today.* W. Junk, The Hague, Holland.

BOMBERGER, E. H., DILL, H. W. *et al.* (1960). Photo-interpretation in agriculture. *Manual of Photo-interpretation*, ch. 11, 561–666.

BONNIER, G. (1895). Les feuilles des plantes de haute altitude. *Annls. Sci. Nat. (Botanique)*, VII series, **20**, 217.

BOON, D. A. (1956). Recent development in photo-interpretation of tropical forests. *Photogrammetria* **12**, 382, 386.

BOON, D. A. (1960a). Aerial photography and forestry. *Rept. ECAFE Seminar on Aerial Survey Methods and Equipment, Bangkok.*

BOON, D. A. (1960b). Report of working group 4. Interpretation of Vegetation, Comm. VII Inter. Soc. of Photogrammetry. *Photogramm. Engng.* **26**, 283–302.

BOON, D. A. (1960c). Results of forest type mapping in Switzerland. *Int. Train. Cent. Delft*, Ser. B 1, 4, 18 pp.

BOON, D. A. (1961a). The evaluation of grouping tendency of tree species in tropical rain forest. IUFRO 25/5.

BOON, D. A. (1961b). Results of forest type mapping in Switzerland. *ITC (Delft)*, Publication 8.

BOON, D. A. (1964). Some aspects of plant ecology in the tropics in connection with the use of aerial photography. U.N. Cartographic Conference, Manila.

BOURNE, R. (1928). Aerial survey in relation to the economic development of new countries, with special reference to an investigation carried out in Northern Rhodesia. *Oxf. For. Mem.* No. 9, 35 pp.

BOURNE, R. (1931a). Regional survey and its relation to stocktaking of the agricultural and forest resources of the British Empire. *Oxf. For. Mem.* No. 13.

BOURNE, R. (1931b). Air survey in relation to soil survey. *Imperial Bur. Soil Sci. Tech. Comm.* **19**.

BOYER, R. E. & MCQUEEN, J. E. (1964). Comparison of mapped rock fractures and air photo linear features. *Photogramm. Engng.* **30** (4), 633.

BRADFORD, J. (1957). Ancient landscapes. *Studies in Field Archaeology*, 296. Bell, London.

BRAITHWAITE, J. D. (1940). Canadian aerial photography in Burma. *Indian Forest Rec.* 96.

BRANCH, M. C. (1948) *Aerial Photography in Urban Planning and Research*, 117. Harvard University Press.

BRAUN, E. L. (1941). The differentiation of deciduous forest of the eastern United States. *Ohio J. Sci.* **41**, 235–41.

BRAUN-BLANQUET, J. (1928). *Pflanzensoziologie Grundzuge der Vegetationskunde.* Springer, Berlin.

BRAUN-BLANQUET, J. (1951). *Pflanzensoziologie*, 126–33. Vienna.

BRAY, J. R. (1956). A study of mutual occurrence of plant species. *Ecology*, **37**, 21–8.

BRAY, J. R. (1962). Use of non-area analytic data to determine species dispersion. *Ecology*, **43**, 328–33.

BIBLIOGRAPHY

BRAY, J. R. & CURTIS, J. T. (1957). An ordination of the upland forest communities of southern Wisconsin. *Ecol. Monogr.* **27**, 325–49.

BRENCHLEY, G. H. & DADD, C. V. (1962). Potato blight recording on aerial photographs. *N.A.A.S. Q. Rev.* 21–5. London.

BRINK, A. B., MABBUTT, J. A., WEBSTER, R. & BECKETT, P. H. (1966). Report of the working group on land classification and data storage, M.E.X.E. No. 940, Christchurch, U.K.

BROCK, G. C. (1952). *Physical Aspects of Aerial Photography*, 7, 9, 29–32, 39, 45, 52–4, 78–9, 80, 89–90. 96, 103, 114–15, 117, 193–221, 233–5, 238–45. Longmans, London.

BROCK, G. C. (1953). Negative exposure in air photography. *Photogramm. Rec.* **1** (1).

BROCK, G. C. (1959). The quality of the photographic image. *Photogramm. Rec.* **16** (3).

BROCK, G. C. (1961). A note on exposure in air photography of desert areas. *Photogramm. Rec.* **3** (18), 530–8.

BROWN, R. T. & CURTIS, J. T. (1952). The upland conifer–hardwood forests of northern Wisconsin. *Ecol. Monogr.* **22**, 218–30.

BROWNE, F. G. (1958). Study of tropical vegetation, Sarawak. *Proc. Kandy Symposium, UNESCO*, 166–7.

BRUCE, D. & REINEKE, L. H. (1931). Correlation alignment charts in forest research; a method of solving problems in curvilinear multiple correlation. *U.S. Dept. Agric. Techn. Bull.* **210**, 87 pp.

BRUNNSCHWEILER, D. H. (1957). Seasonal changes of the agricultural pattern. *Photogramm. Engng.* **23**, 131–9.

BRUNT, M. (1966). Methods employed by the Directorate of Overseas Surveys in the assessment of land resources. *Symposium, Commission VII*, I.S.P. Paris.

BUCKINGHAM, F. M. (1959). Area measurement by weight. *Tech. Rep. Fac. For. Toronto*, No. 2, 449.

BUCKMASTER, J. L. (1950). The camera lucida for aero-mapping. *Photogramm. Engng.* **12**, 235–45.

BUDYKO, M. I. (1958). *The Heat Balance of the Earth's Surface* (trans. N. A. Stepanoua). U.S.D.C., Weather Bureau, Washington.

BURBRIDGE, N. T. (1960). The phytogeography of the Australian region. *Aust. J. Bot.* **8**, 75–212.

BURINGH, P. (1960). Application of aerial surveys in soil surveys. *Manual of Photo-interpretation, U.S.A.*

BURNS, G. R. (1934). Long and short wavelength limits of photosynthesis. *Pl. Physiol.* **9**, 645–52.

BURRELL, A. J. (1963). Mapping the forests of Western Australia from aerial photos. Typescript, Forest Dept., Western Australia.

BURTT-DAVY, D. (1942). Some East African vegatation communities *Journ. Ecology*, **30**, 67–146.

BUSHNELL, T. M. (1951). Use of aerial photographs for Indiana land studies. *Photogramm. Engng.* **17**, 725–38.

BYLES, B. U. (1932). A reconnaissance of the mountainous part of the river Murray catchment in N.S.W. *Comm. For. Bureau*, **13**. Canberra.

CABLE, D. R. (1958). Estimating surface area of ponderosa pine foliage in central Arizona. *Forest Sci.* **4**, 45–9.

CAHUSAC, A. B. (1957). Forest mapping from aerial photography, Uganda. *7th British Commonwealth Forestry Conference*, 1–5.

BIBLIOGRAPHY

CAJANDER, A. L. (1943). Forest types and their significance. *For. Fenn.* **56,** 71 pp.

CALHOUN, J. M. (1960). A method for studying possible local distortions in aerial films. *Photogramm. Engng.* **26,** 661–72.

CAMERON, H. C. (1965). Radar as a surveying instrument in hydrology and geology. *Proc. 3rd Symposium on Remote Sensing,* 440–52. Report 4864-3-x. Univ. Michigan, Ann Arbor.

CAMERON, H. L. (1952). The measurement of water current velocities by parallax methods. *Photogramm. Engng.* **18,** No. 3, June.

CAMERON, H. L. (1961). Interpretation of high altitude small-scale photography. *Can. Sci.* **15** (10), 567–73.

CAMPBELL, C. J. (1964). Compact field kit for photo-interpretation. *J. For.* **62,** 266–7.

CAMPBELL, J. D. (1935). Air survey in Tripura State, Bengal. *Report, Conference of Empire Survey Officers,* 170. H.M.S.O., London, 1936.

CARMAN, P. D. & BROWN, H. (1961). Camera calibration in Canada. *Can. Surveyor,* **15** (8), 425–49.

CARRON, L. T. & HALL, N. (1954). National forest inventory: Beech Forest. *Aust. For.* **18** (2), 128–40.

CATINOT, R. & SAINT-AUBIN, G. DE (1960). Utilisation des photographies aériennes sans point au sol en cartographie forestière. *Bois Forêts Trop.* **69,** 17–25.

CHAMBERS, T. C. & HALLAM, N. D. (1964). Structure of the surface waxes of the eucalypts. Section M, *A.N.Z.A.A.S.,* Australia.

CHAMPION, H. G. (1936). A preliminary survey of the forest types of India and Burma. *Indian Forest Rec.* **1,** 1–286.

CHAPMAN, V. J. (1947). The application of aerial photography to ecology as exemplified by the natural vegetation of Ceylon. *Indian Forester,* **73,** 287–314.

CHASE, C. D. & SPURR, S. H. (1955). Photo-interpretation aids. *U.S. For. Serv. Rep. Lake St. For. Exp. Sta.* **38.**

CHATTIN, J. E. (1952). Appraisal of Californian wild fowl concentrations by aerial photography. *17th North American Wildlife Conference,* 421–6.

CHITTENDEN, H. M. (1959). Differential elevation by adaption of parallax correction graphs to parallax measurements on aerial photographs. *Photogramm. Engng.* **25,** 144.

CHRISTIAN, C. S. (1958). The concept of land units and land systems. *Proc. 9th Pacific Congress,* **20,** 74–81.

CHRISTIAN, C. S. & STEWART, G. A. (1953). *Survey of the Katherine–Darwin Region.* Land Research Series, C.S.I.R.O., Melbourne.

CHRISTIAN, C. S. & STEWART, G. A. (1964). Methodology of integrated surveys. *UNESCO Conference, Toulouse,* 1–146.

CHRISTIAN, C. S., STEWART, G. A. & PERRY, R. A. (1960). Land in northern Australia. *Austr. Geographer,* **7,** 217.

CHROSTHWAIT, H. L. (1930). Air surveys of Central and East African territories. *J. Agric. Soc.*

CHURCHILL, E. A. & STITT, R. L. (1955). Association analysis applied to the interpretation of aerial photographs. *Photogramm. Engng.* **21,** 598–602.

CLAPHAM, A. R. (1932). The form of the observational unit in quantitative ecology. *J. Ecol.* **20,** 192–7.

CLARK, W. (1949). *Photography by Infrared,* 2nd edition, p. 255. Wiley, New York.

CLARKE, A. B. (1962). A photographic edge-isolation technique. *Photogramm. Engng.* **28,** 393–9.

CLEMENTS, F. E. (1936). Nature and structure of the climax. *J. Ecol.* **24,** 253.

CLUFF, N. A. (1952). Aerial camera shutters. *Manual of Photogrammetry,* 180, 186.

BIBLIOGRAPHY

COBLENTZ, C. S. (1939). Restitution of aerial enlargements for area determination. *Photogramm. Engng.* **5**, 138–46.

COBLENTZ, W. W. (1912). Diffuse reflection of various substances. *Bull. Bur. Stand. 9.*

COBLENTZ, W. W. & STAIR, R. (1929). Absorption for chlorophyll and xanthophyll in the infrared. *Phys. Rev.* **33**, 1092.

COLLIER, C. W. (1936). The use of aerial maps in soil conservation studies. *News. Notes. Am. Soc. Photogramm.* **II** (2).

COLNER, B. J. (1962). Vertical accuracy analysis. *Photogramm. Engng.* **28**, 415.

COLWELL, R. N. (1950). Determining relative illumination for overlapping photographs for maximum stereoscopic effect. *J. For.* **48**, 369.

COLWELL, R. N. (1952). Photographic interpretation for civil purposes. *Manual of Photogrammetry*, 536–45, 547, 553–5, 560–7.

COLWELL, R. N. (1954). A systematic analysis of some factors affecting photographic interpretation. *Photogramm. Engng.* **20** (3) 433–54.

COLWELL, R. N. (1956). Determining the prevalence of certain cereal crop diseases by means of aerial photography. *Hilgardia*, **26** (5), 223–86.

COLWELL, R. N. (1959). The future of photogrammetry and photo-interpretation. *Photogramm. Engng.* **25**, 713–36.

COLWELL, R. N. (1960a). Some uses and limitations of aerial colour photography in agriculture. *Photogramm. Engng.* **26**, 220–2.

COLWELL, R. N. (1960b). Some uses of infrared aerial photography in the management of wildland areas. *Photogramm. Engng.* **26**, 774–85.

COLWELL, R. N. (1961a). Some practical applications of multiband spectral reconnaissance. *Am. Scient.* **49**, 9–36.

COLWELL, R. N. (1961b). Aerial photographs show range conditions. *Calif. Agric.* **15** (12), 12–13.

COLWELL, R. N. (1963a). Basic matter and energy relationships involved in remote reconnaissance. Report of Sub-Committee I (several authors). *Photogramm. Engng.* **29**(5), 761–99.

COLWELL, R. N. (1963b). To measure is to know. *Photogramm. Engng.* **29** (1), 70–83.

COLWELL, R. N. (1964). Pictures don't lie. *Photogramm. Engng.* **30** (2), 266–72.

COLWELL, R. N. & OLSON, D. L. (1965). Thermal infrared imagery and its use in vegetation analysis by remote aerial reconnaissance. *Proc. 3rd Symposium on Remote Sensing*, 607–21, Report 4864-9-x. Univ. Michigan, Ann. Arbor.

CORTEN, F. L. (1963). Rational planning and execution of aerial photography. *Archives Int. de Photogramm.* **14**, 39–44. Delft.

COSMA, D. (1953). Preliminary forest area determination by use of narrow tonal scale aerial photographs (in Italian). *Monte e Boschi*, **3**, 124–6.

COTTAM, G. & CURTIS, J. T. (1948). The use of the punched card method in phytosociological research. *Ecology* **29**, 516–19.

COTTAM, G. & CURTIS, J. T. (1949). A method for making rapid surveys of woodlands by means of pairs of randomly selected trees. *Ecology*, **30**, 101–4.

COTTAM, G. & CURTIS, J. T. (1956). The use of distance measures in phytosociological sampling. *Ecology* **57**, 451–60.

CRAIG, R. D. (1920). An aerial survey of the forests in northern Ontario. *Canadian Forestry Magazine*, **16**, 516–18.

CRANDALL, C. J. (1963). Advanced radar map compilation. *Photogramm. Engng.* **29** (6), 947–55.

CRITTENDEN, E. C. (1923). The measurement of light. *J. Wash. Acad. Sci.* **113**, 69–90.

CROMER, D. A. N. (1960). Surveys applicable to extensive forest areas in the Asia–Pacific region. *5th World Forestry Conference*, Washington, GA/7/1 Australia (1p.).

BIBLIOGRAPHY

CROMER, D. A. N. & AITKEN, J. D. (1948). An aerial inventory of Norfolk Island pine. *Aust. For.* **12,** 82–7.

CROMER, D. A. N. & BROWN, A. G. (1954). Plantation inventories with aerial photographs and angle count sampling. *Bull. For. Timb. Bur. Aust.* **34,** 23.

CRONE, D. R. (1951). The Rand method of heighting from near vertical aerial photographs. *Emp. Surv. Rev.* **9** (82).

CROSBY, P. & KOERBER, B. W. (1963). Scattering of light in the lower atmosphere. *J. Opt. Soc. Am.* **53** (3), 358–61.

CROSSLEY, D. I. (1966). Application of scientific discoveries and modern techniques in silviculture. World Forestry Congress, Madrid.

CROW, A. B. (1945). A method of determining forest areas from aerial photo index sheets. *J. For.* **43,** Nov. 812–13.

CUMMINGS, W. H. (1941). A method for sampling the foliage of a silver maple tree. *J. For.* **39,** 382–4.

CUNNINGHAM, R. N. (1952). Land use and forest survey of the Ryukyus. *FAO Asia–Pacif. For. Comm.* No. 52, 15.

CURTIN, R. C. (1964). Stand density and the relationship of crown width to diameter and height in *Eucalyptus obliqua. Aust. For.* **28,** 91–105.

CURTIS, J. T. (1959). *The Vegetation of Wisconsin: An Ordination of Plant-communities.* Univ. Wisconsin Press, Madison, Wis., U.S.A.

CURTIS, L. F. (1966). Air photography and air photo interpretation in the United Kingdom. Report to Working Group V, Commission VII, I.S.P.

DAEHN, R. E. (1949). A standardized tone scale as an aid in photo-interpretation. *Photogramm. Engng.* **15,** 287.

DAGNELIE, P. (1960). Contribution à l'étude des communautés végétales par l'analyse factorielle. *Bull. Serv. Carte Phytogéogr. Series B,* **5,** 7–71, 93–195.

DAHL, B. (1954). The assessment of standing timber volumes from aerial photographs. *Aust. For.* **18** (1).

DANSEREAU, P. (1957). *Biogeography: An Ecological Perspective.* Ronald Press, New York.

DARGAVEL, J. (1965). A.P.M. Ltd, Melbourne, Victoria, personal communication.

DAVIES, N. (1954). The assessment of Murray River red gum forests by aerial survey methods. N.S.W. Forestry Commission.

DAWKINS, H. C. (1963). Crown diameters: their relation to bole diameter in tropical forest trees. *Commonw. For. Rev.,* **42,** 318–33.

DESJARDINS, L. (1943a). A rapid method of drafting an accurate map from vertical aerial photographs. *Photogramm. Engng.* **9,** 172–5.

DESJARDINS, L. (1943b). Contouring and elevation measurement on vertical aerial photographs. *Photogramm. Engng.* **9,** 214–24.

DEVAUX, H. (1939). Une différence physique marquée entre les portions extrêmes des mousses. *Compte Rendu,* 1260. Paris.

DEVILLE, E. (1895). *Photographic Surveying.* Govt. Printer, Ottawa.

DILL, H. W. (1959). Use of the comparison method in agricultural air photo-interpretation. *Photogramm. Engng.* **25,** 44–9.

DILL, H. W. (1963). Airphoto analysis in outdoor recreation: site inventory and planning. *Photogramm. Engng.* **29** (1), 67–70.

DILLEWIJN, F. J. VAN (1957). Sleutel voor de interpretatie van begroeingsvormen uit luchtfotos 1/40,000 van het Noordelijk deel van Suriname (Key for the interpretation of vegetative types on aerial photographs in North Surinam). *Landsbosbeheer.* Paramaribo, Surinam.

BIBLIOGRAPHY

DILWORTH, J. R. (1959). Aerial photo mensuration tables. *Res. Note Ore. Agric. Exp. Sta.* No. 46.

DINGER, J. E. (1941). The absorption of radiant energy in plants. *Iowa St. Coll. J. Sci.* **16**, 44–5.

DODGE, H. F. (1954). Mapping evergreen timber areas. *Photogramm. Engng.* **20**, 796–802.

DOMMERGUES, Y. (1952). La prospection des peuplement forestiers tropicaux par application des méthodes statistiques. *Bois Forêts Trop.* **23** (May–June).

DONALDSON, E. M. (1962). Fifty years of air power. *Daily Telegraph*, London, 9th May.

DOVERSPIKE, G. E. & HELLER, R. C. (1962). Identification of tree species on large scale panchromatic and colour photographs. *Proc. Intern. Symposium, Delft, Holland.*

DOWNES, R. G., GIBBONS, F., ROWAN, J. & SIBLEY, G. (1957). Principles and methods of ecological surveys for land use purposes. Soil Sci. Conference, C.S.I.R.O., Melbourne.

DOYLES, F. J. (1964). The historical development of analytical photogrammetry. *Photogramm. Engng.* **30** (2), 259–65 (includes extensive bibliography).

DRAKE, H. L. (1963). A spectral reflectance study using a wedge spectrograph. *Tech. paper, 29th meeting, Am. Soc. Photogramm.*

DUNTLEY, S. Q. (1948). Reduction of apparent contrast by the atmosphere. *J. Opt. Soc. Am.* **38** (2), 179–90.

DU RIETZ, G. E. (1954). Vegetation analysis in relation to homogeneity and size of sample. Typescript, University of Uppsala, 14 pp.

EDEN, J. A. (1957). Edge sharpness technique for assessing photographic quality. *Photogramm. Rec.* **2**, 279–88.

ELIEL, L. (1939). Analysis of error present in planimetric prints. *Photogramm. Engng.* **5**, 94–9.

EMMONS, W. H., THIEL, G. A., STAUFFER, C. R. & ALLISON, I. S. (1939). *Geology*, 24–30, 330, 316, 317. McGraw-Hill.

ENGELN, O. D. VON (1942). *Geomorphology.* Macmillan, New York.

ENGLISH, J. S. (1965). Vertical colour: air survey's new weapon. *Hunting Group Review*, **3**, 4–9.

ERIKSSON, H. (1960). Concerning accuracy in measuring tree stand heights. *Svensk Lantmät. Tidskr.* **3**.

EULE, H. W. (1959). Tree crown measurement and relationships between crown size, stem diameter and growth for beech in north-west German thinning trials. *Allg. Forst.-u. Jagdztg.* **30**, 185–201.

FAGERHOLM, P. O. (1959). The application of photogrammetry to land use planning. *Photogramm. Engng.* **25**, 523–33.

FAIRBAIRN, W. A. (1954). Difficulties in the measurement of light intensity. *Emp. For. Rev.* **33**, 262–9.

FARRER, R. P. (1953). Crown diameter in relation to girth and height for three tree species in Tanganyika. Typescript, Library, Comm. Fort. Inst., Oxford.

FAULDS, A. H. & BROCK, R. H. (1964). Atmospheric refraction and its distortion of photographs. *Photogramm. Engng.* **30**, (2), 296.

FAULDS, H. (1959). Some notes on the displacement of photographic images caused by tilt and relief. *Photogramm. Engng.* **25**, 110–15.

FEREE, M. J. (1953). Estimating timber volumes from aerial photographs. *Tech. Publ. N.Y. St. Coll. For.* **75**.

BIBLIOGRAPHY

FICHTER, H. J. (1953). Geometry of the imaginary stereoscopic model. *Photogrammetria*, **10** (4), 134–9.

FINNEY, D. J. (1948). Random and systematic sampling in timber surveys. *Forestry*, **22**, 64–99.

FISCHER, W. A. (1960). Reflectance measurements as a basis for film filter selection for photographic differentiation of rock types. *U.S.G.S. Professional Paper* 400B, 61 pp., B-136 to B-138.

FISCHER, W. A. (1962). Colour aerial photography in geologic investigations. *Photogramm. Engng.* **28** (1), 133–9.

FISCHER, W. A. & GRAY, R. G. (1962). Are aerial photographs obsolete? *Photogramm. Engng.* **28** (1), 94–6.

FISH, R. W. (1953). Lens and film resolution. *Photogramm. Rec.* **1** (2), 21–32.

FLEMING, E. A. (1961). Recognition of air survey lens types. *Can. Surveyor*, **15** (2), 107–12.

FLEMING, E. A. (1963). Aberrations of the aerial photographic process. *Can. Surveyor*, **17** (2), 68–79.

FLEMING, E. A. (1964). Solar nomograms. *Bull.* **64** (3). Mines and Techology, Ottawa.

FLEMING, G. (1960). Can tilted photographs be assumed to be vertical for the purposes of calculating point elevations? *Photogramm. Engng.* **26**, 50–4.

FLETCHER, R. J. (1964). The use of aerial phtographs for engineering soil reconnaissance in arctic Canada. *Photogramm. Engng.* **30** (2), 211–19.

FLORET, C. (1966). Cartographie phyto-écologique à petite échelle et photo-interprétation en Tunisie du nord. Symposium, Commission VII, I.S.P. Paris.

FORBES, A. (1938). Northernmost Labrador mapped from the air. *Bull. Am. Geogr. Soc.* **22**. N.Y.C.

FOSTER, F. W. (1951). Some aspects of the field use of aerial photographs by geographers. *Photogramm. Engng.* **17**, 771–6.

FOULDS, A. H. & BROCK, R. H. (1964). Atmospheric refraction and its distortion of aerial photographs. *Photogramm. Engng.* **30** (2), 211–19.

FOURCADE, H. G. (1935). The rectification of air photographs. *Emp. Surv. Rev.* **3** (19), 972.

FOURCADE, H. G. (1962). On a stereoscopic method of photographic surveying. *S. Afr. Photogr. Soc.*

FRANCIS, D. A. (1957a). Use of aerial survey methods in Scandinavian forestry. *Emp. For. Rev.* **38** (3), 266–76.

FRANCIS, D. A. (1957b). The use of aerial photographs in tropical forests. *Unasylva*, **11**, 103–9.

FRANCIS, D. A. (1959). Forest inventory in the Sudan. *Sudan Sylva*, 49–55.

FRANCIS, D. A. (1962). Aerial survey methods for forestry and forestry industry preinvestment surveys in developing countries. *Proc. Intern. Symposium, Delft, Holland*.

FRANCIS, E. C. (1955). Interpretation of mangrove species from aerial photographs in North Borneo. Typescript, Directorate of Overseas, Surveys, Tolworth, U.K.

FRANCIS, E. C. & WOOD, G. H. (1954). Classification of vegetation in North Borneo from aerial photographs. *Proc. 4th World For. Conf., Dehra Dun*, **3**, 623–9.

FRANCIS, E. C. & WOOD, G. H. (1955). Classification of vegetation of North Borneo from aerial photographs. *Malay For.* **18**, 38–44.

FREESE, F. (1961). Relation of plot size to variability. *J. For.* **59**, 679.

GATES, D. M. (1962). *Energy Exchange in the Biosphere*. Harper & Row, New York.

GATES, D. M. (1963). The energy environment in which we live. *Am. Scient.* **51**, (3), 327–47.

BIBLIOGRAPHY

GATES, D. M. (1965). Characteristics of soil and vegetational surfaces to reflected and emitted radiation. *Proc. 3rd Symposium of Remote Sensing*, Report 4864–9-x. Univ. Michigan, Ann Arbor.

GATES, D. M., KEEGAN, H. J., SCHELTER, J. C. & WEINER, V. R. (1965). Spectral properties of plants. *Appl. Optics*, **4**, 11–20.

GATES, D. M. & TANTRAPORN, W. (1952). The reflectivity of deciduous trees and herbaceous plants in the infra-red to 25 microns. *Sc.* **115**, 613–16.

GAY, S. P. (1957). Measurement of vertical heights from single oblique aerial photographs. *Photogramm. Engng.* **23**, 900–8.

GETCHELL, W. A. & YOUNG, H. E. (1953). Length of time necessary to obtain proficiency with height finders on air photos. *Note Univ. Maine. For. Dept.* **26**.

GIBBONS, F. & DOWNS, R. G. (1965). A study of the land in south-western Victoria. *Soil Conservation Authority, Victoria.*

GIER, J. T., DUNKLE, R. V. & BEVANS, J. (1954). Measurement of absolute spectral reflectivity from 1.0 to 15 microns. *J. Opt. Soc. Am.* **44**, 560.

GILL, C. B. (1950). A study of the relations between number of trees per acre and dispersion. *For. Chron.* **26**, 186–96.

GIMBAZEVSKY, P. (1964). Making full use of photography in the pulp and paper industry. *Pulp Paper Mag. Can.* May.

GIMBAZEVSKY, P. (1964). The significance of landforms in the evaluation of forest land. *Woodlands Review*, July.

GINGRICH, S. F. & MEYER, H. A. (1955). Construction of an aerial stand volume table for upland oak. *Forest Sci.* **1**, 140–7.

GLEASON, H. A. (1926). The individualistic concept of the plant association. *Bull. Torrey Bot. Club*, **53**, 7–27.

GOGUEY, R. (1963). Avion, stéréoscope ou projecteur. Archéologie Aérienne, Colloque International, Paris.

GOGUEY, R. (1966a). Recherches aériennes sur les structure archéologiques rurales et urbaines. Symposium, Commission VII, I.S.P. Paris.

GOGUEY, R. (1966b). Recherches sur l'influence des dates et des conditions techniques des prises dans la detection aérienne des indices d'origine archéologique. Intern. Symposium, Commission VII, Paris.

GOODALL, D. W. (1952a). Some considerations in the use of point quadrats for the analysis of vegetation. *Aust. J. Sci. Res.* **B.5**, 1–41.

GOODALL, D. W. (1952b). Quantitative aspects of plant distribution. *Biol. Rev. Cam. Phil. Soc.* **27**, (2), 194–243.

GOODALL, D. W. (1953a). Objective methods for the classification of vegetation. 1. The use of positive interspecific correlation. *Aust. J. Bot.* **1**, 39–63.

GOODALL, D. W. (1953b). Objective methods for the classification of vegetation. 2. Fidelity and indicator value. *Aust. J. Bot.* **1**, 434–56.

GOODALL, D. W. (1954). Vegetational classification and vegetational continua. *Agnew Pflanzsoz*, **1**, 168–82.

GOODALL, D. W. (1961). Objective methods for the classification of vegetation. *Aust. J. Bot.* **9**, 162–96.

GOODALL, D. W. (1962). Bibliography of statistical plant sociology. *Excerpta Bot. Sec. B*, **4**, 253–322.

GOODMAN, M. S. (1959). A technique for the identification of farm crops on aerial photographs. *Photogramm. Engng.* **25**, 131–7.

GOOSEN, D. (1961). A study of geomorphology and soils in the middle Magdalena Valley, Columbia. *I.T.C. Series B*, No. 9.

U

BIBLIOGRAPHY

GOUDALE, E. R. (1957). The measurement of elevation differences by photogrammetry where no elevation data exists. *Photogramm. Engng.* **23**, 774–8.

GRADMAN, R. (1909). Über Begriffsbildung in der Lehre von Pflanzen Formationen. *Bot. Jb. Beibl.* **99**, 91–103.

GRANT, K. (1964). Terrain classification. C.S.I.R.O. Mobile Field Conference, Australia.

GREEN, J. (1964). N.S.W.F.C., Sydney, personal communication.

GREENWAY, P. J. (1943). Second draft report on vegetation classification. Typescript, Pasture Research Conference, Amani, Tanzania.

GREGORY, J. M. (1946). The effect of the angle of incidence of the exposing light on the resolving power of photographic materials. *Proc. Phys. Soc.* **58**, 769.

GREIG-SMITH, P. (1964). *Quantitative Ecology*, 13, 36, 95, 105–69, 203. Methuen, London.

GREIG-SMITH, P., KERSHAW, K. & ANDERSON, D. (1963). The analysis of pattern in vegetation. *J. Ecol.* **51**, 223–9.

GROSENBAUGH, L. R. (1952). Plotless timber estimates. *J. For.* **50**, 32–7.

GROSENBAUGH, L. R. (1958). The elusive formula of best fit: a comprehensive new machine programme. *U.S. For. Serv. Sth. For. Exp. Sta. Occ. Pap.* No. 158, 9 pp.

GROSSENBAUGH, L. R. (1963). Optical dendrometers for out of reach diameters. *Forest Sci.* Mon. 4.

GRUNER, H. (1964). Super wide-angle projection mapping instrumentation. *Photogramm. Engng.* **30**, 745–9.

HAACK, P. M. (1962). Evaluating colour, infrared and panchromatic aerial photographs for the forest survey of interior Alaska. *Photogramm. Engng.* **28**, 592–8.

HAACK, P. M. (1963). Aerial photo volume tables for interior Alaska. *U.S. For. Serv. Res. Note Nth. For. Exp. Sta.* No. Nor-3.

HACKMAN, R. J. (1956). The stereo slope comparator—an instrument for measuring angles of slope in stereoscopic models. *Photogramm. Engng.* **22**, 893–8.

HACKMAN, R. J. (1957). The flying carpet—a stereoscopic grid used in photo-interpretation. *Photogramm. Engng.* **23**, 108–15.

HAGBERG, N. (1956). Skoggsvattning pa flybilder. *Norrlands SkogsvFörb. Tidskr.* **3**, 369–410.

HAGBERG, N. (1956). Aerial photographs for mapping purposes in forests in Sweden. Intern. Congress Photogrammetry, Stockholm.

HAGBERG, E. (1960). The new national forest survey. Typescript, Dept. of Forest Survey, Stockholm, 19 pp.

HAGGETT, P. (1964). Regional and local components in the distribution of forested areas in S.E. Brazil. *Geogr. Journ.* **130**, 365–80.

HAGSTROM, B. (1953). Three dimensional forest map. *Norrlands SkogsvFörb. Tidskr.* **2**, 257–84.

HAIG, I. T. (1958). Tropical silviculture. *Unasylva*, **12**, 158–63.

HALL, R. (1954). The effect of haze and high solar altitude on the density of air survey negatives. *Photogramm. Rec.* **1** (4), 20–37.

HALL, W. (1959). Application of photogrammetry to forest inventory in British Columbia. *Can. Surveyor*, **14** (7), 300–5.

HALLERT, B. (1966). Notes on the calibration of cameras and photographs in photogrammetry. Commission I, Symposium, I.S.P. London.

HANNIBAL, L. W. (1952). Aerial photo-interpretation in Indonesia. *2nd Asia–Pacific Forestry Commission of F.A.O.*, **75**, 1–23.

BIBLIOGRAPHY

HARE, F. K. (1957). *Photo Reconnaissance of Labrador–Ungava.* Queen's Printer, Ottawa.

HARRIS, D. E. & WOODBRIDGE, C. L. (1964). Terrain mapping by use of infrared radiation. *Photogramm. Engng.* **30** (1), 134–9.

HARRIS, R. W. (1951). The use of aerial photographs and sub-sampling in range inventory. *Journ. Range Management,* **4** (4), 270–8.

HARRISON, G. B. (1933). The infra-red content of daylight. *Photogr. J.* (72), Sci. Tech. Suppl. 1–3.

HARRISON, J. D. & SPURR, S. H. (1955). Planning a national forest inventory. F. A. O., Rome (Mimeo).

HARRY, H. (1951). De l'importance de la vue aérienne et de la mensuration aéro-photo-grammétrique. *J. For. Suisse,* Jan., 20–34.

HART, C. A. (1943). *Air Photography Applied to Surveying,* 27, 109, 136–42, 196, 251–6, 257. Longmans, London.

HARTMAN, F. J. (1947). A simplified method of localizing sample plots on aerial photographs. *U.S. For. Serv. Stat. Note Ntheast For. Exp. Sta.* **3.**

HEDDLE, D. W. (1962). Observations in the southern hemisphere of ultra-violet light from celestial objects. Letter to *Nature,* **4818,** 861.

HEINSDIJK, D. (1952). Forest interpretation in Surinam. *Photogramm. Engng.* **18,** 158–62.

HEINSDIJK, D. (1955). Forest type mapping with the help of aerial photographs in the tropics. *Trop. Woods,* **102,** 27–46.

HEINSDIJK, D. (1957). The upper storey of tropical forests. *Trop. Woods,* **107,** 66–84; **108,** 31–45.

HEINSDIJK, D. (1960). Surveys particularly applicable to extensive forest areas (South America). F.A.O. 5th World Forestry Conference, 1960.

HEINSDIJK, D (1961) Forest survey in the Amazon valley *Unasylva* **15,** 167–74.

HELLER, R. C., ALDRICH, R. C. & BAILEY, W. F. (1959a). Evaluation of several camera systems for sampling forest insect damage at low altitude. *Photogramm. Engng.* **25,** 137–44.

HELLER, R. C., ALDRICH, R. C. & BAILEY, W. F. (1959b). An evaluation of aerial photography for detecting southern pine beetle damage. *Photogramm. Engng.* **25,** 595–606.

HELLER, R. C., ALDRICH, R. C. & BAILEY, W. F. (1963). Identification of tree species on large scale panchromatic and colour photographs. *Tech. Paper, 29th meeting Am. Soc. Photogramm.*

HEMMINGS, W. D. (1949). The use of aerial photographs in forestry in Tasmania. *Aust. Timb. J.* **15** (4).

HEMPENIUS, S. A. (1964). Aspects of photographic systems in engineering. *Appl. Optics,* **3,** 45–53.

HENDLEY, C. D. & HECHT, S. (1949). The colours of natural objects and terrains and their relation to colour deficiency. *J. Opt. Soc. Am.* **39,** 870–3.

HIGGINS, G. C. & JAMES, T. H. (1960). *Fundamentals of Photographic Theory,* 266–306. Morgan & Morgan, New York.

HILDEBRANDT, G. (1960). Ein Zeitvergleich von Wald-taxationen mit und ohne Luftbildern. 9th Int. Congress Photogrammetry, London, 1960.

HILDEBRANDT, G. (1961). Zur Frage der radialen Punktversetzungen auf Luftbildern. *Forst u. Jagd.* **132** (6), 143–8.

HINDLEY, E. & SMITH, J. H. (1957). Spectrometric analysis of foliage of some British Columbia conifers. *Photogramm. Engng.* **23,** 894–5.

HOCKING, D. (1964). Department of National Development, Australia, personal communication.

BIBLIOGRAPHY

HOLLERWOGER, F. (1954). Is there a correlation in teak forests between crown diameter and the height of the trees with regard to the diameter at breast height? *J. Scient. Res. Indonesia*, **3**, 3–20.

HOLLOWAY, J. T. (1949). Ecological investigations in the *Nothofagus* forests in New Zealand. *N.Z. Jl. F.* **5**, 401–10.

HOLLOWAY, J. T. (1954). Forests and climates in South Island of New Zealand. *Trans. R. Soc. N.Z.* **82** (2), 329–40.

HOLMES, A. T. (1961). The geodimeter. *Can. Surveyor*, **15** (9), 445–54.

HOPKINS, B. (1955). The species-area relations of plant communities. *J. Ecol.* **43**, 409–26.

HORTON, R. E. (1945). Erosional development of streams and their drainage basins. *Bull. Geol. Soc. Am.* **56**, 275–370.

HOTHMER, J. (1958). Possibilities and limitations for elimination of distortion in aerial photographs. *Photogramm. Rec.* **2, 3** (1958–59).

HOTINE, M. (1929). Professional papers of the Air Survey Committee, No. 6, War Office, London. Extension of the Arundel method.

HOUGH, A. F. (1954). Control method of forest management in an age of aerial photography. *J. For.* Aug., **52**, 568–71.

HOWARD, J. A. (1956). Aerial photogrammetry in American forestry. Forestry School, University of Minnestoa.

HOWARD, J. A. (1957). An assessment of the value of woodland from aerial photographs. Typescript, Forest Department, Tanganyika.

HOWARD, J. A. (1959). The classification of woodland in western Tanganyika for type mapping from aerial photographs. *Emp. For. Rev.* **38**, 348–64.

HOWARD, J. A. (1965). Small scale photographs and land resources in Nyamweziland, East Africa. *Photogramm. Engng.* **32**, 287–93.

HOWARD, J. A. (1966a). Total interpretation in relation to evergreen broad-leaved forest. World Forestry Congress, Madrid.

HOWARD, J. A. (1966b). Spectral energy relations of isobilateral leaves. *Aust. J. Biol. Sci.* **19**, 757–66.

HOWARD, J. A. (1966c). Ecological analysis in photo-interpretation. Symposium, Commission VII, I.S.P., Paris.

HOWARD, J. A. & KOSMER, H. (1967). Monocular mapping by micro-film. *Photogramm. Engng.* November, 1299–1302.

HOWE, R. H. L. (1960). The application of aerial photographic interpretation to the investigation of hydrologic problems. *Photogramm. Engng.* **26**, 85–95.

HOWIE, E. L. (1956). Forest mapping for the forest industry in New Brunswick. *Can. Surveyor*, **13** (1), 23–5.

HOWROYD, C. S. (1954). An investigation into commercial volume estimation from air photographs. Typescript, Comm. For. Inst., Oxford.

HUGERSHOFF, R. (1939). Die Bildmessung und ihre förstlichen Andwendungen. *Der Deutsche Forstwirt*, **21**, 612–15.

HUGON, M. (1930). Variation de la brilliance des lointains avec la distance. *Sci. Inds. Photogr.* **1**, 161–8, 201–12.

HUGUET, L. (1961). Des Eaux et des Forêts, Bordeaux, personal communication.

HUGUET, L. *et al.* (1958) A forest inventory in Mexico. *Unasylva*, **12**, 55–62.

HULBURT, E. O. (1941). Optics of atmospheric haze. *J. Opt. Soc. Am.* **31**, (7), 467–74.

HURAULT, JEAN (1949). *Elements de Photogrammétrie.* L'Institut de Géographie National, Paris.

HUSCH, B. (1963). *Forest Mensuration and Statistics*, 125, 203–5, 350–70. Ronald Press, New York.

BIBLIOGRAPHY

ILVESSALO, Y. (1950). On the correlation between crown diameter and the stem of trees. *Comm. Inst. Forestalis Fenn.* **38** (2), 5–32.

IVES, R. L. (1939). Infra-red photography as an aid in ecological surveys. *Ecology*, **20**, 433–9.

JACKSON, K. B. (1960). Factors affecting the interpretability of air photos. 9th Congress, I.S.P.

JACOBS, M. R. (1932). Die Luftaufnahme im Dienste der Forsteinrichtung mit Vorschlägen zu ihrer Weiterentwicklung, insbesondere in unentwickelten Ländern (Ideas for the use of aerial photography in forestry, especially in underdeveloped countries.) Tharandt Library & Forestry and Timber Bureau, Canberra.

JENSEN, H. & COLWELL, R. (1949). Panchromatic versus infrared minus blue aerial photography for forestry purposes in California. *Photogramm. Engng.* **15**, 201–23.

JERIE, H. G. (1957). Errors in height and planimetry in the Stereotop due to its approximate solution of the double resection in space. *Photogrammetria* **13** (3), 117–27, (4), 137–42.

JOHNSON, E. W. (1954). Ground control for planimetric base maps. *J. For.* **52**, 89–95.

JOHNSON, E. W. (1957). The limit of parallax perception. *Photogramm. Engng.* **23**, 933–7.

JOHNSON, E. W. (1958a). A training programme for measuring tree heights with parallax instruments. *Photogramm. Engng.* **24**, 50–5.

JOHNSON, E. W. (1958b). The effect of photographic scale on precision of individual height measurement. *Photogramm. Engng.* **24**, 142–52.

JOHNSON, E. W. (1962a) The effect of tilt on the measurement of spot heights using parallax methods. *Photogramm. Engng.* **28** (3), 492–508.

JOHNSON, E. W. (1962b). Aerial photographic site evaluation for long-leaf pine. *Agricultural Experimentation Station*, Auburn University, Alabama.

JONES, E. W. (1945). The structure and reproduction of the virgin forests of the North Temperate Zone. *New Phytol.* **44**, 130–48.

JONES, L. A. & NELSON, C. N. (1943). Control of photographic printing by measured characteristics of the negative. *J. Opt. Soc. Am.* **32**, 558–619.

JONSSON, A. (1960). On the accuracy of stereo-plotting of convergent aerial photographs. *Photogrammetria* **17** (3), 84–98.

JULESZ, B. (1965). Texture and visual perception. *Scient. Am.* Feb., 38–48.

KEITZ, H. A. (1955). *Light calculations and measurements. Philips Tech. Library*, 23–36, 137–77.

KELEZ, G. (1947). Measurement of salmon spawning by means of aerial photography. *Pacif. Fisherm.* **45**, 49–51.

KELLER, M. (1963). Tidal current surveys by photogrammetric methods. *Photogramm. Engng.* **29** (5), 827.

KELSH, H. T. (1940). The slotted templet method for controlling maps made from aerial photographs. *Misc. Publ. U.S. Dept. Agric.* **404**, 29 pp.

KELSH, H. T. (1952). Radial triangulation. *Manual of Photogrammetry*, 409.

KEMP, R. C., LEWIS, C. G., SCOTT, C. W. & ROBINSON, C. R. (1925). Aero photo survey and mapping of the forests of the Irrawaddy delta. *Burma Forest Bulletin*, No. 11, 42 pp.

KENDALL, R. H. & SAYN-WITTGENSTEIN, L. (1961). A test of effectiveness of air photo stratification. *For. Chron.* **37**, 350–5.

KING, C. W. B. (1959). Computations of ground control for air surveys at large scale. *Emp. Surv. Rev.* **15**, 128–38.

BIBLIOGRAPHY

KIPPEN, F. W. & SAYN-WITTGENSTEIN, L. (1964). Tree measurements on large-scale, vertical 70 mm air photographs. *Forest Research Branch, Canada, Tech. Note* 1053.

KITTREDGE, J. (1938). The interrelations of habitat, growth rate and associated vegetation in the aspen community of Minnesota and Wisconsin. *Ecol. Monogr.* **8**, 151–144.

KITTREDGE, J. (1944). Estimation of the amount of foliage of trees and stands. *J. For.* **42**, 393–404.

KLESHNIN, A. F., SHULGIN, I. A. & VERBDOVA, M. I. (1959). The optical properties of plant leaves. *Bot. Sci. Sect. Transl.* **125**, 108–10. (Translation from *Doklady Akad. Nauk. S.S.R.* **125**, 1158–60.)

KNUDSON, G. J. (1951). Wisconsin fur research project 15-R. *Wis. Wild.* **10** (1), 168–84.

KOETCHLEY, C. W. (1960), Simple applications of photographs in geographic analysis of rural settlements. *Photogramm. Engng.* **17**, 759–71.

KRAMER, P. J. & KOZLOWSKI, F. T. (1960). *Physiology of Trees*, 76–80. McGraw Hill, New York.

KRAMER, P. & STURGEON, E. (1942). Transect method of estimating forest area from aerial photograph index sheets. *J. For.* **40**, 693–6.

KRINOV, E. L. (1947). *Spectral Reflectance Properties of Natural Formations*. Academy of Science, U.S.S.R. Translated by G. Belkov (1953), Natural Resources Council of Canada, T-439, 267 pp.

KRUTZSCH, H. (1925). Das Luftbild im Dienste der Forsteinrichtung (The aerial photograph in the service of forest management). *Tharandt Forstl. Hb.* **76**, 97–150.

KÜCHLER, A. W. (1960). Mapping tropical forest vegetation. Symposium, Helsinki, *Silva Fenn.* 60–64.

KUMMER, R. H. (1964). A use for small scale photography in forest management. *Photogramm. Engng.* **30** (5), 727–32.

KUUSELA, K. (1957). Outlines of cartographical and timber surveying units. *Silva Fenn.* **97**, 11 pp.

LACKNER, H. (1964). Untersuchungen von 9 Filmmaterial-Masstab-Kombinationen für Zwecke der Holzarteninterpretation. *9th Congress, I.S.P. Lisbon.*

LACMANN, O. (1950). *Die Photogrammetrie im ihrer Andwendung auf nichttopographischen Gebieten.* Herzel-Verlag, Leipzig.

LAER, W. VON (1960). Das Luftbild im Forstwesen. *9th Intern. Congress Photogramm.,* London.

LAMBERT, J. H. (1759). *Freie Perspective.* Zürich.

LAMBERT, J. M. & DALE, M. B. (1964). The use of statistics in phytosociology. *Advances in Ecological Research,* **3**, 59–28.

LAMPRECHT, H. (1954). Über Strukturuntersuchungen im Tropenwald. *Weltforstwirtsch.,* **17** (5), 161–8.

LANCASTER, C. W. (1965). The multisensor mission. *Photogramm. Engng.* **31** (3), 504.

LANCIE, R. DE et al. (1957). Quantitative evaluation of photo interpretation keys. *Photogramm. Engng.* **23**, 599–600.

LANDEN, D. (1962). Photo-interpretation of ice and snow features in the Antarctic mapping programme. *Proc. Intern. Symposium, Delft, Holland.*

LANE-POOLE, C. E. (1936). Crown ratio. *Aust. For.* **1**, 5–11.

LAUSSEDAT, A. (1898). *Recherches sur les instruments, les méthodes et le dessin topographique,* **1**, **2**. Gauthier-Villars, Paris.

LAWRENCE, A. O. (1939). Notes on *Eucalyptus obliqua* regeneration in central Victoria. *Aust. For.* **4**, 4–10.

BIBLIOGRAPHY

LAWRENCE, J. C. (1960). A pilot scheme for a grant of land titles in Uganda. *J. African Admin.* **12** (3).

LAWRENCE, P. R. (1957). Testing the efficiency of photo-interpretation as an aid to forest inventory. *7th British Commonwealth Forestry Conference,* 22.

LAWRENCE, P. R. & WALKER, B. B. (1954). Methods and results of forest assessment using random sampling units in photo-interpretated strata. *Aust. For.* **18**, 107–27.

LEE, H. C. (1941). Aerial photography, a method for final type mapping. *J. For.* **39**, 531–3.

LEE, R. (1963). Evaluation of solar beam irradiation. *Hydrology Papers,* Colorado State University.

LEE YAM (1959). A comparison of some 12 inch and 6 inch focal length photographs for photo mensuration and forest typing. M. F. thesis, Univ. Br. Columbia, 124 pp.

LEEDY, D. L. (1948). Aerial photographs, their interpretation and suggested uses in wildlife management. *Journal of Wildlife Management,* **12**, 191–210.

LEEDY, D. L. (1953). Aerial photo use in the fields of wild life and recreation. *Photogramm. Engng.* **19**, 127–37.

LEIGHTON, E. (1941). How many tones? *Photo. Technique,* Dec., 37–9.

LEYDOLPH, W. K. (1954). Stereophotogrammetry in animal husbandry. *Photogramm. Engng.* **20**, 804–8.

LIANG, T. *et al.* (1951). A photo analysis key for the determination of ground conditions. U.S.N., Office of Naval Research, N.R. 25700.

LINSLEY, R. K. *et al.* (1958). *Hydrology for Engineers,* McGraw Hill. 1–3.

LITTLETON, J. W. (1944). Detail mapping by the use of enlarged aerial photographs. *Photogramm. Engng.* **10**, 214–16.

LOETSCH, F. G. (1957a). A forest inventory in Thailand. *Unasylva,* **11**, (4), 174–80.

LOETSCH, F. G. (1957b). Report to the government of Thailand on inventory methods for tropical forests. Pt. *E.T.A.P. Report* 545. F. A. O., Rome.

LOETSCH, F. G. (1960). Aspects of forestry in relation to the rayon plant project in south Sumatra. *F.A.O. Rep.* No. 1278.

LOETSCH, F. & HALLER, K. (1962). The adjustment of area computation from sampling devices on aerial photographs. *Photogramm. Engng.* **28**, 789–810.

LOETSCH, F. & HALLER, K. (1964). *Forest Inventory. Vol. I. Statistics of Forest Inventory and Information from Aerial Photographs.* Verlagsgesellschaft, Munich, 436 pp.

LOPIK, J. R. VAN & KOLB, C. R. (1959). A technique for preparing desert terrain analogs. *Technical Report* 3-506, U.S. Army.

LOSEE, S. T. (1942). Air photographs and forest sites. *For. Chron.* **18**, 129–44, 169–81.

LOSEE, S. T. (1951). Photographic tone in forest interpretation. *Photogramm. Engng.* **17**, 785–99.

LOSEE, S. T. (1953). Timber estimates from large scale photographs. *Photogramm. Engng.* **19**, 752–62.

LOSEE, S. T. (1955). *Forest Inventory in Eastern Canada.* K. B. Wood & Associates Inc., Portland, Oregon, U.S.A.

LOSEE, S. T. (1956). The measurement of stand density in forest photogrammetry. *Can. Surveyor,* **13** (2), 81–90.

LOUCKS, O. L. (1962). Ordinating forest communities by means of environmental scalars and phytosociological indices. *Ecol. Mongr.* **32** (2), 137–65.

LOWDERMILK, W. C. (1938). Use of aerial mapping in soil conservation studies. *C.E. (Aust).* **8** (9), 605.

LOWMAN, P. D. (1965). Photography of the earth from sounding rockets and satellites. *Photogramm. Engng.* **31** (1), 76.

BIBLIOGRAPHY

LUEDER, D. R. (1950). A system for designating map-units on engineering soil maps. *Highway Research Board, Bull.* **28**. U.S.A.

LUEDER, D. R. (1960). *Aerial Photographic Interpretation*, 390–400. McGraw Hill, N.Y.

LUTZ, H. & CAPORASO, A. (1958). Indicators of forest land classes in air-photo interpretation of the Alaska interior. *U.S. For. Serv. Stat. Pap. Alaska For. Res. Cen.* **10**, 31 pp.

LYON, D. (1964). Let's optimize stereo-plotting. *Photogramm. Engng.* **30**, 897–911.

LYONS, E. H. (1960). Preliminary studies of two camera, low elevation stereo-photography from helicopters. *Photogramm. Engng.* **27** (1).

LYONS, E. H. (1964). Recent developments in 70 mm stereo-photography from helicopters. *Photogramm. Engng.* **30** (5), 750–6.

LYONS, E. H. (1966). Fixed air base 70 mm photography, a new tool for forest sampling. *For. Chron.* **42** (4), 420–37.

MACANDREWS, F. D. (1955). Average height weighted by volume in airphoto interpretation. *Depth. of North. Affairs, Canada, For. Res. Div. Tech. Note* **17**.

McBETH, F. H. (1961). Aerial photographic investigation of leaching and sapping as an erosion process. *Photogramm. Engng.* **27** (1), 154–5.

McCURDY, P. (1950). Coastal delineation from aerial photographs. *Photogramm. Engng.* **16**, 550–5.

McDANIEL, J. F. & ARNTZ, J. F. (1959). Aerial colour film in military photo-interpretation. *Photogramm. Engng.* **25**, 529–33.

MACDONALD, D. E. (1953). Interpretability. *Photogramm. Engng.* **19**, 102–7.

MACDONALD, D. E. (1958). Resolution as a measure of interpretability. *Photogramm. Engng.* **24**, 58–62.

MACDONALD, D. E. *et al.* (1956). Detection and recognition of photographic detail. *J. Opt. Soc. Am.* **46**, 715–20.

MACDONALD, H. (1963). Forestry Commission, New South Wales, Australia, personal communication.

McINTOSH, R. P. (1962). Pattern in a forest community. *Ecology* **43**, 25–33.

McINTOSH, R. P. (1963). Ecosystems, evolution and relational patterns of living organisms. *Am. Scient.* June, 246–67.

McNAIR, A. J. (1957). General review of aero-triangulation. *Photogramm. Engng.* **23**, 573–8.

McNAMARA, P. J. (1959). Air photo mapping of the western Australian eucalypt forests. A.N.Z.A.A.S. meeting, 1959.

McNEIL, G. T. (1954). *Photographic Measurements. Problems and Solutions.* Pitman.

MACPHERSON, A. F. (1962). Testing the efficiency of air photo stratification. *For. Chron.* **38**, 450–8.

McVAY, D. (1963). The use by the forest service of photogrammetry in cadastral surveys. *Photogramm. Engng.* **29** (5). 867–9.

McVEAN, D. N. & RATCLIFFE, D. A. (1962). Plant communities of the Scottish Highlands. Monograph, Nature Conservancy, No. 1, 445 pp., London.

MADOW, W. G. (1949) The theory of systematic sampling. *Ann. Math. Statist.* **20**, 333–45.

MAHAN, R. O. (1958). The photo contour map. *Photogramm. Engng.* **24**, 451–7.

MAKAREYSKII, N. I. (1938). Reflectance, transmittance and absorption of solar radiation by plant leaves. *Transl. Transactions of the Phytophysiological Laboratory of the Physico-Agronom. Inst.* **1**, 111–19.

MALING, D. H. (1960). A review of some Russian map projections. *Emp. Sur. Rev.* **15**, 203–15, 255–66.

BIBLIOGRAPHY

ANON. *Manual of Photogrammetry* (1952). American Society of Photogrammetry, 875 pp. Reference 56, 464–99, 506–8, 628, 839 (stereocomparagraph), 623-62 (photogrammetric plotting instrument).

MARTIN, D. B. (1966). A new scanning stereoscope with optical means of measuring parallax. *Intern. Symposium*, Commission VII, I.S.P. Paris.

MARTIN, R. (1948). *Cours de Photo-Topographie*, 2. Paris.

MARUYASU, T. & NISHIO, M. (1960). On the study and application of infrared aerial photography. *Report, Institute of Industrial Science*, **10** (1), 1–16.

MARUYASU, T. & NISHIO, M. (1962). Experimental studies on colour aerial photographs in Japan. *Photogrammetria*, **18** (3), 87–106.

MASON, B. (1953). The hotspot in wide angle photographs. *Photogramm. Engng.* **19**, 619–25.

MASTERS, S. E., HOLLOWAY, J. T. & MCKELVEY, P. J. (1957). *The National Forest Survey of New Zealand. Vol. I. The Indigenous Forest Resources.* Government Printer, Wellington.

MATERN, B. (1960). Forest surveys and the statistical theory of sampling. Some recent developments. *5th World Forestry Conference*, GA/75/1/B/Sweden, 12 pp.

MAY, J. R. (1960). A stand aerial volume table for *E. obliqua*. Typescript, School of Forestry, Melbourne Univ.

MAYCOCK, P. F. & CURTIS, J. T. (1960). The phytosociology of boreal conifer–hardwood forests of the Great Lakes region. *Ecol. Monogr.* **30**, 1–35.

MAYER, I. (1959). The aerial survey camera as an entity. *Cartography*, **3** (3), 119–22.

MEES, C. (1953). *The Theory of the Photographic Process.* Macmillan, New York.

MEESTER, T. DE (1961). The application of airphoto analysis in a detailed soil survey in Tanganyika. *I.T.C.* Series B, No. 10.

MEJORADA, N. S., ABRAHAM, E. H. & HUGUET, L. (1958). A forest inventory in Mexico. *Unasylva*, **12**, 55–62.

MEKEL, J. E., VAVAGE, J. F. & ZORN, H. C. (1964). Slope measurements and estimates from aerial photographs. *I.T.C.* Series B, No. 6.

MERRILL, D. F. (1956). The economics of forest aerial photography. *Can. Surveyor*, **13**, 149–55.

MERRIT, E. L. (1963). Image aberration. *Photogramm. Engng.* **29** (1), 119–26.

MERRITT, V. G. & RANATUNGA, M. S. (1958). Report on aerial photographic survey of Sinharaja forest (Ceylon). *Ceylon Forester*, **4**, 103–56.

MEYER, D. (1961). A reflecting projector you can build. *Photogramm. Engng.* **27**, 76–8.

MEYER, H. A. & WORLEY, D. P. (1957). Volume determination from aerial stand volume tables and their accuracy. *Journ. Forestry*, **55**, 368–72.

MEYER, M. P. (1955). Photogrammetric training for the technical forester. *Photogramm. Engng.* **21**, 741–6.

MEYER, M. P. (1957). A preliminary study of the influence of photopaper characteristics on a stereo image perception. *Photogramm. Engng.* **23**, 149–55.

MEYER, M. P. (1963). Quantitative methods in forest aerial photo-interpretation research. *Tech. paper, 29th meeting Am. Soc. Photogramm.*

MEYER, M. P. (1964). Relationship of aerial photo measurements to stand diameter classes of a Minnesota hardwood forest. *Photogramm. Engng.* **30** (1), 142–4.

MEYER, M. P. & HUGO, H. J. (1961). Comparative forest aerial photo-interpretation results from variable-contrast and single-contrast paper prints. *Photogramm. Engng.* **27**, 697–703.

MEYER, M. P. & MYHRE, D. W. (1961). Variations in aerial photo image recovery, resulting from differences in film and printing technique. *Photogramm. Engng.* **27**, 595–9.

BIBLIOGRAPHY

MEYER, M. P. & TRANLOW, L. H. (1957). Some observed effects of variations in photo-paper emulsion and tone upon stereo-perception of tree crowns. *Photogramm. Engng.* **23**, 896–9.

MEYER, M. P. & TRANLOW, L. H. (1961). A test of polaroid variable-colour filters for forest aerial photography. *Photogramm. Engng.* **27**, 703–5.

MIDDLETON, C. E. (1955). Aspects and trends of photogrammetry in Australia. *Cartography*, **1**, 56–65.

MIDDLETON, W. E. K. (1935). How far can I see? *Scient. Mon.* **41**, 343–6.

MIDDLETON, W. E. K. (1950). The attenuation of light by the atmosphere. *Photogramm. Engng.* **16**, 663–72.

MIDDLETON, W. E. K. (1952). *Vision through the Atmosphere*, 7, 9–12, 18–59, 60–8, 71–3, 86, 104–19, 122–8, 132–6, 145–72, 200–2. Univ. Toronto Press.

MIGNERY, A. L. (1957). Use of low altitude continuous strip aerial photography in forestry. *U.S. For. Serv. Sth. For. Exp. Sta. Paper* **178**.

MIKHAILOV, V. Y. (1961). The use of colour sensitive films in aerial photography in the U.S.S.R. *Photogrammetria* **17** (3).

MIKODA, P. M. (1960). A brief history of film development by Ansco. *Photogramm. Engng.* **26**, 116–19.

MILLER, C. I. (1958). The stereoscopic space image. *Photogramm. Engng.* **24**, 810–15.

MILLER, C. I. (1960). Vertical exaggeration in the stereo space-image and its use. *Photogramm. Engng.* **26**, 815–18.

MILLER, O. M. & SUMMEROON, C. H. (1960). Slope zone maps. *Geogrl. Rev.* **1**, 194–202.

MILLER, R. G. (1952). Progress in modern map making. *Emp. For. Rev.* **31**.

MILLER, R. G. (1957). The use of aerial photographs in forestry in British Colonies. *7th British Commonwealth Forestry Conference*, 1–7. Colonial Office, London.

MILLER, R. G. (1960). The interpretation of tropical vegetation and crops on aerial photographs. *Photogrammetria*, **16** (3) 230–40.

MILLER, V. C. (1953). Some factors causing vertical exaggeration and slope distortions. *Photogramm. Engng.* **19**, 592–607.

MILLER, V. C. (1961). *Photo-geology*, 35–55. McGraw-Hill, N.Y.C.

MILLER, W. L. (1946). Aerial survey of upland game. *N. Dak. Outdoors*, **8**, 11.

MILNE, G. (1936). A provisional soil map of East Africa (Kenya, Uganda, Tanganyika) with explanatory memoir. *Amani Memoirs*, Amani.

MILLS, H. L. *et al.* (1963). Quantitative physiognomic analysis of the vegetation of the Florida everglades. *Rep. Waterways Exp. Sta. Vicksburg.*

MINOR, C. O. (1951). Stem crown diameter relations in southern pine. *J. For.* **49**, 490–3.

MINOR, C. O. (1960). Estimating tree diameter of Arizona Ponderosa pine from aerial photographs. *U.S. For. Serv. Res. Note Rocky Mt. For. Range Exp. Sta.* **46**, 2 pp.

MOESSNER, K. E. (1948). Photo classification of forest sites. *Soc. Amer. Foresters Proc.* 278–91.

MOESSNER, K. E. (1954). A simple test for stereoscopic perception. *U.S. For Serv. Pap. Cent. St. For. Exp. Sta.* **144**.

MOESSNER, K. E. (1957a). Preliminary aerial volume tables for conifer stands in the Rocky Mountains. *U.S. For. Serv. Res. Pap. Intermt. For. Range Exp. Sta.* **41**, 17 pp.

MOESSNER, K. E. (1957b). How important is relief in area estimates from dot sampling in aerial photographs? *U.S. For. Serv. Pap. Intermt. For. Range Exp. Sta.* **42**.

MOESSNER, K. E. (1957c). Relation of minimum area standards to proportions obtained by dot samples of aerial photographs. *U.S. For. Serv. Res. Pap. Intermt. For. Range Exp. Sta.* **44**.

BIBLIOGRAPHY

MOESSNER, K. E. (1960a). Estimating timber volume by direct photogrammetric methods. *Proc. Soc. Am. For.* 1959, 148–51.

MOESSNER, K. E. (1960b). Estimating the area in logging roads by dot sampling on aerial photographs. *U.S. For Serv. Res. Note Intermt. For. Range Exp. Sta.* **77**.

MOESSNER, K. E. (1960c). *Training Handbook*. Basic techniques in forest photo-interpretation. *Intermt. For. Range Exp. Sta.*, Ogden, Utah, 73 pp.

MOESSNER, K. E. (1963a). Estimating depths of small mountain lakes by photo measurement techniques. *Photogramm. Engng.* **29** (4), 580–8.

MOESSNER, K. E. (1963b). A test of aerial photo classifications in forest management—volume inventories. *U.S. For. Serv. Res. Pap. Intermt. For. Range Exp. Sta.* **3**.

MOESSNER, K. E., BRUNSEN, J. & JENSEN, C. (1951). The accuracy of stand height measurements in air-photos. *Stn. Notes Ohio. Exp. Stn.*

MOESSNER, K. E. & ROGERS, E. J. (1957). Parallax wedge procedures in forest surveys. *Intermt. For. Range Exp. Sta.*, *Misc. Pub.* **15**.

MOFFAT, R. C. (1964). Recent developments with surveying aids. *APPITA*, **8** (3), XVII–XII.

MOFFIT, F. H. (1961). *Photogrammetry*. International Textbook Company.

MOLLINEUX, C. E. (1965). Aerial reconnaissance of surface features with the multiband spectral system. *Proc. 3rd Symposium on Remote Sensing*, Report 4864–9–x. Univ. Michigan, Ann Arbor.

MONTEITH, J. L. (1959). The reflectance of short wave radiation by vegetation. *Quart. Jl. R. Met. Soc.* **85**, 386–92.

MORGAN, J. O. & PRENTICE, V. L. (1966). Third symposium on remote sensing. *Photogramm. Engng.* **32**, 98–107.

MORISAWA, M. & TAIRA, K. (1957). The amount of foliage of beech of Gunwa district. *Jap. Min. Agric. For. Sta. Bull.* **95**, 121–8.

MOTT, P. G. (1956). Contouring of a tropical forest area in Ceylon *Emp. For. Rev.* **35** (1), 36–41.

MUELLER, W. A. (1959). A floating mark scans the Australian Alps. *Aust. Surv.* **17**, 405–16.

MUIR, A. (1955). The use of air photographs in soil survey. *Photogramm. Rec.* **1** (6), 50–7.

MUMBOWER, L. E. & RICHARDS, T. W. (1962). Image information processing for photo-interpretation operations. *Photogramm. Engng.* **28**, 569–78.

NAKAJIMA, I. (1962). Forest type mapping in combination with a forest sampling project. *Bull. Forest Exp. Stn. Meguro, Tokyo*, No. 129 (61)—English edition.

NAKAJIMA, I., HIWATASKI, Y. & HASEGAWA, K. (1962). Forest type mapping and volume estimation on a natural forest in Japan. *Bull. Forest Exp. Stn. Meguro, Tokyo*, No. 146.

NAKANO, T. M. & KANUKUBO, T. (1962). Land form classification in aid of flood prevention using aerial photographs. *Proc. Intern. Symposium. Delft, Holland.*

NAKAYAMA, H. (1958). A study of forest survey from aerial photographs, Nagoya University. *Bull. Forests*, **1**.

NASH, A. J. (1948). Some volume tables for use in aerial survey. *For. Chron.* **24**, 4–14.

NASH, A. J. (1949). Some tests of the determination of tree heights from aerial photographs. *For. Chron.* **25**, 243–9.

NAYLOR, R. (1956). The determination of area by weight. *N.Z. Jl. F.* **7**, 109–11.

NEBLETTE, C. B. (1927). Aerial photography for the study of plant diseases. *Photo. Era Mag.* **58**, 346.

NIELSEN, J. N. & GOODWIN, F. K. (1961). Environmental effects of supersonic and hypersonic speeds on aerial photography. *Photogramm. Engng.* **27**, 427–35.

NOWICKI, A. L. (1952). Elements of stereoscopy. *Manual of Photogrammetry*, 521–7.

BIBLIOGRAPHY

NUMATA, M. (1949). The basis of sampling in the statistics of plant communities. Studies on the structure of plant communities. III. *Bot. Mag. Tokyo*, **62**, 35–8.

NYYSSONEN, A. (1955). On the estimate of growing stock from aerial photographs. *Comm. Inst. Forestalis Fenn.* **46** (1), 57 pp.

NYYSSONEN, A. (1961). Survey methods of tropical forests. F. A. O., Rome.

NYYSSONEN, A. & POSO, S. (1962a). Aerial photographs of tropical forests. *Unasylva*, **16**, 3–12.

NYYSSONEN, A. & POSO, S. (1962b). Tree stand classification from aerial photographs—an experiment. *Silva Fenn.* **112**.

NYYSSONEN, A., POSO, S. & KEIL, E. M. (1966). The use of aerial photographs in the estimation of some forest characteristics. *Paper No. 3, Institute of Forest Mensuration and Management*, University of Helsinki.

OBATON, F. (1941). Sur la réflexion du proche infrarouge par les surfaces vegetales. *Compte Rendu*, **212**, 621–3.

OBATON, F. (1944). La réflexion des radiations de grande longeur d'onde par les plantes haute montagne. *Compte Rendu*, 721–3.

OGLE, K. N. (1952). Limits of stereoscopic vision. *J. Exper. Psych.* **44**, 253–9.

OLSON, C. E. (1960). Elements of photographic interpretation common to several sensors. *Photogramm. Engng.* **26**, 651–6.

OLSON, C. E. (1963). Photographic interpretation in the earth sciences. *Photogramm. Engng.* **29** (6), 968–78.

OLSON, C. E. (1963). Seasonal trends in light reflectance from tree foliage. *Archives Int. Photogramm.* **14**, 226–32.

OLSON, C. E. (1964). *Spectral Reflectance Measurements Compared with Panchromatic and Infrared Aerial Photographs*. Inst. of Science and Technology, Univ. Michigan, U.S.A., U.S.A., 20 pp.

OLSON, C. E. & GOOD, R. E. (1962). Seasonal changes in light reflectance from forest vegetation. *Photogramm. Engng.* **28** (1), 107–14.

OLSON, D. L. & CANTRELL J. L. (1965). Comparison of airborne conventional photography and scanned ultra-violet imagery. *Photogramm. Engng.* **31** (3), 506.

OLSON, D. P. (1964). The use of aerial photographs in studies of marsh vegetation. *Bull.* **13**, *Maine Agricultural Exp. Sta.*

O'NEIL, H. T. & NAGEL, W. J. (1957). The Diachromoscope; an instrument for increasing contrast between a coloured object and a different coloured background on colour photographs. *Photogramm. Engng.* **23**, 180–5.

OTTOSON, L. (1956). Comparative tests of the accuracy of panchromatic, infrared and colour pictures from the air for photogrammetry purposes. International Congress for Photogrammetry, Stockholm.

OVINGTON, J. D. (1957). Dry matter production by *P. sylvestris. Ann. Bot.* **21**, 287–314.

OVINGTON, J. J. (1957). Stereo-templates and stereo-plotting. *Cartography*, **2**, 21–2.

PAELINCK, P. (1958). Note sur l'estimation du volume des peuplements à limba à l'aide des photos aériennes. *Bull. agric. Congo belge*, **49**, 1045–54.

PAIJMANS, K. (1951). Interpretation of aerial photographs in a virgin forest complex, Celebes, Indonesia. *Tectona*, **41**, 111–35 (English summary).

PAIJMANS, K. (1966). Typing of tropical vegetation by aerial photographs and field sampling in northern Papua. *Photogrammetria*, **21** (1), 1–25.

PALLEY, M. N. & LEAH, G. H. (1961). Properties of some random and systematic point sampling estimators. *For. Sci.* **7**, 52–63.

BIBLIOGRAPHY

PARKER, D. C. & WOLFF, M. F. (1965). Remote sensing. *Sci. Technol.* July. New York.

PARSONS, R. F. (1965). School of Botany, Melbourne University, personal communication.

PASTORELLI, A. (1956). Large scale photogrammetry and economy of precision photogrammetry in Switzerland (in German). *Photogrammetria,* **12** (1), 36–8.

PATTON, R. T. (1955)—*see* ANON. (1955).

PECHANEC, J. F. & STEWART, G. (1945). Sagebrush-grass range sampling studies: size and structure of sampling unit. *J. Am. Soc. Agron.* **32,** 669–82.

PERVIS, M. (1950). Drainage pattern significance in air photo identification of soils and bedrocks. *Photogramm. Engng.* **16** (3), 380–5.

PHILLIPS, J. (1934). Succession, developments, the climax and the complex organism: an analysis of concepts. *J. Ecol.* **22,** 554–71.

PHILLIPS, J. (1935). Ibid. **23,** 218–46, 488–508.

PIDGEON, I. M. (1942). Ecological studies in New South Wales. Types of primary succession and forest ecology in the central coastlands, with a classification of Eucalyptus formations in N.S.W. D.Sc. thesis, Botany Dept., Univ. of Sydney.

PLATONENKO, M. A. (1963). Quantitative description of airphoto interpretation indices for soils using correlation co-efficients. Kirov Agricultural Institute, Omsk (English translation *Soviet Soil Science,* Univ. of Nebraska, No. 1, 1964), 54–60.

PLOCHMANN, R. (1956). Bestockungs Aufbau und Baumartenwandel der nordischer Urwälder. *Beih. Forstw. Cent. Forstw. Forsch.* **6,** 1–96.

PLUMMER, B. F. (1959). An evaluation of the S.O.M. Stereoflex as a third order plotting instrument. *Cartography,* **3,** 82–7.

POISSONET, P. (1966). Place de photo-interprétation dans un programme d'étude détaillé de la flore, de la végétation et du milieu. Symposium Commission VII, I.S.P. Paris.

POKROWSKI, G. I. (1925). Über die Lichtabsorption von Blättern einiger Bäume. *Biochem. Z.* **165,** 420–6.

POLYCARPOU, A. (1957). The scope of aerial photography in stratification for sampling purposes in *Pinus brutia* forest of Cyprus. *7th British Commonwealth Forestry Conference,* 1957 (*Cyprus*), 1–20.

POORE, M. E. D. (1955). The use of phytosociological methods in ecological investigations. I. The Braun Blanquet system. *Ecology,* **43,** 14, 3 refs.

POORE, M. E. D. (1956). The use of phytosociological methods in ecological investigations. IV. General discussion of phytosociological problems. *J. Ecol.* **44,** 28–50.

POPE, R. B. (1957). The effect of photo-scale on the accuracy of forest measurements. *Photogramm. Engng.* **23,** 869–73.

POPE, R. B. (1960). Ocular estimation of crown density on aerial photographs. *For. Chron.* **36** (1).

POPE, R. B. (1962). Constructing aerial photo volume tables. *U.S. For. Serv. Res. Pap. For. Range Exp. Sta.* **49,** 25 pp.

PRIEST, R. B. (1959). Harbour survey in the Cook Islands. *N.Z. Draughtsman,* **1** (3), 55–7.

PRONIN, A. K. (1949). The investigation of vegetation by means of aerial photography in different zones of the spectrum (Russian only). *Trudy Laer,* **1,** 69–91. IzAN, Moscow.

PRYOR, L. D. (1939). The vegetation of the A.C.T.: A study of the syncology. M.Sc. thesis, Univ. of Adelaide.

PRYOR, L. (1959). Species distribution and association in Eucalyptus. Biogeography and ecology in Australia. Monograph. Biol. 8, 291–302, W. Junk, Hague, Holland.

QUINN, A. O. (1963). Proyecto aerofotogrametrico. *Chile Tech. paper, 29th meeting. Am. Soc. Photogramm.*

BIBLIOGRAPHY

RAASVELDT, H. C. (1956). The stereo-model, how it is formed and deformed. *Photogramm. Engng.* **22,** 708–26.

RABBEN, E. L. *et al.* (1955). The eyes have it. *Photogramm. Engng.* **21,** 573–8.

RABBEN, E. L. *et al.* (1960). Fundamentals of photo interpretation. *Manual of Photo-Interpretation,,* 99–169.

RABIDEAU, G. S., FRENCH, C. S. & HOLT, A. S. (1946). The absorption and reflection spectra of leaves, chloroplast suspensions and chloroplast fragments. *Am. J. Bot.* **33,** 769.

RABINOWITCH, E. I. (1951). *Photosynthesis and Related Processes,* **2** (1), ch. 22. Interscience Publishers Inc., N.Y.C.

RANTZIEN, HORN AF (1951). Macrophyte vegetation in lakes and temporary pools of the Alvar of Öland, south Sweden. I. *Svensk Bot. Tidskr.* **45,** 75–120.

RAUNKIAER, C. (1934). *Life Forms of Plants and Statistical Plant Geography,* English translation, Chs. 9, 15, 16, 17. Oxford.

RAY, R. G. & FISCHER, W. A. (1960). Quantitative photography—a geologic research tool. *Photogramm. Engng.* **26,** 143–50.

RAYLEIGH, LORD (1910). Incident of light upon a transparent sphere of dimensions small compared with the wavelength. *Proc. Roy. Soc.* (*A*), **84,** 25.

RAYLEIGH, LORD (1914). Diffusion of light by spheres of small refractive index. *Proc. Roy. Soc.* (*A*), **90,** 219.

RAYNAR, T. (1959). Frost problems and photo-interpretation. *Photogramm. Engng.* **25,** 779–86.

READING, H. G. (1956). A method of supplying contours from four suitably placed spot heights using parallax law observations. *Photogramm. Rec.* **2** (7), 65–75.

REVERTERA, K. (1961). An example of geological interpretation of small scale verticals. *I.T.C.* Series B, No. 11.

REVERTERA, K. (1963). Notes on mapping of ground features in mountainous regions. *I.T.C.* Series B, No. 21.

REY, P. (1947). Quelques exemples d'interprétation botanique et agricole de photographies aériennes. *Bull. technique des ingénieurs des Service Agricole,* 65–81.

REY, P. (1957). *L'interprétation des photographies aériennes.* C.N.R.S. Paris.

REY, P. (1962). Les perspectives fondamentals de la cartographie de la végétation. *Ser. Carte Veg.* **1.** C.N.R.S. Toulouse.

REY, P. (1964). *Photographie aérienne et végétation.* UNESCO Conference, Toulouse, 31 pp.

RHODY, B. (1962). An approximate method of rectification for photographs of mountain forests (in German). *Archives Internationales de Photogrammetrie, Delft,* **14,** 245–51; *J. For. Suisse,* **114** (5/6), 314/332.

RICHARD, P. (1962). Photogeology survey of the Sahara. *Photogramm. Rec.* **4,** 163–70.

RICHARDS, P. W. (1952). *The Tropical Rain Forest,* 1–25, 372–4, 399. Cambridge University Press.

ROBBINS, A. R. (1949). A method of finding gradients from air photographs with no control. *Photogramm. Engng.* **15,** 636.

ROBBINS, C. R. (1929). Air survey and forestry. *Can. For. J.* **8** (2).

ROBBINS, C. R. (1931). An economic aspect of regional survey. *J. Ecol.* **19** (1).

ROBBINS, C. R. (1934). Northern Rhodesia: An experiment in classification of lands of use of aerial photographs. *J. Ecol.* **22** (1).

ROBBINS, R. (1957). The status and classification of New Zealand forest vegetation. Ph.D. thesis, Auckland University.

ROBERTSON, V. C. (1955). Aerial photography and proper land utilization. *Photogramm. Rec.* **1** (6), 5–9.

BIBLIOGRAPHY

ROBINSON, M. W. (1947). An instrument to measure crown cover. *For. Chron.* 222–5.

ROCARD, Y. (1932). Distribution of light scattered by the atmosphere at ground level. *Revue Opt* **2**, 193.

ROGERS, E. J. (1949) Estimating tree heights from shadows on vertical aerial photographs. *J. For.* **47**, 182–91.

ROGERS, E. J. (1949). Estimating tree heights from shadows on vertical aerial photographs. *U.S. For. Serv., Northeast For. Exp. Sta. Pap.* 12.

ROGERS, E. J. (1960). Forest survey design applying aerial photographs and regression techniques for the Caspian forests of Iran. *Photogramm. Engng.* **26**, 441–3.

ROGERS, E. J. (1961). Application of aerial photographs and regression technique for surveying Caspian forests of Iran. *Photogramm. Engng.* **27**, 811–16.

ROGERS, E. J. (1963). A mathematical scheme for checking the bias of relief in determining areas from photo points. *Photogramm. Engng.* **29** (2), 272–4.

ROGERS, E. J., AVERY, G. & CHAPMAN, R. A. (1959). Three scales of aerial photography compared for making standard measurements. *For. Res. Notes U.S. Dept. Agric.*, No. 8. Pennsylvania.

ROGERS, G. J. (1953). The natural occurrence of the Eucalypts. *Aust. For. Timb. Leaflet* 65. Canberra.

ROGERS, J. (1948). A short cut for scaling aerial photographs. *N.W. Forest Exp. Sta. Paper*, No. 20.

ROLLET, B. (1960). Emploi de photographies aériennes au 1/40,000 pour l'interprétation de la végétation (Cambodia and Vietnam). *Bois Forêts Trop.* **74**, 16–24.

ROMEL, L. G. (1930). Comments on Raunkiaer's and similar methods of vegetation analysis and the law of frequency. *Ecology* **11**, 589–96.

ROSAYRO, R. A. DE (1954). A reconnaissance of Sinharaja rain forest. *Ceylon Forester*, **1** (3), 68–74.

ROSAYRO, R. A. DE (1958). Tropical ecological studies in Ceylon. *Proceedings of Kandy Symposium, UNESCO*, 33–9.

ROSAYRO, R. A. DE (1959). The application of aerial photography to stockmapping and inventories in rainforest in Ceylon. *Emp. For. Rev.* **38** (2), 141–74.

ROSAYRO, R. A. DE (1960). Surveys particularly applicable to extensive areas. *5th World Forestry Conference*, GA/157/1/Ceylon (2 pp.).

ROSCOE, J. H. (1954) Photogeography, photo-geology and the photo-interpretation key. *A.A.A.S.* Section E.

ROSCOE, J. H. (1960). Photo interpretation in geography. *Manual of Photographic Interpretation*, ch. 14.

ROSENZU, M. D. (1966). Photographic image quality: Prediction and analysis. Symposium, Commission I, I.S.P. London.

ROSSETTI, C. (1962). Un dispositif de prises de vues aériennes à basse altitude et ses applications pour l'étude de la physiognomie de végétations ouvertes. *Bulletin de service*, **7** (2), 219–38. C.R.N.S. Paris.

ROSSETTI, C. (1965). A propos des images photographiques aériennes de la végétation. *Photo-interpretation*, **5**.

ROSSETTI, C. (1966). Reflections sur l'utilization des photographies aériennes pour létude du couvert vegetal. Intern. Symposium, Commission VII, I.S.P. Paris.

ROSSETTI, C. & KOWALISKI, P. (1966). Relation entre les caractéristiques de réflexion spectral de quelques espèces et leur images sur des photographies en couleur. Symposium, Commission VII, I.S.P. Paris.

ROURKE, J. D. & MORRIS, E. A. (1951). The use of air photos for soil classification and mapping in the field. *Photogramm. Engng.* **17**, 738–44.

BIBLIOGRAPHY

ROUSSEL, L. (1962). Etat actuel de la photologie forestière. *Revue For. Fr.* (1), 8–13.

ROWE, J. S. (1956). Uses of undergrowth plant species in forestry. *Ecology* 37, 461–73.

ROWE, J. S. (1960). Can we find a common platform for the different schools of forest type classification? *Silva Fenn.* 105, 82–8.

ROWE, J. S. (1961). Critique of some vegetational concepts as applied to forests of north-western Alberta. *Can. J. Bot.* 39, 1007–17.

ROWE, J. S. (1962). Soil site, and land classification. *For. Chron.* 38, 420–32.

RYKER, H. C. (1933). Aerial photography. Method of determining timber species. *Timberman* 24 (5), 11–17.

SAASTAMOINEN, J. (1959). Tacheometers and their use in surveying. *Can. Surveyor,* 14 (10), 444–53.

SABET, A. H. (1962). An example of photo-interpretation of crystallic rocks. *I.T.C.* Series B, No. 15/15.

SALGUEIRO, R. (1956). Photogrammetry in Latin America. *Photogrammtria,* 13 (2), 47–8.

SAMOJLOVIC, G. G. (1958). Visual characteristics of forests as seen from the air in the Abakan river basin, Khakass (Siberia) (in Russian). *Bot. Z.* 43, 1304–10.

SANKTJOHANSER, L. (1960). Forstliche Luftbildmessung im Hochgebirge. Mitt. *Staatsforstverw. Bayerns,* 31, 208–13.

SARALEGUI, A. M. (1960). Sur la préparation rapide du plan de vol photographique (aero-photogrammes horizontaux), Institute Fota-Topografico, Argentino. *Photogrammetria,* 17, (3), 105–12.

SARUAS, R. (1953). Measurement of the crown closure of a stand. *Comm. Inst. Forestalis Fenn.* 41 (6).

SATOO, T., KUNUGI, R. & KUMEKAWA, A. (1956). Amount of leaves and production of wood in aspen second growth in Hokkaido. *Bull. Tokyo Univ. For.* 52, 33–51.

SAUNDERS, R. G. (1945). Stereoscopy, its history and uses. *Photogramm. Engng.* 11, 101–13.

SAYN-WITTGENSTEIN, L. (1960). Recognition of tree species on air photographs by crown characteristics. *Tech. Note 95, Canadian Forest Research Division.*

SAYN-WITTGENSTEIN, L. (1961a). Recognition of tree species on air photographs by crown characteristics. *Photogramm. Engng.* 27, 798–809.

SAYN-WITTGENSTEIN, L. (1961b). Phenological aids to species identification on air photographs. *Tech. Note 104, Dept. For. Can.*

SAYN-WITTGENSTEIN, L. (1966). The best season for aerial photography. *Intern. Symposium,* Commission VII, I.S.P. Paris.

SCHIMPER, A. F. (1903). *Plant Geography on a Physiological Basis.* Clarendon Press, Oxford.

SCHIMPER, A. F. (1935). *Pflanzengeographie auf physiologischer Grandlage,* 260. F. van Faber, Jena.

SCHMIEDT, G. (1960). Report from Italy. *Intern. Soc. Photogramm., London, Commission VII Report* (7–13).

SCHNEIDER, S. J. (1960). Report from Germany. Vegetation interpretation. *Intern. Soc. Photogramm., London, Commission VII Report* (1–6).

SCHNEIDER, S. J. (1966). The contribution of geographical airphoto interpretation to problems of land division. *Intern. Symposium, Commission VII, I.S.P. Paris.*

SCHULTE, O. W. (1951). The use of panchromatic, infrared and colour aerial photography in the study of plant distribution, Quebec, Canada. *Photogramm. Engng.* 17, 688–714.

SCHUMACHER, F. X. & CHAPMAN, R. A. (1942). *Sampling Methods in Forestry and Range Management.* Duke Univ. Bull. 213 pp.

SCHWIDEFSKY, K. (1959). (*Grundriss der Photogrammetrie, Wissenschaft und Fachbuch, Germany,* 1950); *Outline of Photogrammetry.* Pitman, New York.

BIBLIOGRAPHY

SEELY, H. E. (1929). Computing tree heights from shadows on the aerial photographs. *For. Chron.* **5**, 24–7.

SEELY, H. E. (1934). Aerial photography in forest survey. *E.F.J.* **13** (2).

SEELEY, H. E. (1942). Determination of tree heights from shadows in air photography. *Dept. Res. Div. (Canada) Research Note;* also *Photogramm. Engng.* **8**, 100.

SEELY, H. E. (1948). The shadow height calculator. *Ser. Dept. For. Can. Leaflet* **2**, 7 pp.

SEELY, H. E. (1959). Some developments in photogrammetry. *Pulp Paper Mag. Can.* June.

SEELY, H. E. (1960). Aerial photogrammetry in forest surveys. *5th World Forestry Conference*, GP/12/1/B—Canada, 11 pp.

SEELY, H. E. (1964). Canadian forest inventory methods. *Dep. For. Can.* 11 pp.

SELWYN, E. W. H. (1948). Photographic and visual resolving power of lenses. *Photogr. J.* **88B**, Jan. and June.

SELWYN, E. W. H. & ROMER, W. (1943). An instrument for the measurement of graininess. *Photogr. J.* Jan.

SEMENOVA, N. N. (1959). A study of soil erosion by aerial photographs. *Soviet-Soil Science, Am. Inst. Biol. Sciences*, **5**, 582–90.

SEYBOLD, A. (1933). Über die optischen Eigenschaften der Laubblätter, Planta III. *Arch. Wiss. Bot.* **20**, 577–601.

SHARONOV, V. V. (1934). Visual luminance co-efficients of some natural objects. *Transl. Tekh. bulleten, Aerophotography IzAN Tsniiga*, **6** and **7**.

SHAW, S. H. (1953). The value of air photographs in the analysis of drainage patterns. *Photogramm. Rec.* **1** (2), 4–15.

SHIELL, C. A. (1929). Reflectivity of leaves. *Bot. Gaz.* **87**, 583.

SHULGIN, I. A. (1960). The optical characteristics of xeromorphy and succulence of plants. *Dokl. Akad. Nauk. S.S.R.* **134** (4), 972–5.

SHULGIN, I. A., KHASANOV, V. S. & KLESHNIN, A. F. (1960). On the reflection of light as related to leaf structure. *Bot. Sci. Sect. Transl.* **134**, 471–4.

SHULL, C. A. (1929). A spectrophotometric study of reflection of light from leaf surfaces. *Bot. Gaz.* **87**, 583–607.

SIMAKOVA, M. S. (1964). *Soil Mapping by Colour Aerial Photography.* Academy of Sciences, U.S.S.R. (Translation Israel Program for Scientific Translations, Jerusalem—over 100 Russian references), 2.

SIMMONSSEN, G. (1956). Device for measurement of parallaxes. Intern. Congress Photogrammetry, Stockholm.

SIMONET, M. & VAN ROOST, J. (1957). La prise de vue aérienne en infra-rouge au Congo Belge. *Institut Geographique du Congo Belge.*

SIMONSON, R. (1950). Use of aerial photography in soil surveys. *Photogramm. Engng.* **16**, 308–15.

SIMS, W. G. (1954). Shadow point. *For. Timb. Bur. Aust.* No. 67.

SIMS, W. G. & HALL, N. (1955). Some requirements for air photo interpretation in Australia. *Aust. For.* **19** (2), 121–35.

SISAM, J. W. (1947). The use of aerial survey in forestry and agriculture. *Imp. Agric. Bur.* No. 9, 59 pp.

SJORS, H. (1955). Remarks on ecosystems. *Svensk Bot. Tidskr.* **49**, 155–69.

SMITH, J. H. (1953). Analysis of crown development can establish biological and economic limits to growth of trees and stands. *Commonw. For. Rev.* **42** (1).

SMITH, J. H. (1957). Problems and potential uses of photo-mensurational techniques for estimation of volume of some immature stands of Douglas fir and western hemlock. *Photogramm. Engng.* **23**, 595–9.

BIBLIOGRAPHY

SMITH, J. H. (1965). Biological principles to guide estimation of stand volumes. *Photogramm. Engng.* **31**, 87–91.

SMITH, J. H. & BAJZAK, D (1961). Photo-interpretation provides a good estimate of site index of fir, hemlock and cedar. *J. For.* **59**, 261–3.

SMITH, J. H., LEW, Y. & DOBIE, J. (1960). Intensive assessment of factors influencing photo-cruising shows that local expressions of photo volumes are best. *Photogramm. Engng.* **26**, 463–9.

SMITH, J. T. (1963). Colour—A new dimension in photogrammetry. *Photogramm. Engng.* **29** (6), 999–1011.

SMITH, J. W. (1958). The effects of the earth's curvature and refraction on the mensuration of vertical photographs. *Photogramm. Engng.* **24**, 751–6.

SNEDECOR, G. W. (1956). *Statistical Methods.* Iowa State Press, Ames.

SONLEY, G. R. (1946). Interim report on experimental air photography for forest-cover classifications. *For. Chron.* **22** (2), 157–8.

SPECHT, R. L. & PERRY, R. A. (1948). Plant ecology of the Mount Lofty Ranges. *Trans. Roy. Soc. South Australia,* **73**, 91–132.

SPECHT, R. L. & RAYSON, P. (1957). Dark Island Heath. I. Definition of the ecosystem. *Aust. J. Bot.* **5**, 52–85.

SPINNER, G. P. (1949). Observations on the greater snow geese. *Auk,* **66**, 197–8.

SPOONER, M. G. (1965). A review of some recent research in techniques for automated photo-analysis. *Photogramm. Engng.* **31** (3), 504.

SPURR, S. H. (1948). *Aerial Photographs in Forestry,* 77–159, 261–80. Ronald Press, New York.

SPURR, S. H. (1952). A further note concerning shadows on aerial photographs. *Photogramm. Engng.* **18**, 833–4.

SPURR, S. H. (1954). History of forest photogrammetry and aerial mapping. *Photogramm. Engng.* **20**, 551–60.

SPURR, S. H. (1960). *Photogrammetry and Photo-interpretation,* 4, 27, 30–2, 47, 50, 72–6, 86–91, 350–2, 357, 365–77, 383–96, 420, 426–32, 443. Ronald Press.

SPURR, S. H. & BROWN, C. T. (1945). The Multiscope. *Photogramm. Engng.* **11**, 171–8.

ST. JOSEPH, J. K. *et al.* (1966). *The Uses of Air Photography.* Cambridge University Press, 196 pp.

STAIR, R. & COBLENTZ, W. W. (1933). Infrared absorption spectra of some plant pigments. *J. Res.* **11**, 703–11.

STAMP, D. (1925). The aerial survey of the Irrawaddy delta forests (Burma). *J. Ecol.* **13**, 515–22.

STEEN, W. W. & LITTLE, J. C. (1959). A new portable reflectance spectrophotometer for the selection of film and filters. *Photogramm. Engng.* **25**, 615–18.

STEIGERWALDT, E. F. (1950). Stereo types for aerial photo interpreters. *J. For.* **48**, 693–6.

STEINER, D. (1963a). Luftaufnahme und Luftbildinterpretation in der Sowjetunion. *Erdkunde Heft,* **1–2**, 17, 77–100.

STEINER, D. (1963b). Technical aspects of air photo interpretation in the Soviet Union (includes bibliography). *Photogramm. Engng.* **29** (6), 988–97.

STEINER, D. & HAEFNER, H. (1965). Tone distortion for automated interpretation. *Photogramm. Engng.* **31**, 269–80.

STELLINGWERF, D. A. (1960). Methods and results of forest photo-interpretation. *I.T.C.* Delft, Series B, **8**, 1–11.

STELLINGWERF, D. A. (1963). Volume determination on aerial photographs. *I.T.C.* Series B, No. 18, pp. 19.

STELLINGWERF, D. A. (1966). Interpretation of tree species and mixtures on aerial photographs. *Intern. Symposium, Commission VII, I.S.P. Paris.*

BIBLIOGRAPHY

STELLINGWERF, D. A. & YEARSLEY, M. (1961). Forest maps from aerial photographs. *I.T.C.* Delft.

STONE, K. H. (1948). Aerial photographic interpretation of natural vegetation in the Anchorage area, Alaska. *Geogrl. Rev.* **38**, 465–74.

STONE, K. H. (1951). Geographical air-photo interpretation. *Photogramm. Engng.* **17**, 754–71.

STONE, K. H. (1954). A selected bibliography for geographic instruction and research. *Photogramm. Engng.* **20**, 561–6.

STRAHLER, A. N. (1952). Hypsometric analysis of erosional topography. *Geol. Soc. Am.* **63**, 117–42.

STRANGEWAY, D. W. & HOLMER, R. C. (1965). Infrared geology. *Proc. 3rd Symposium on Remote Sensing*, Report 4866-9-x. Univ. Michigan, Ann Arbor.

STRIDSBERG, E. (1957). The air photo as a forest map (in Swedish). *Norrlands SkogsvForb. Tidskr. Stockholm*, **3**, 401–7.

STRIDSBERG, E. (1966). Department of Forest Economy, Stockholm, personal communication.

STRIDSBERG, E., ERIKSSON, H. & SKARBY, R. (1961). An example for forest land use planning and inventory in south Sweden (in Swedish). *Svenska SkogsvForen. Tidskr.* **59**, 267–306.

STRINGER, K. V. (1953). Air photography in the mapping of sedimentary, metamorphic and igneous rocks, with special reference to areas in Burma and Central Africa. Ph.D. thesis, London University.

STUBNER, K. (1954). Aerial photographs in geomorphic micro analysis. *Wiss. Z. Friedr. Schiller-Univ. Jena*, **3** (German only).

SUITS, G. H. (1946). The nature of infrared radiation and ways to photograph it. *Photogramm. Engng.* **36**, 763, 772.

SWANSON, L. W. (1954). Aerial photography requirements for soil survey field operations. *Photogramm. Engng.* **20**, 709–11.

SWANSON, L. W. (1964). Aerial photography and photogrammetry in the coast and geodetic survey. *Photogramm. Engng.* **30** (5), 699–726.

SWELLENGREBEL, E. J. (1959). On the value of large scale aerial photographs in British Guiana. *Emp. For. Rev.* **38**, 56–64.

TAGEVEVA, S. V., BRANDT, A. B. & DEREVYANKO, V. G. (1960). Changes in the optical properties of leaves in the course of the growing season. *Bot. Sci. Transl.* **135**, 226–8. *Transl. Dokl. Akad. Nauk S.S.S.R.* **135**, 1270–3.

TANIGUCHI, S. (1961). Forest inventory by aerial photographs. *Res. Bull. Hokkaido Exp. For. Stat.* **21** (1), English ed.

TANSLEY, A. G. (1920). The classification of vegetation and the concept of development. *J. Ecol.* **8**, 118–49.

TANSLEY, A. G. (1935). The use and abuse of vegetational concepts and terms. *Ecology*, **16**, 284–307.

TANSLEY, A. G. (1939). *The British Isles and their Vegetation*. Cambridge Univ. Press.

TANSLEY, A. G. (1949). *Introduction to Plant Ecology*. Allen & Unwin, London, 266 pp.

TARDIVO, CAPT. (1913). International Society of Photogrammetry, Vienna.

TARKINGTON, R. G. (1959). Kodak panchromatic negative films for aerial photography. *Photogramm. Engng.* **25**, 695–99.

TARKINGTON, R. G. & SOREM, A. L. (1963). Color and false-color films for aerial photography. *Photogramm. Engng.* **29**.

BIBLIOGRAPHY

TAUBENHAUS, J. J. *et al.* (1929). Aeroplane photography in the study of cotton root-rot fungus. *Phytopathology*, **19**, 1025–9.

TAYLOR, B. W. & STEWART, G. A. (1958). Vegetation mapping in Australian territories of Papua and New Guinea. *Proc. Kandy Symposium, UNESCO*, 127–36.

TEWINKEL, G. C. (1952). Basic mathematics. *Manual of Photogrammetry*, 409.

TEWINKEL, G. C. (1962). Kelsh plotter notes. *Photogramm. Engng.* **28**, 485–91.

THEIS, J. B. (1958). Increased base height ratio. *Photogramm. Engng.* **24**, 127–32.

THOMAS, H. H. (1920a). Geological reconnaissance by aeroplane photography, with special reference to work on the Palestine front. *Geogrl. J.* **55** (5), 349–76.

THOMAS, H. H. (1920b). Aircraft photography in the service of science. *Nature*, **105** (No. 2641), 257–9.

THOMPSON, E. H. (1954). Heights from parallax measurements. *Photogramm. Rec.* **1**, (4), 38–49.

THOMPSON, E. H. (1955). Note on the accumulation of random error in a long strip of photographs. *Photogramm. Rec.* **1** (5), 71–2.

THOMPSON, E. H. (1957). The geometric theory of the camera and its application to photogrammetry. *Photogramm. Rec.* **2**, 241–63.

THOMPSON, E. H. (1962). The theory of the method of least squares. *Photogramm. Rec.* **4**, 53–65.

THOMSON, A. P. (1947). The national forest survey in New Zealand. *Proc. 5th Empire For. Conf., London.*

THOMSON, A. P. (1949). Technical developments in air survey and the interpretation of forest data. *N.Z. Jl. F.* **6**, 39–44.

TOMASEGOVIC, Z. (1961). Stereophotogrammetric lines surveys. English translation from *Sum. List*, **1**, 36–45. N.Z.F.S., Wellington, New Zealand.

TOMLINSON, R. F. & BROWN, W. G. E. (1962). The use of vegetation analysis in photo-interpretation of surface material. *Photogramm. Engng.* **28**, 584–93.

TOOMING, H. (1958). On the spectral reflectivity properties of maize leaves. *Akad. Nauk*, Est., S.S.R. *Investigations of the Physics of the Atmosphere*, **1**, 68–82.

TOTEL, R. (1960). Einige Ergebnisse des Luftbildeinsatzes in der forstlichen Standortserkundigung der D.D.R. Int. Congr. Photogramm. London.

TRAPNELL, C. G. (1952). The use of aerial photography in vegetation, soil and land use surveys. *Proc. Nairobi Sci. Philos. Soc.* **6**, 17–20.

TREECE, W. (1955). Estimation of vertical exaggeration in stereoscopic viewing. *Photogramm. Engng.* **21**, 518–27.

TROLL, C. (1947). Der asymmetrische Aufbau der Vegetationszonen und Vegetationsstufen auf der Nord- und Sud-Halbkugel. *Ber. Jeobot., Inst. Rubel*, 46–83.

TROLL, C. (1939) Luftbildplan und ökologische Bodenforschung. *Zeitschrift des Gessellschaft, Berlin*, **710**, 277–311.

TROREY, L. G. (1950). *Handbook of Aerial Mapping and Photogrammetry*, 178. Cambridge University Press.

TRUESDELL, P. E. (1953). Report of unclassified military terrain studies. *Photogramm. Engng.* **19**, No. 3.

TUOMIKOSKI, R. (1942). Untersüchinngen über die vegetation der Bruchmoore in Ostfinnland. Zur Methodik der pflanzensoziologischen Systematik. *Ann. Bot. Vanamo*, **17** (1), 1–203.

TUPPER, J. L. & NELSON, C. N. (1955). The effect of atmospheric haze. *Photogramm. Engng.* **21**, 116–26.

TUPPER, J. P. *et al.* (1952). *Manual of Photogrammetry*, 271–4, 277, 279, 280, 282. 288.

BIBLIOGRAPHY

UHL, A. (1937). Untersuchungen über die Assimilation Verhaltnisse und die Ursachen ihrer Unterschiede in der Gattung Pinus. *J. b. Wiss. Bot.* **75**, 368–421.

ANON. U.S. Forest Service (1959). Third progress report on the status of aerial survey and techniques in research. *Forest Economics and Marketing Memorandum*, 14 pp.

VERR, R. (1962). *Light and Plant Growth.* Phillips Technical Library, Eindhoven, Holland.

VERBOOM, W. C. (1961). Planned rural development in the Northern Province of Northern Rhodesia. *I.T.C.* Series B, Nos. 12/13.

VERHOEVEN, H. J. (1963). A study of soil survey and photo-interpretation in the Grand Duchy of Luxembourg. *I.T.C.* Series B, Nos. 19/20.

VERSTAPPEN, H. T. (1964). Geomorphology as an essential element in aerial survey. U.N. Cartographic Conference, Manila.

VERSTAPPEN, H. T. (1966). The role of landform classification in integrated surveys. *Intern. Symposium, Commission VII, I.S.A. Paris.*

VESTAL, A. G. (1949). Minimum areas for different vegetations: Their determination from species-area curves. *Univ. Ill. Biol. Mongr.* **20**, (3), 1–129.

VESTAL, A. G. & HEERMANS, M. F. (1954). Size requirements for reference areas in mixed forest. *Ecology*, **26**, 122–34.

VIAL, A. J. (1959). A brief history of the aerial survey unit, South Australia. *Cartography*, **3**, 165–71.

VIKTROV, S. V. (1947). A study of the distribution and dispersion of plants by aerial photographs. *Bull. Soc. Nat. Moscow* (Sect. biol.), N.S. **52** (4), 71–8. (Transl. from the Russian.)

VINK, A. P. (1964). Practical soil surveys and their interpretation for agricultural purposes. *I.T.C.* Series B. No. 16.

VINK, A. P. *et al.* (1965). Some methodological problems in interpretation of aerial photographs for natural resource surveys. *I.T.C.* Series B, No. 32.

VINOGRADOV, B. V. (1962). *Objective Methods of Vegetation Analysis from Aerial Photographs* (in Russian). Acad. Sci. U.S.S.R. Inst. Bot. Moscow, 12 pp.

VINOGRADOV, B. V. (1964). Les enquêtes générales de l'élaboration et de l'application des aéromethodes d'intégration des études des zones arides. Photo-interpretation Congress, Toulouse.

VINOGRADOVA, A. I. (1955). Surveying vegetation and soils (Russian only). *Geograficheskii, Sbornikz*, 59–74 IzAN, Moscow.

VRIES, D. M. (1948). Method and survey of the characterization of Dutch grasslands. *Vegetatio*, **1**, 51–7.

WALDHAUSL, P. (1959). Height adjustment by the I.T.C. Jerie method. *Photogrammetria*, **16** (1).

WALDRAM, J. M. (1945a). Measurement of the photogrammetric properties of the upper atmosphere. *Transl. Illum. Engng. Soc.* (London), **10**, (8), 147–88.

WALDRAM, J. M. (1945b). Measurement of the photometric properties of the upper atmosphere. *Q. Jl. R. Met. Soc.* **71**, 319–36.

WALKER, F. (1953). *Geography from the Air.* Dutton, New York, 111 pp.

WALKER, J. E. (1961). Progress in spectral reflectance film filter research applicable to engineering and geological studies. *Photogramm. Engng.* **27**, 445–50.

WALLIS, I. (1943). How reflection changes appearance. *Am. Photogr.* Aug., 20–24.

WARDROP, T. N. (1964). Forest Research Institute, Rotorua, N.Z., personal communication.

BIBLIOGRAPHY

WARDROP, T. N. & DUFF, C. E. (1952). An account of reconnaissance and stock-mapping surveys in tropical deciduous woodlands in a part of central Africa. *6th Br. Comm. For. Conf., Rhodesia*, 1–21.

WASHBURTON, J. (1951). Log block piles inventorized by air. *Pulp Paper Mag. Can.* Nov.

WASHER, F. E. (1951). Characteristics of the wide-angle aeroplane lens. *Nat. Bur. Standards Journ. Research*, **29**, 234–43.

WASHER, F. E. (1963a). The precise evaluation of lens distortion. *Photogramm. Engng.* **29**, (2) 327–32.

WASHER, F. E. (1963b). Calibration of photogrammetric lens and cameras at the national bureau of standards (extensive bibliography). *Photogramm. Engng.* **29** (1), 113–19.

WASHER, F. E. & TAYMAN, W. P. (1960). Location of the plane of best average definition for aeroplane camera lens. *Photogramma. Engng.* **26**, 475–88.

WASSINK, E. C. (1956). On the mechanism of photosynthesis. *Radiat. Biol.* **3**, 293–342.

WATT, A. S. (1947). Pattern and process in the plant community. *Ecology*, **35**, 1–22.

WEAR, J. F. & BONGBERG, J. W. (1951). Uses of aerial photographs in control of insects. *J. For.* **49**, 632–3.

WEAR, J. F. & DILWORTH, J. R. (1955). Colour photos and salvage. *Lumberman*.

WEBB, L. J. (1959). A physiognomic classification of Australia rainforests. *J. Ecol.* **47**, 551–69.

WEBB, D. A. (1954). Is the classification of plant communities either possible or desirable? *Bot. Tidsskr.* **51**, 362–70.

WEBSTER, R. (1963). The use of basic photographic units in air photo interpretation. *Archives Internationales de Photogrammetrie*, **14**, 143–8.

WEDE, A. J. VAN DER, (1962). The accuracy of radial triangulation. *I.T.C.*

WEIGL, J. W. & LIVINGSTONE, R. (1963). Infra red spectra of chlorophyll and related compounds. *J. Am. Chem. Soc.* **75**, 2173–6.

WEIR, I. C. (1959). A crown ratio study in even-aged stands of *E. Obliqua.* Typescript, School of Forestry, Melbourne University.

WELANDER, E. (1952). Estimation of mixture of species, tree height, density and volume with aerial photographs (Swedish only). *Norrlands SkogsvFörb. Tidskr.* 2, 209–41, Stockholm.

WHEELER, P. R. (1959). Preliminary plan. *Forest Survey of Cambodia.* USOM, Cambodia.

WHITMORE, G. D. (1949). *Advanced Surveying and Mapping.* 609, ref. 344–512. International Textbook Co.

WHITMORE, G. D. (1952). The development of photogrammetry. *Manual of Photogrammetry*, 1–10.

WHITTAKER, B. B. (1900). Photogrammetry and land tenure surveys in Uganda. *Photogramm. Rec*, **4**, 53–64.

WHITTAKER, R. H. (1951). A criticism of the plant association and the climatic climax concepts. *New. Sci.* **25**, 17–31.

WHITTAKER, R. H. (1953). A consideration of the climax theory: the climax as a population and pattern. *Ecol. Mongr.* **23**, 41–78.

WHITTAKER, R. H. (1960). The vegetation of Sisiyou Mountains. *Ecol. Monogr.* **30**, 279–335.

WHITTAKER, R. H. (1962). Classification of natural communities. *Bot. Rev.* **28**, 1–239.

WIJK, W. R. VAN & UBING, D. W. (1963). *Physics of Plant Environment*, North Holland Publishing Co., Amsterdam.

WILCOX, F. R. (1936). Gatineau scale ratio for use with vertical photographs. *J. For.* **34**, 1049–57.

WILLIAMS, R. J. (1955). *Vegetation Regions.* Dept. of National Development, Canberra, 23 pp.

BIBLIOGRAPHY

WILLIAMS, W. T. & LAMBERT, J. M. (1959). Multivariate methods in plant ecology. *J. Ecol.* **47,** 83–101.

WILLIAMS, W. T. & LAMBERT, J. M. (1960). Ibid. **48,** 689–710.

WILLIAMS, W. T. & LAMBERT, J. M. (1961). Ibid. **49,** 717–29.

WILLIAMS, W. T. & LANCE, G. N. (1958). Automatic subdivision of associated populations. *Nature,* **182,** 1755.

WILLINGHAM, J. W. (1957). The indirect determination of forest stand variables from vertical aerial photographs. *Photogramm. Engng.* **23,** 892–3.

WILLSTATTER, R. & STOLL, A. (1913). *Untersuchungen über Chlorophyll.* Springer, Berlin. English translation by F. M. Schertz and A. R. Mertz.

WILSON, E. (1920). Use of aircraft in forestry and logging. *Canadian Forestry Mag.* **16,** Oct., 439, 444.

WILSON, R. C. (1949). The relief displacement factor in forest area estimates by dot templets on aerial photographs. *Photogramm. Engng.* **15,** 225–36.

WILSON, R. C. (1950). Controlled forest inventory by aerial photographs. *Timberman,* **50** (4)—reprint.

WILSON, R. C. (1962). Surveys particularly applicable to extensive forest areas. 5th World Forestry Congress, Washington, D.C.

WINKLER, E. M. (1962). Relationship of air photo tone control and moisture content in glacial soils. *Proc. 2nd Symposium on Remote Sensing,* 107–13. Univ. Michigan.

WOLFF, G. (1966). Schwerz-Weiss und falschfarbige Luftbilder als diagnostiche Hilfsmittel für operative Arbeitenim Forstschutz. Symposium, Commission VII, I.S.P. Paris.

WOOD, E. S. (1946). The maximum values of tilt and relief when compiling radial line plots using the principal point. *Photogramm. Engng.* **12,** 205–11.

WOOD, E. S. (1950). The uses of high altitude photography for mapping and reconnaissance: A discussion. *Photogramm. Engng.* **16,** 613–18.

WOOD, J. G. (1937). *The Vegetation of South Australia.* Govt. Printer, Adelaide.

WOOD, J. G. (1959). The phytogeography of Australia (in relation to Eucalyptus, Acacia, etc.). Biogeography and Ecology in Australia. *Monogr. Biol.* **8,** 291–302. W. Junk, Haag.

WOOD, J. G. & WILLIAMS, R. J. (1960). *Vegetation in the Australian Environment,* Ch. 6, 66–84. Melbourne University Press.

WOOD, R. W. (1959). *Physical Optics,* 423–48. Macmillan & Co., N.Y.

WOODWARD, L. (1952). Multiplex aerial projectors. *Photogramm. Engng.* **18,** 655.

WOOLRIDGE, S. W. (1932). The cycle of erosion and representation of relief. *Scottish Geogr. Mag.* **48,** 30.

WORLEY, D. P. & LANDIS, G. H. (1954). The accuracy of height measurement with parallax instruments on 1/12,000 photographs. *Photogramm. Engng.* **20** (5), 823–9.

WORLEY, D. P. & MEYER, H. A. (1955). Measurement of crown diameter and crown cover and their accuracy on 1/12,000 scale photographs. *Photogramm. Engng.* **21,** 372–5.

WRIGHT, W. D. (1954). Stereoscopic vision applied to photogrammetry. *Photogramm. Rec.* **1** (3), 29–45.

WUNDERLICH, W. (1962). *Numerical Techniques in Photogrammetry.* 1960/62 Year Book, Hanover Technical University, 11 pp.

YORDANSKY, A. N. (1957). The uses and principles of spectrozonal photography. *J. Scient. Appl. Photogr.* **2** (1).

YORDANSKY, A. N. (1958). Chromatic and achromatic details of darkening as a measure of spectrozonal images. Ibid. 3.

BIBLIOGRAPHY

Yost, E. F. (1960). Resolution and sinewave response as measures of photo-optical quality. *Photogramm. Engng.* **26**, 489–97.

Young, H. E. (1963). Use of air photos for location of truck roads. *University of Maine, Tech. Note* 20.

Young, H. E., Call, F. M. & Tryon, T. C. (1963). Multimillion acre forest inventories. *Photogramm. Engng.* **29**, 641–4.

Young, H. E., Tryon, T. T. & Hale, G. A. (1955). Dot gridding air photos and maps. *Photogramm. Engng.* **21**, 737.

Zeller, M. (1952). *Textbook of Photogrammetry*, 11. Lewis, London.

Zernitz, E. R. (1932). Drainage patterns and their significance. *J. Geol.* **40** (6).

Zieger, E. (1928). Ermittlung von Bestandmassen aus Flugbildern mit Hilfe des Hugershoff-Heydeschen Autokartographen. *Versuchsanstalt zu Tharandt*, 3, 97–127.

Zinger, A. (1964). Systematic sampling in forestry. *Biometrics*, **20**, 553–65.

Zinke, P. J. (1960). The soil vegetation survey as a means of classifying land for multiple use forestry. 5th World Forestry Conference, SP/56/1—U.S.A., 10 pp.

Zinke, P. J. *et al.* (1960). Photo-interpretation in hydrology and watershed management. *Manual of Photo-interpretation*, 549–56.

Zorn, H. C. (1965). Instruments for testing stereoscopic acuity. *Photogrammetria*, **20**, 229–207.

Zorn, H. C. (1966). Requirements and compilation of base maps for photo-interpretation. Symposium, Commission VII, I.S.P. Paris.

Zsilinsky, V. G. (1963). *Photographic Interpretation of Tree Species in Ontario*. Department of Lands and Forests, Toronto, 73 pp. Also in *Photogrammetria* (1964), **14**, 192–207.

Zsilinsky, V. G. (1964). The practice of photo-interpretation for a forest inventory. 10th Intern. Congress, I.S.P.

Zukov, A. J. (1960). Measuring the diameters of the crown projection of *Larix siberica* trees on large scale photographs (in Russian). *Lesn Z.* **3**, 6–9.

FURTHER RELEVANT PUBLICATIONS (1965–68)

Aldred, A. H. & Blake, G. R. (1967). Photo mosaics for forest stand mapping. Forest Management Research and Services Institute, Ottawa Report FMR-X-3.

Anon. (1968). Land evaluation (ed. G. A. Stewart). Papers of a CSIRO symposium, Canberra. Macmillan, Melbourne.

Anson, A. (1966). Colour photo comparison. *Photogramm. Engng.* **32**, 286–97.

Benson, M. C. & Sims, W. G. (1967). False colour film fails in practice. *J. For.* **65**, 904.

Bernstein, D. A. (1968). Constructing stereograms. *Photogramm. Engng.* **34**, 370–4.

Carneggie, D. M. & Lauer, D. T. (1967). Use of multiband remote sensing in forest and range inventory. *Photogrammetria*, **21**, 115–41.

Chevallier, R. (1968). Application du filtrage optique à l'étude des photographies aériennes. Commision VII, 11th Congress, I.S.P. Lausanne.

Corten, F. L. (1968). Performance and contract flight functions. Commission VII, 11th Congress, I.S.P. Lausanne.

Croney, W. F. (1967). Some vegetative and soil aspects in the use of vertical Kodak Ektachrome MS aerographic photography, Cornell University, pp. 19.

Duddeck, M. (1967). Practical experiences with aerial color photography. *Photogramm. Engng.* **33**, 1117–25.

BIBLIOGRAPHY

FRITZ, N. L. (1967). Optimum methods for using infrared sensitive color films. *Photogramm. Engng* **33**, 1128–38.

GIMBAZEVSKY, P. (1966). Land inventory interpretation. *Photogramm. Engng.* **32**, 967–76.

HEMPENIUS, S. A. (1968). Physiological and phsychological aspects of photo-interpretation. Commission VII, 11th Congress, I.S.P. Lausanne.

LANGLEY, P. G. (1965). Automating aerial photo-interpretation in forestry—How it works and what it will do for you. *Proc. American Foresters.*

LEIBOWITZ, T. H. (1966). Techniques required to build up a library of annotated aerial photographs. Soil Science Laboratory, Oxford University.

LOPIK, J. VAN, PRESSMAN, A. E. & LUDLUM, R. L. (1968). Mapping pollution with infrared. *Photogramm. Engng.* **34**, 561–4.

LYONS, E. H. (1967). Forest sampling with 70 mm. fixed air-base photography from heli-copters. *Photogrammetria*, **22**, 213–31.

MALILA, W. A. (1968). Multispectral techniques for image enchancement and discrimination. *Photogramm. Engng.* **34**, 490–501.

MEYER, M. P. (1967). Detection of diseased trees. *Photogramm. Engng.* **33**, 1128–38.

MORAIN, S. A. & SIMONETT, D. S. (1967). K-band radar in vegetation mapping *Photogramm. Engng.* **33**, 730–30.

MULLINS, L. (1966). Some important characteristics of photographic materials for air photographs. *Photogramm. Rec.* **5**, 240–70

NORMAN, G. G. & FRITZ, N. L. (1965). Infrared photography as an indicator of disease and decline in citrus trees. *Proc. Florida State Hort. Soc.* **78**, 59–63.

NYYSSONEN, A., POSO, S. & KEIL, C. (1968). The use of aerial photographs in the estimation of some forest characteristics. *Acta for. fenn* **82**, 32 pp.

PHIPPS, M. (1968). Contribution à l'analyse et la classification des types de paysage. Commission VII, 11th Congress I.S.P. Lausanne.

REINHOLD, A. (1967) Large scale aerial photographs as an aid in assessing the silvicultural condition of pine plantations and thickets (German). *Arch Forster*, **16**, 905–10.

SAYN-WITTGENSTEIN, L. & ALDRED, A. H. (1967). Tree volumes from large-scale photos. *Photogramm. Emgng.* **33**, 69–73.

SCHINDLER, C. (1968). Reduction of photogrammetric errors of area estimates from aerial photographs by use of radial lines as sampling units (English summary) *Allg. Forst- u. Jagdztg.* **139**,(2), 39–43.

SCHNEIDER, W. J. (1968). Colour photographs for water resource studies. *Photogramm. Engng.* **34**, 257–63.

SOEHNGEN, H. F. (1968). The development of a programming language for photogrammetry. *Photogrammetria*, **23**, 45–54.

SORE, A. L. (1967). Principles of aerial colour photography. *Photgramm. Engng.* **33**, 1008–18.

STALLARD, A. D. (1965). An evaluation of colour aerial photography for engineering purposes. *State Hwy. Comm. Kansas. Report* No. 1,

STELLINGWERF, D. A. (1968). The usefulness of Kodak Ektachrome infrared and aerofilm for forestry purposes. Commission VII. 11th Congress I.S.P. Lausanne.

TOMLINSON, R. F. (1968). Data handling and interpretation, pp. 200–10. Paper in *Land Evaluation* (ed. G. A. Stewart), Macmillan.

WEAR, J. F., POPE, R. B. & ORR, P. W. (1966). Aerial photographic techniques for estimating damage by insects in Western forests. *U.S.D.A. Pacific NW Forest and Range Ex. Sta., Oregon.*

WELCH, R. (1968). Film transparencies vs. paper prints. *Photogramm. Engng.* **34**, 490–501.

WERT, S. L. & ROETTGERING, B. (1967). Aerial survey of insect caused mortality. *U.S. For. Serv. Res. Note No. PSW*-150.

INDEX

(Page references to photogrammetric definitions are given in bold type)

INDEX

INDEX

INDEX

INDEX

INDEX

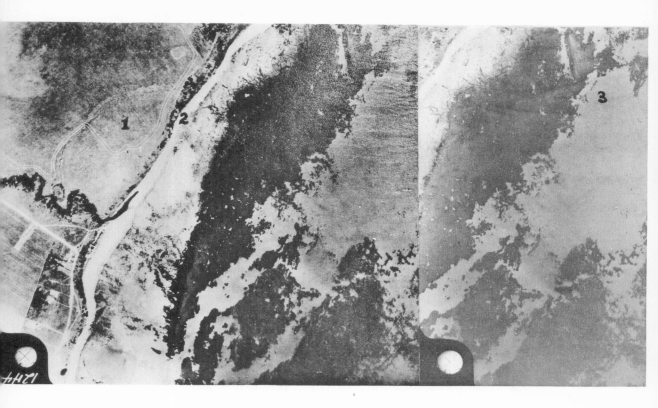

FIG. 6.3a. The stereo-photographs show farm-land (1), coast line and dunes (2), and clear coastal waters (3), of Port Philip Bay, Melbourne, (panchromatic film; scale: 1/11000). The reflex reflection zone and aircraft's shadow on the left photograph is located 1.4″ from the left margin and 2.2″ from the bottom margin. On the right-hand photograph, only the shadow of the aircraft (dark spot) can be seen in the same location. This indicates that the reflex reflection zone is not primarily a phenomenon of the troposphere.

Fig. 6.3b. The stereo-triplet shows (sclerophyll) woodland and (dry sclerophyll) forest near Eden, south-eastern Australia (panchromatic film; scale: 1/43,000). With summer photography the dark tree crowns of the woodland are often conspicuous against the (dry) light-toned herb layer. The reflex reflection zone occupies the same geometric position on each of the three photographs. For the right and centre photographs, this can be checked by measuring from the black margin and the fiducial mark (X). On the right photograph, the centre of reflection zone is absent over the deep water channel, which has recorded black, but the zone is present on the forested banks on either side of the channel. Within the reflection zone, the trees are without shadows. On the centre photograph, the shadow of the 'plane cannot be seen in the reflection zone, as was observed over land in fig. 6.3a, but this is normal for photography at higher altitudes (i.e. 16,000 feet at Eden).

Fig. 7.4. Stereo-triplet. On the centre panchromatic photograph, the zones of reflex reflection and specular reflection (on water surface) are indicated by a straight black line passing through the principal point (cross). On the left-hand photograph, part of the water surface appears white due to the recording of reflected light from the waves. At the top of the photograph, two plant formations can be observed (dry sclerophyll forest and sclerophyll woodland—see Chapter 22.) For the region of the photography, the weathering pattern suggests sedimentary rocks (see Chapters 17 and 18).

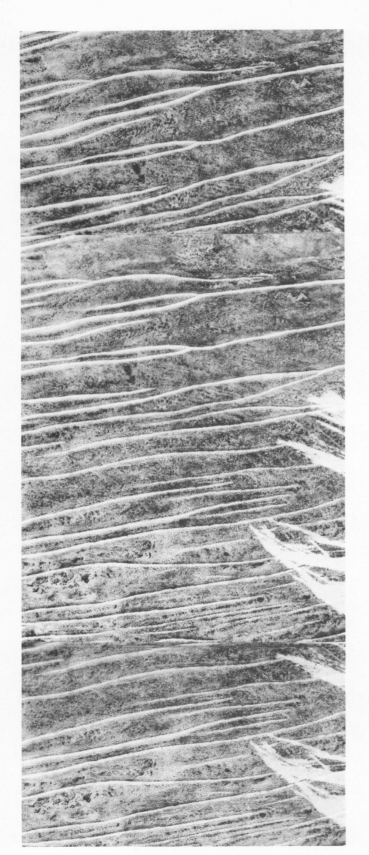

FIG. 13.4. Stereo-triplet. Examine the centre photograph without a stereoscope and decide whether or not the white lines are water courses or similar features.

(If the conclusion was in favour of the white lines being depressions, then it is incorrect and is an optical illusion. Under the stereoscope the lines will be found to be crests of sand dunes: Tuning Fork Dunes, Simpson Desert, Central Australia).

FIG.14.1. Low oblique photographs, Genlis, Cote d'Or, France. An Interesting example of vegetation providing a distinctive pattern and so revealing an ancient defensive system comprising a double ditch and palustrade. The site is flat, the sub-stratum is gravel and the land has been cultivated for hundreds of years. (a) Within the field of lucerne, the ancient individual post-holes are conspicuous as 'tufts' of lucerne inside the lucerne growing on the site of the double ditch (dry summer, June 1964). (b) The same area in May 1966, after being ploughed and left to fallow in 1965. Again the pattern is conspicuous and is now provided by cereals. The pattern was not present when fallow and the location of the post-holes of the palustrade cannot now be seen.

FIG. 14.2. A stereo-pair of terrestrial photographs, which records the profile of a young unthinned beech stand in north-east West Germany. The reader should examine a single photograph before examining the stereo-pair, as this will help him to appreciate better the value of stereo-pairs as an aid to field studies. The photographs were taken with an amateur's 35 mm. camera at two stations approximately three yards apart.

Fig. 17.4. *Alpine Woodland and Sclerophyll Forest.* Stereo-triplet of a high rainfall near Buckland Valley, Victoria, showing granites in the left-hand photograph in the vicinity of the lake and Silurian sedimentary rocks in the right-hand photograph. It is recommended that these photographs should be studied in conjunction with the description of granite and sedimentary rocks in the text.

Fig. 17.5. *Sclerophyll Woodland*. Basaltic plains are present in the bottom left-hand corner of the centre photograph and Silurian sediments in the right-hand photograph. In the latter, the marked differences in tone and texture result from clearing the woodland (dark) for agriculture (light-toned grassland); (rainfall: c.20″; elevation: 1,000—1,500 feet 1/32/000; west of Mount Disappointment, Victoria).

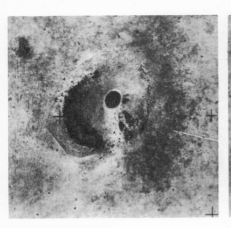
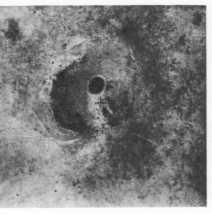

FIG. 17.6. A weathered lava cone near Mount Hamilton, Victoria. The cone is probably 10,000 years old. Note the small lake in the centre of the old crater.

FIG. 17.7. Conspicuous sedimentary rocks in an arid region (Flinders Range, South Australia). The angle of dip is from left to right. A ridge with this profile is termed a hogback. Alluvial fans are present towards the lower left of the photographs. The small black dots along the water courses are *Eucalyptus microtheca* (koolabah) and *Acacia aneura* (mulga). The culines of a few of the strata are emphasised by lines of *Acacia aneura* (Scale: 1/000000).

BELOW, TOP

FIG. 17.8. Sink-holes (1) are characteristic is some limestone areas (2) (see text) and are conspicuous on photographs by providing a pock-marked pattern. This stereo-triplet also illustrates the influence that a rock-type can have on land use. The limestone ends abruptly at the river (3) and gives way to sedimentary rocks (4), which are still heavily forested and not used for grazing as the limestone areas (scale: 1/31,000; Murrindal, Australia).

BELOW, BOTTOM

FIG. 17.9. The recognition of lineations is helpful in identifying rock types (see text). Faulting results from a fracture in the earth's surface and is accompanied by movement. This is evidenced in the fault line (1) along the arch of the anticline. Joints (2), as distinct from a fault, can also be observed in the stereo-triplet (scale 1/50,000; Hodgson Downs, Australia).

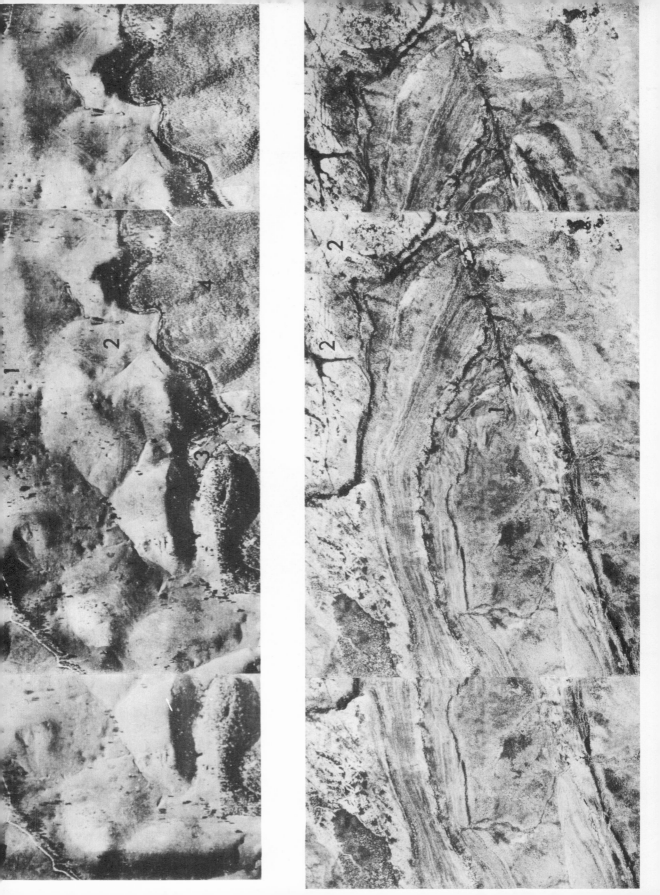

Description of Land System as a whole.

Climate : 1000 – 1300 mm. rainfall, bimodal ; mild dry season.
Rock : Pre-Cambrian basement complex, mainly schists and gneisses mainly deeply weathered and lateritised.
Morphogenesis: Dissected old land surface in which massive laterite is preserved as level caps to major interfluves.
Below these are long hill slopes leading to wide aggraded and frequently swampy valleys.
Soils : A variety of red loam lateritic (ferrallitic) type. (Buganda catena Kifu and Kaku series)
Vegetation : Forest/savanna mosaic with forest dominant along valleys.
Altitude : 1300 m. approx.
Relief : 120 – 150 m.

Diagram

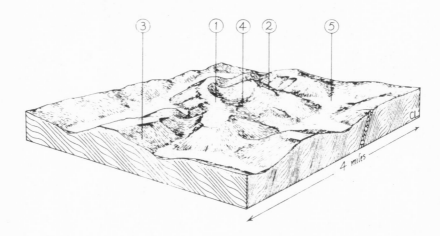

Land facet	Form	Soils, materials and hydrology	Land cover
1	Plateau crest, flat usually several hundred metres across. Abrupt short steeply sloping margins.	Thick massive laterite over weathered Pre-Cambrian metamorphic rock, mainly schist. Above laterite is dark brown sandy soil, humus stained and frequently containing murram to surface. Above ground water influence.	Themeda - Cymbopogon grass savanna.
2	Quartzite ridge, steep sided, with rounded crest. Usually occur on interfluve summits but may also occur on mid slopes.	Quartzite, bare rock or with shallow stony soil. Above ground water influence.	Themeda - Cymbopogon grass savanna.
3	Convex interfluve and slope steepening to ca 7° lower and mid slopes mainly straight. Where adjacent to laterite margins (land facet 1) uppermost part of slope steeper. May possess steeper portion on lower slope immediately above valley floor.	Two variants a) Reddish yellow loam, 1m. or more deep, over stone line, over weathered schist ; little or no concretionary iron material. b) Red clay loam or clay with little quartzite over murram, partially indurated laterite or both over weathered schist. Drainage free, above ground water influence.	Cultivated, or Penniset um fallow. Moist deciduous forest on steeper sites.
4	Small valley. Narrow drainage lines with rounded bottoms, occasionally broadening to 100m. with flat bottoms.	As 3 above but grading to mixed alluvial deposits in flatter occurrences within the influence of ground water.	Moist deciduous forest.
5	Main valley floor. Wide, 100–500m. across, and flat.	Sand with occasional layers of gravel or mottled clay. Humus stained but little peat. High ground water table.	Papyrus and Miscanthidium

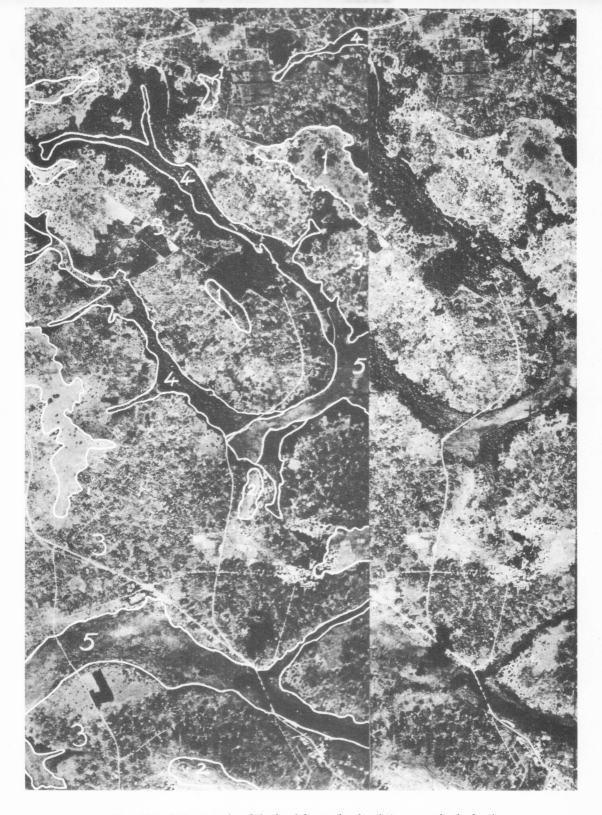

FIG. 18.1. A stereo-strip of the land facets (land units) present in the land system in south-western Uganda (Masaka Land System). The block diagram opposite portrays, as an abstraction, all the facets occurring in this land system. Below the block diagram, the facets have been defined or described.

Fig. 18.5. Stereo-triplet of glaciated terrain (Alberta) showing glacial till (1), Morainic ridges (2), alluvium plain (3), an esker (4), a kame and a drumlin. Lodgepole pine (P) occurs as pure stands in the vicinity of the ridges (lighter tone, coarse texture). In younger, heavily stocked stands the pine is light toned and smooth textured. White Muskeg along the valley shows up clearly towards the bottom of the centre photograph in very light tones (M). (Summer panchromatic photography 1/15,840.) For the same latitude in North America as in Europe, the number of tree species is usually greater in a region. However, in boreal forests the tree species remain relatively few in number and this makes identification easier than in temperate forests (for Europe, see Hagberg, 1957; Kuusela, 1957; Lutz and Caporaso, 1958; Samojlovic, 1958; Itsch and Hallert, 1964).

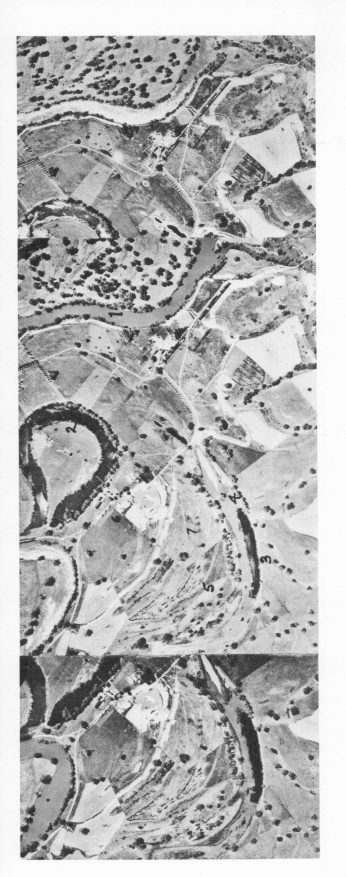

Fig. 18.6. Stereo-triplet of a fluvial landform (River Murray, Victoria). (1) Tributary of River Murray; (2) Ox-bow lake; (3) Lateral lake of old loop which is now partly swamp; (4) Natural levees formed previously by flood waters; (5) & (6) Farm buildings and small orchard. Most of the large trees are *Eucalyptus camaldulensis* (Murray red gum). Several dead red gums are conspicuous in the paddock (7).

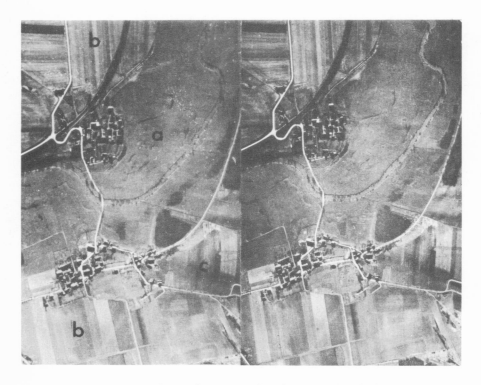

FIG. 18.7(c). A panchromatic stereo-pair (1/16,000) of a section of the Leine valley, south of Hanover. The alluvium soil (a) of the valley has recorded in an even grey tone on the black-and-white photographs and in green on the colour photograph; the brown earths (c) on Bunter differ in tone and colour and are conspicuously mottled; and the grey-brown podsolic soil on Loess (b) are somewhat lighter in tone and colour.

FIG.19.1. Contemporary high quality photographs at scales as small as 1/100,000 contain valuable information about the terrain and land-use. The above effective area of a photograph, taken with a $3\frac{1}{2}''$ focal length lens at a flying height of 25,000 feet, is at a scale of approximately 1/90,000 and covers an area of nearly 90 square miles. Using a magnifying glass, individual trees will be easily identified on the photograph. Note the complete removal of trees from the paddocks towards the centre of the photograph.

19.2. (1) Fruit orchard (citrus). Newly planted orchard. Abandoned orchard. Water storage. Ridge not cleared of native etation. Cool stores for fruit. Farm house. (8) Poultry house.

FIG.19.3. Eucalypt forest damaged by the phasmids (reddish).

FIG. 19.4a. Part of one of a series of single vertical photographs of wildebeest herd in the Serengeti National Park, East Africa, which indicated the calf percentage was 14 per cent. The photographs were taken to obtain a sample of 5000 animals from which the percentage of yearling animals can be determined as an index of the trend of the population. The calves (2) can be identified from the older animals if the photograph is examined with a hand-lens.

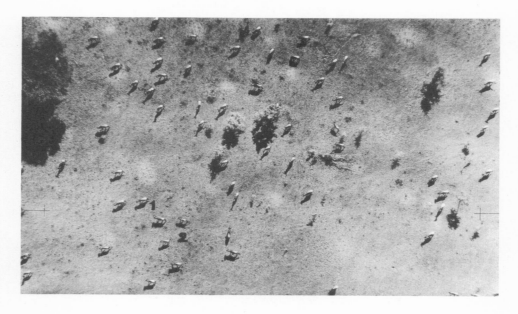

FIG. 19.4b. Part of one of a series of vertical single photographs of elephant herds in the Serengeti National Park, which was taken to obtain a sample of 1000 and from which the population structure can be obtained. The sizes of the elephants can be used to determine their age with considerable accuracy for young animals. The picture shows a typical cow herd (1) with calves (2) of varying ages. (Kodak Super XX, F.5·6, 1/1000 secs.).

FIG. 19.4c. In the United Kingdom, single and stereo-pairs of vertical and oblique photographs have been used for consensuses of black-headed gulls and oystercatchers. This photograph, at a scale of about 1/750, was used for counting the individual oystercatchers confined to a narrow littoral (2) between the high tide (1) and the hinterland (3).

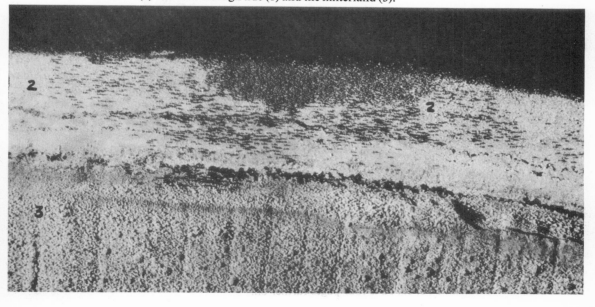

FIG. 19.5. The detection of disease ('Die-back' of *Eucalyptus marginata* near Teasdale, Western Australia). (a) Terrestrial photograph taken from area marked (1) in the stereo-pair. Tree height about 80 feet. Note the sharp edge to the die-back zone. (b) Stereo-pair (1/15,840) showing (1) area of deaths; healthy forest (dark toned) and early stage of the disease (light toned). In the early stages of the attack, the understorey is killed and results in the forest recording light toned on aerial photographs.

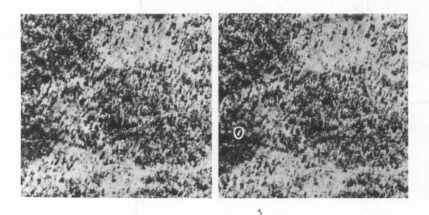

Erratum. FIG. 19. 5a is four pages further on.

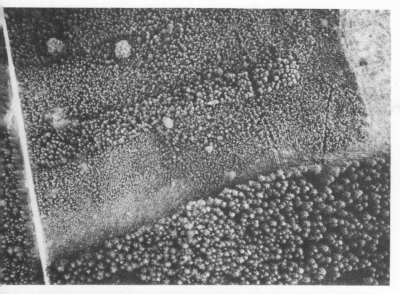

Fɪɢ. 19.6. In addition to aerial photography being used to detect disease (Fig. 19.5), photographs may be helpful in evaluating the physical condition of crops. The figures illustrate the value of carefully selecting film-filter combinations. The photographs are of stands of Scots pine and Norway spruce near Eberswalde, East Germany. Figures (a), (b) and (c) show healthy trees, trees dying from the effect of air pollution and dead trees. (a) was taken using panchromatic film and a minus blue filter, and the three classes cannot be satisfactorily separated. (b) was taken using Russian two-layer colour film (Spectrozonal SN-2M) and a red filter; now the dead trees are conspicuous due to their bright green colour (3), the healthy trees are greenish brown (1) and dying trees are yellowish green (2). Finally, in (c), part of the area is shown in (normal) three-layer colour; dead trees are brown, healthy trees are green to yellowish green.

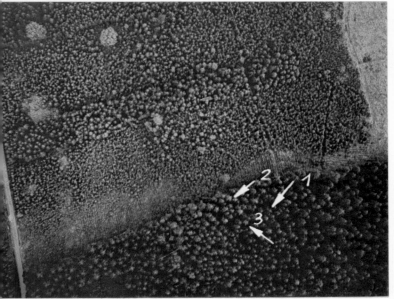

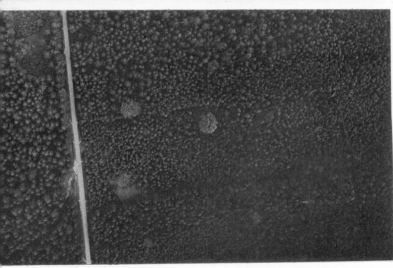

FIG. 19.6 (d) and (e) are photographs at a scale of 1/4000 of an area, part of which was dressed with nitrogen fertiliser. On the panchromatic photograph (minus blue filter), (fig d), there is no evidence to show where the fertiliser was applied; but in (e), infra-red film (minus blue filter), the plots to which the fertiliser was applied are conspicuous by their light tone.

Fig. 20.1. *Temperate rain forest*. The large emergent crowns of *Agathis australis* (1) are conspicuous against a 'back-ground' provided by a closed canopy of angiospermous species and several *Podocarpus species* (2). (Scale: 1/15,840) (Road. Basaltic plateau, North Island, New Zealand). The heights of the emergents probably attain 180 feet.

Fig. 20.2. *Tropical Montane Forest*. The large emergent crowns of *Araucaria* app. (1) are conspicuous as small light toned dots, against a 'background' provided by a closed canopy of the many other species. (2) Note the waterfall towards the top of the stereo-pair, the river (3), shifting cultivation (taungya) (4) and regeneration of the forest following shifting cultivation (5). *Araucaria* app. are probably about 150 feet in height and have cone-shaped crowns. (Scale: 1/44,000; Dagia river, Papua, New Guinea).

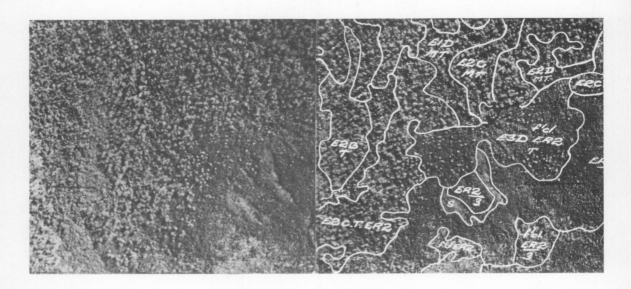

FIG. 20.3. *Wet sclerophyll Forest*, River Styx, Tasmania. On the right-hand photograph of the stereo-pair, the area has been delineated stereoscopically into forest types (production classes) for management on the basis of maturity, mean dominant height of eucalypts and their stocking. This is commonly referred to as forest typing. The eucalypts, El, have top heights up to 300 feet and a second storey of myrtle (M) and sassafras (T) up to 100 feet. The undelineated area at the foot of the photograph contains ti-tree up to 30 feet high. The crowns of large dead trees are conspicuous in 'ED3' (Scale: 1/24,000).

Abbreviations. E: eucalypt forest; ER2: eucalypt regrowth, 50-90 feet; E2, E3: mature forest over 180 feet, 135-180 feet, 90-135 feet; f'd: Fire damaged mature trees counted per acre: B(11-15), C(6-10), D(1-5).

FIG. 19. 5a. This photograph (infrared film, deep red filter) is of a potato field in the Fenlands (United Kingdom) and shows several well-established blight foci. Many developing 'daughter' blight foci can be observed along the wheelings made by a Sprayer. Probably only the largest foci would be recorded on a panchromatic photograph.

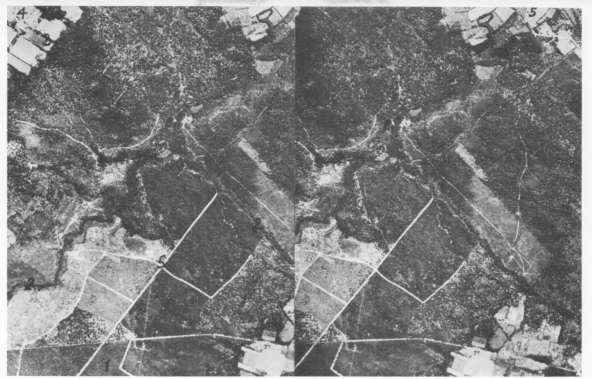

PANCHROMATIC PHOTOGRAPHY

FIG. 21.1. *Hardwoods and conifers in the Northern Hemisphere*. These can usually be separated using (pure) infra-red photography. Both stereo-pairs are of the same area (Landes, France), and show plantations of pine (P. pinaster) of several age classes (1); hardwoods (e.g. alder, poplar) along the streams (2); several fields surrounded by hardwood hedges (3); a house (4) and roads (5). On the infra-red photographs, the pines are conspicuously dark-toned and the hardwoods light-toned. Other contrasts can also be made (e.g. tone of house and roads).

INFRA-RED PHOTOGRAPHY

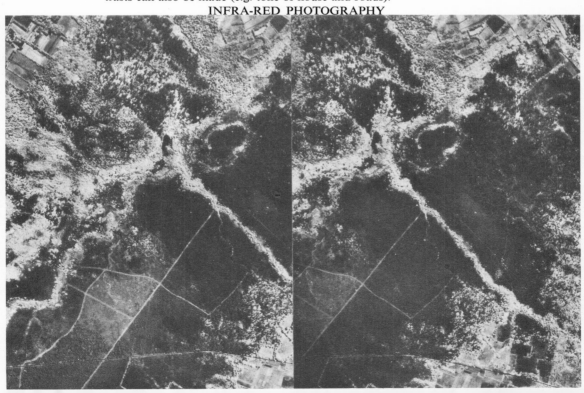

FIG. 22.2a. A stereo-pair of part of Wilson's Promontory, Victoria. (1/15,840). This stereo-pair should be studied in conjunction with the text and figures 22.2b. and 22.2c. *Woodland* (mesophanerophytes), *thicket* (microphanerophytes) and *heath* (nanophanerophytes) and grassland are present.

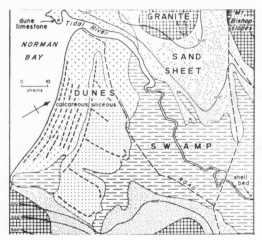

FIG. 22b.

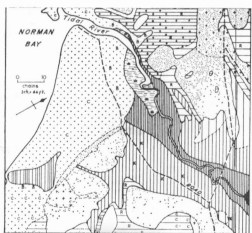

FIG. 22c.

EUCALYPTUS BAXTERI TALL SCLEROPHYLL SHRUB WOODLAND	LEPTOSPERMUM LAEVIGATUM THICKET	MELALEUCA ERICIFOLIA THICKET
E. BAXTERI LOW SCLEROPHYLL SHRUB WOODLAND	JUNCUS MARITIMUS– SAMOLUS REPENS SALT–MARSH	M. SQUARROSA THICKET
E. BAXTERI– CASUARINA PUSILLA HEATH	CYPERUS LUCIDUS– PHRAGMITES COMMUNIS GRASSLAND	M. SQUARROSA HEATH
C. PUSILLA– LEPTOSPERMUM MYRSINOIDES HEATH	KUNZEA AMBIGUA THICKET	B BANKSIA INTEGRIFOLIA
R E. RADIATA	C C. STRICTA	K E. KITSONIANA
		● E. OBLIQUA

22. 1.

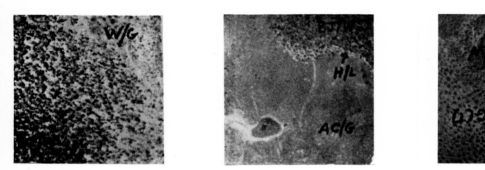

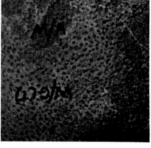

22. 1a.   22. 1b.

TROPICAL WOODLAND

FIGS. 22. 1. Over vast areas of Southern Africa *Brachystegia-Julbernardia* woodland (BJ/W fig. 22. 1a, left) is an important formation and is recognisable on the single photograph even at scale 1/31,000. The bottom land, however, is commonly *Acacia-Combretum* grassland (AC/G fig. 22. 1b, centre).

The herb line (Chapter 20) is also shown in fig. 22. 1b, centre. Fig. 22. 1b, right shows wooded grassland with termitaria (WGT) and valley woodland (V/W).

Erratum. These plates should have preceded Nos. 22. 2. a, b and c.